Energy Storage Technologies

Energy Storage Technologies

Pathways to a Sustainable Future

Edited by

Deepanraj Balakrishnan

Prince Mohammad Bin Fahd University, Dhahran, Saudi Arabia

Prabhakar Sharma

Delhi Skill and Entrepreneurship University, New Delhi, India

Aman Maung Than Oo

University of the Sunshine Coast, Sippy Downs, QLD, Australia

Bachir Benhala

Sidi Mohamed Ben Abdellah University, Fez, Morocco

ELSEVIER

Elsevier
Radarweg 29, PO Box 211, 1000 AE Amsterdam, Netherlands
125 London Wall, London EC2Y 5AS, United Kingdom
50 Hampshire Street, 5th Floor, Cambridge, MA 02139, United States

ISBN: 978-0-443-45858-3

For information on all Elsevier publications visit our website at
https://www.elsevier.com/books-and-journals

Publisher: Priscilla Braglia
Acquisitions Editor: Ambika Kapoor
Editorial Project Manager: Avni Gupta
Production Project Manager: Anitha Sivaraj
Cover Designer: Raman Kumar

Typeset by TNQ Tech

Contents

Contributors

Mona O. Albalawi Department of Chemistry, Faculty of Science, University of Tabuk, Tabuk, Saudi Arabia

Mohammed Alenezi Wolfson Centre for Magnetics, School of Engineering, Cardiff University, Cardiff, United Kingdom

Gaydaa AlZohbi Department of Mechanical Engineering, College of Engineering, Prince Mohammad Bin Fahd University, Al-Khobar, Saudi Arabia

Tanzeela Asghar College of Science, National Taiwan University (NTU), Taipei, Taiwan; Institute of Physics, Academia Sinica, Taipei, Taiwan

Abul Kalam Azad School of Engineering and Technology, Central Queensland University, Melbourne, VIC, Australia

Deepanraj Balakrishnan Department of Mechanical Engineering, College of Engineering, Prince Mohammad Bin Fahd University, Al-Khobar, Saudi Arabia

Bachir Benhala Sidi Mohammed Ben Abdellah University, Fez, Morocco

Yassine Benkirane Sidi Mohammed Ben Abdellah University, Fez, Morocco

Sachin Lotan Borse Department of Mechanical Engineering, JSPM Narhe Technical Campus, Pune, Maharashtra, India

Badre Bossoufi LIMAS Laboratory, Faculty of Sciences Dhar El Mahraz, Sidi Mohammed Ben Abdellah University, Fez, Morocco

Ajcharapa Chuanchai Department of Aeronautics and Astronautics, National Cheng Kung University, Tainan, Taiwan

Arun Teja Doppalapudi School of Engineering and Technology, Central Queensland University, Mackay, QLD, Australia

Natthawud Dussadee School of Renewable Energy, Maejo University, Chiang Mai, Thailand

Mourad Fakhfakh University of Sfax, Sfax, Tunisia

I.M. Rizwanul Fattah University of Technology Sydney, Ultimo, NSW, Australia

Mohammed Amine Hoummadi LIMAS Laboratory, Faculty of Sciences Dhar El Mahraz, Sidi Mohammed Ben Abdellah University, Fez, Morocco

Shamsul Huda Faculty of Science, Engineering and Built Environment, School of Information, Deakin University, Burwood, VIC, Australia

Amal Kadri ENSET of Mohammedia, Hassan II University of Casablanca, Mohammedia, Morocco

Onur Karaman Akdeniz University, Antalya, Turkey

Ceren Karaman Akdeniz University, Antalya, Turkey

Mohammed Karim LIMAS Laboratory, Faculty of Sciences Dhar El Mahraz, Sidi Mohammed Ben Abdellah University, Fez, Morocco

Tajeddine Khalili ENSA of Beni Mellal, Sultan Moulay Slimane University, Beni-Mellal, Morocco

Taimoor Ali Khan College of Electrical Engineering and Computer Science, National Taiwan University (NTU), Taipei, Taiwan; Research Center for Information Technology Innovation, Academia Sinica, Taipei, Taiwan

Mohammed Kissaoui ENSET of Mohammedia, Hassan II University of Casablanca, Mohammedia, Morocco

Ali Ihsan Komur Akdeniz University, Antalya, Turkey

Muninathan Kota St. Joseph's College of Engineering, Chennai, Tamil Nadu, India

H. Leena Catherine St. Joseph's College of Engineering, Chennai, Tamil Nadu, India

G.M. Lionus Leo St. Joseph's College of Engineering, Chennai, Tamil Nadu, India

Hiren Mewada Prince Mohammad Bin Fahd University, Al Khobar, Saudi Arabia

Yosra M'zabiri ENSET of Mohammedia, Hassan II University of Casablanca, Mohammedia, Morocco

S. Murugapoopathi PSNA College of Engineering and Technology, Dindigul, Tamil Nadu, India

Jayashri Narayanan Nair Department of Mechanical Engineering, VNR Vignana Jyothi Institute of Engineering and Technology, Hyderabad, Telangana, India

Abdelhadi Raihani ENSET of Mohammedia, Hassan II University of Casablanca, Mohammedia, Morocco

Rajeswaran Ramaraj Institute of Green Products, Feng Chia University, Taichung, Taiwan

Rameshprabu Ramaraj School of Renewable Energy, Maejo University, Chiang Mai, Thailand

Sajeeb Saha School of Science, Technology and Engineering, University of the Sunshine Coast, Moreton Bay, QLD, Australia

Mohamed Abdel Salam Department of Chemistry, College of Science and Technology, Florida A&M University, Tallahassee, FL, United States

Prabhakar Sharma Delhi Skill and Entrepreneurship University, Delhi, New Delhi, India

Amanullah Maung Than Oo School of Science, Technology and Engineering, University of the Sunshine Coast, Moreton Bay, QLD, Australia

Yuwalee Unpaprom Program in Biotechnology, Maejo University, Chiang Mai, Thailand

Dhana Raju Vallapudi Department of Mechanical Engineering, Lakireddy Bali Reddy College of Engineering, Mylavaram, Andhra Pradesh, India

Rohitha Weerasinghe School of Computing, Engineering and Creative Industries, University of Bedfordshire, Luton, United Kingdom

Preface

In the present era, the world is hugely dependent on energy for comfort in our daily lives. The intermittency in renewable energy systems and supply interruptions of fossil fuels owing to global political challenges necessitate a stable energy storage system. The challenge, however, is to develop a reliable and efficient energy storage system for a variety of energy needs. The present book presents a comprehensive framework for different domains and the need for energy storage systems.

The book opens with a brief introduction to different types of energy storage systems and their functioning. Next, the book discusses the key technologies: batteries and electrochemical storage, hydrogen and metal hydride, thermal and thermochemical storage, and mechanical and electrical storage. The chapters build on one another by covering artificial intelligence, machine learning, and optimization methods—methods that are transforming the way storage systems are designed and operated. Smart grids, decentralized energy, global policy, investment trends, and materials innovation all find their own space as well.

Two chapters are particularly worthy of mention. One considers battery storage in modern power systems—with emphasis on reliability and economics. The other focuses on cybersecurity in smart energy networks, which is becoming an ever more important issue. Both are real-life issues that can no longer be regarded as secondary ones.

The last chapter provides a glimpse into future research directions. The profession is evolving rapidly, and we would like readers to not only leave informed but also be prepared to contribute to the future.

The editors owe a debt of gratitude to all the researchers and colleagues whose work is mirrored in these pages. We hope that this book can be used as a practical guide as well as a reliable reference book for anyone working toward a more sustainable energy future.

The Editors

Introduction to energy storage systems

1

Deepanraj Balakrishnan[1], *Prabhakar Sharma*[2], *Amanullah Maung Than Oo*[3] *and Bachir Benhala*[4]
[1]Department of Mechanical Engineering, College of Engineering, Prince Mohammad Bin Fahd University, Al-Khobar, Saudi Arabia; [2]Delhi Skill and Entrepreneurship University, Delhi, New Delhi, India; [3]School of Science, Technology and Engineering, University of the Sunshine Coast, Moreton Bay, QLD, Australia; [4]Sidi Mohammed Ben Abdellah University, Fez, Morocco

1.1 Introduction

There is a profound transformation of the global energy sector as a result of increasing deployment of renewable energy (RE), electrification of end-use sectors, and the urgent need to mitigate global warming. In the past, dispatchable generation systems have provided energy based on fossil fuels that can be employed to meet demand. However, the rapid deployment of different sources of RE generation (like wind and solar) has introduced uncertainty and variability into the energy system, thus challenging conventional generation-demand paradigms (Chen et al., 2009; Senthilkumar et al., 2024). The emerging field of energy storage is being considered an enabling technology in this new landscape to bridge the temporal and spatial gaps between the supply of energy and growing demand. The growing attention on decarbonization and sustainability has also brought an increased requirement for advanced energy storage systems (ESSs). With the electrification of transportation, industrial, and residential systems, there exists an increasing requirement for efficient, economic, and scalable EST. Energy storage is no longer regarded as a supporting factor but an integral part of the energy system. It allows for higher penetrations of RE, reduces fossil fuel consumption, and allows for the development of smart and distributed energy networks (Chougale et al., 2023; Thirumalvalavan et al., 2025).

1.1.1 Background and motivation

The motivation to build energy storage is technology and environment-driven. Traditionally, energy storage has been used on a small scale (e.g., pumped hydro storage) to provide grid stability and peak load shaving facilities. However, these conventional systems are location-dependent and can be associated with high capital costs and environmental impacts. The enhanced variability of RE has shown the limitations of such traditional storage technologies and the need to explore a range of options and sustainable solutions. The wind and solar resources are also intermittent and are affected by

Energy Storage Technologies. https://doi.org/10.1016/B978-0-443-45858-3.00003-6

the weather and day-night cycles. This variable nature of resource availability can be an issue with the continuous energy supply. Lacking adequate storage, the surplus RE generated during high production periods can be reduced, resulting in inefficiencies and losses. Similarly, during low production, there is an increased use of backup systems that are fueled by fossil fuels, which defeats the purpose of sustainability. Energy storage overcomes this problem by enabling the spatial and temporal shifting of energy, thus increasing resource utilization (NREL, 2024).

The other key driver is transportation electrification, especially the explosive growth in electric vehicles (EVs). Widespread adoption of EVs demands high-performance batteries that offer higher energy densities, rapid charging, and longer life. This has driven research on electrochemical EST, in particular, lithium-ion, and new battery chemistries (Kumar Jhaet al., 2025;Srinivasaraoand Mohamed, 2024). In addition, connecting EVs to the grid enables new vehicle-to-grid (V2G) technologies, where EV batteries can be used as distributed energy storage. In addition, energy storage plays an important role in the reduction of greenhouse gas emissions (GHGs) (Serarslan, 2025). Energy storage supports the generation of clean energy by increasing the integration of RES and reducing the fossil fuel consumption in power plants. Furthermore, the storage technologies are also helping to develop off-grid and microgrid systems, particularly in remote and marginalised communities, which improves access to energy and energy equity (Kardenet al., 2007).

Economic factors are also a strong driver. The cost reduction of RE technologies and the advancement of battery technology and materials science have improved the economic viability. As the cost of storage continues to decline, it is becoming more attractive than conventional generation technologies. Moreover, energy storage has multiple streams of revenue, including frequency regulation, demand response, and energy arbitrage, which makes it an attractive investment (Liu and Zhang, 2025).

1.1.2 Energy storage in the modern energy systems

ESSs have a multifaceted role in modern energy systems and are a key enabling technology for the transformation toward sustainable and resilient energy systems. The prime applications of energy storage are to enhance grid stability. For power systems, it is critical to have a balance between generation and demand to obtain a stable frequency and voltage. Energy storage can respond quickly to fluctuations and offers supplementary support services like voltage control, frequency control, and spinning reserve. Such a response time is especially important in systems with higher participation of renewable sources, where fluctuations can occur with short time scales (Wang and Zhao, 2023).

The second important function is to support RE integration. As RES continue to grow, energy storage of the RE surplus is required to increase the RE penetration. Energy storage allows load shifting, where energy generated when there is surplus RE production can be stored and employed in case of higher demand. This not only helps to reduce curtailment but also improves the reliability as well as efficiency of energy systems (Shafie and Nu'man, 2024). The energy storage also enables decentralization of energy systems. Energy systems are being increasingly decentralized with distributed energy resources (DERs). There is an increase in the employment of rooftop PVs and smaller wind turbines. ESS enables distributed resources to operate more effectively by smoothing variations in production and providing back-up in the case of power failure. This is particularly important in the case of microgrids, where storage can offer energy availability and reliability in both grid-connected and islanded modes. In the case of electric mobility, energy storage is the key to electrified transportation. Battery technologies are essential for the development of EVs with longer range and fast charging. In addition, the interaction of EVs with the grid offers the opportunity of demand side management and grid support because of the bi-directional energy flow (Matos et al., 2024).

Energy storage is also critical in enhancing energy security and resilience. Energy storage can be used to power backup power and critical facilities during disasters, cyber-attacks, and other events. Data centers, hospitals, and emergency systems are particularly important, as they need a continuous power supply. And ESS helps to optimize energy markets. Energy storage can be used for storing energy at low prices and discharging it at high prices through means such as energy arbitrage, which improves economic efficiency. They also enable them to participate in the ancillary service markets, which provide them with additional revenues (Njemaet al., 2024).

1.1.3 Scope and objectives

The present chapter will provide an overview of the ESS, its principles, technologies, and applications. The focus is on a broad range of EST, including mechanical, electrochemical, electrical, thermal, and chemical energy storage. Each type is described in terms of its principles, characteristics, advantages, and disadvantages. One of the goals is to set a picture of the fundamental concepts and measurements of energy storage. Energy density, efficiency, power density, response time, and cycle life are some of the characteristics that are crucial to decide the appropriateness of different technologies for a particular application. The chapter is expected to guide researchers and practitioners to choose the right storage technologies by providing a comparative basis.

1.2 Fundamentals of energy storage systems

ESS is one of the important aspects of modern flexible energy systems, helping in the temporal decoupling of energy supply and demand. At its core, these systems convert energy into one form, store it for a time, and convert it back to the original or another form of energy. This energy conversion, storage, and reconversion is associated with some physical processes and efficiency losses. The range of EST is due to the different physical and chemical processes of energy conversion, such as electrochemical, mechanical, electromagnetic, and thermal processes (May et al., 2018). To put energy storage into perspective, it is required to explore energy storage capacity, power capability, and efficiency of energy recycling. These characteristics determine the suitability of technology for a specific application, like power quality regulation over a shorter period or energy balancing over a longer time. The balance of energy, power, and efficiency is the basis of the evaluation and design of the storage technologies in modern energy systems.

1.2.1 Definitions and classification

The ESS can be generally defined as the technologies which capture, store, and release energy in a stable manner. The energy stored can be electrical, mechanical, thermal, or chemical energy, depending on the storage mechanism. Based on these mechanisms, EST can be broadly classified into five categories: electrochemical, mechanical, electrical, thermal, and chemical ESS. Mechanical EST are used to store energy as potential and kinetic energy. Examples include the pumped hydroelectric energy storage using gravitational potential energy, in which water is pumped uphill or downhill between reservoirs of different elevations, and kinetic energy storage in flywheels (Balali and Stegen, 2021).

Electrochemical storage (also referred to as batteries) stores energy in reversible chemical reactions. These batteries (lithium-ion, lead-acid, flow batteries, etc.) are extensively employed owing to their high energy density and scalability. Supercapacitors, which are sometimes included in this category, are used to store energy electrostatically and have high power density and short charge/discharge times (Njemaet al., 2024). Electrical ESSs are based on the use of electromagnetic fields to store energy. An electric field is used to store energy in a capacitor and superconducting magnetic energy storage (SMES) systems store energy in a magnetic field generated by a current. They are highly efficient with fast response time and for this reason can be used for power quality. Thermal energy can be stored as hot or cold. This can be achieved by sensible heat storage, latent heat storage system using phase change materials or thermochemical processes. Thermal storage is particularly useful for heating and cooling and concentrated solar power systems. Chemical energy storage is the storage of energy in chemical bonds, for example, hydrogen or synthetic fuels. This type of energy storage offers long-term and large-scale storage, hence attractive for seasonal storage and decarbonization of hard-to-electrify industries. This typology thus provides a framework to understand the different EST and their use (He and Zhang, 2022).

1.2.2 Performance indicators

There are several important parameters that are used to evaluate the performance of ESS in storing and supplying energy. Efficiency, power density, and energy are the most important of these. Energy density is the amount of energy stored per unit volume. High energy densities allow for long run times without increasing the size or weight of the system. Electrochemical systems, such as Li-ion batteries, possess higher energy densities in comparison to other systems. However, power density is the energy that is delivered per unit volume (W/L) or mass (W/kg). High power density technologies also include supercapacitors and flywheels that are used in domains needing a sudden power release, like frequency regulation and regenerative braking (Lakshmi and Vedhanarayanan, 2023). Another measure that describes the amount of energy that can be recovered during discharge is efficiency. This is also known as round-trip efficiency and accounts for the energy losses that occur during discharging and charging. The objective is to have a high efficiency to minimize energy losses and improve system performance. The efficiencies of electrical storage systems are typically quite high (more than 90%), but thermal and chemical storage systems can have marginally low efficiency owing to losses during energy conversion (Raman and Barooah, 2020).

There are other important measures, such as cycle life, which is the count of the system charging and discharging before its performance is degraded; response time, which is the energy release rate from the system; and self-discharge rate, which is the rate at which the energy degrades when the system is not in use. There are tradeoffs between these metrics that are critical in the technology choice. For instance, systems with higher energy density can have lower power density and vice versa. Therefore, the choice of storage technology is determined by the application-specific requirements in terms of time, size, and conditions (Zhang et al., 2017).

1.2.3 Charge-discharge characteristics

The charge-discharge characteristics of ESS are the foundation of its operation. These are determined by the physical and chemical processes that are associated with energy storage and release. These are important to understand how to improve system performance, safety, and lifetime. Energy is supplied to the storage system during the course of the charging process, and this leads to the accumulation of energy. In electrochemical systems, the flow of ions takes place between two electrodes by an electrolyte and with redox reactions. In mechanical systems, charging can be performed by lifting a weight or spinning a flywheel and in thermal systems by increasing the temperature or phase change. Discharge is the reverse process, where the energy is released to perform work. This process is dependent on several key factors including internal resistance, reaction kinetics, and design of the system. For instance, the internal resistance of batteries is a source of heat loss and has an effect on efficiency and heat management (Wu et al., 2024).

Another critical point of the charge-discharge process is the rate capability, which is usually expressed in the form of C-rate for batteries. By raising the C-rate, the battery can be charged or discharged faster, but more degradation and reduced efficiency can also occur (Mishra et al., 2022). Therefore, the charge-discharge rates are critical in maintaining the battery system. Another important feature is the charging and discharging rate. Most battery systems have a voltage related to the state of charge (SoC), and this provides an indication of the internal state and capacity of the system. This is used to track and manage operation for safety and reliability of battery management systems (Eftekhari, 2017). The other factor is hysteresis and degradation. ESS can experience capacity loss, internal resistance and efficiency loss with cycling. These effects are determined by the temperature, SoC, and operating conditions of the storage system. Additionally, the self-discharge rate affects the long-term storage capacity. For example, systems such as capacitors have a high self-discharge rate while others, such as some chemical storage systems, can store energy for a long time (Ovejas and Cuadras, 2019).

1.3 Types of energy storage technologies

There are a variety of EST that are based on different physical and chemical processes. These differ in terms of power and energy, speed of response, and versatility. The appropriate storage system selection depends on the application, which could be short-term power support, long-term energy storage, or grid-scale energy storage.

1.3.1 Mechanical storage systems

Mechanical EST stores energy in the form of potential or kinetic energy and is the most mature technology. They are ideal for larger-scale and longer-duration storage applications owing to their robustness, lifespan, and low operating costs. The oldest form of mechanical EST is pumped hydro storage (PHS). This type of ETS stores energy by first supplying water to a higher level and then generating electricity by releasing the water through turbines. PHS has high efficiency (70–85%) and can provide large storage capacities (up to gigawatt-hours). However, they require suitable topography and substantial water resources (Haas et al., 2022).

A different mechanical EST is compressed air energy storage (CAES). CAES is used to store energy by making use of compressed. CAES employs air using off-peak electricity for air compression and storing it in caverns or high-pressure tanks. The air is expanded and heated and then used to generate electricity. Recent adiabatic CAES designs aim to improve efficiency by capturing and reusing heat from the compression stage. CAES can be used to store energy on a large scale, but its efficiency and environmental impact depend on system design and thermal management strategies (Qiu, 2023).

A flywheel ESS stores kinetic energy in a rotating mass. It is used to convert electrical energy to kinetic energy for charging. Subsequently, the kinetic energy is converted into electrical energy for discharging. The high-power density, rapid response, and long lifespan of flywheels make them suitable for energy storage for frequency regulation and power quality control. However, they have low energy density, limiting their use in energy storage. So, mechanical ESSs are highly reliable and can be employed at the grid level, but their use can be constrained by size and geography (Xu et al., 2023).

1.3.2 Electrochemical storage systems

Batteries, also referred to as electrochemical ESS, are a very versatile and widely used storage technology. They use reversible chemical reactions to store energy and convert it to electrical energy. Li-ion batteries are the most common owing to their superior energy density, high efficiency (greater than 90%), and longer life. They have been extensively used in electric transportation, portable devices, and energy storage. Advances in materials science, including new active materials and electrolytes, are enhancing their performance and safety (Ramaraj et al., 2025; Ren et al., 2023).

Lead-acid batteries are the earliest batteries and are widely employed owing to their lower cost and durability. The cycle life and energy density of lead-acid batteries are inferior to those of Li-ion batteries, limiting their applications. Vanadium redox flow batteries, also termed as flow batteries, are also gaining traction for energy storage purposes. Flow batteries have a longer cycle life and enhanced safety but generally inferior energy density and higher cost. Supercapacitors (also known as ultracapacitors) store energy as static electrical energy. They have higher-power density, rapid charging and discharging, and can be cycled many times. However, their energy density is low and can only be employed for short-term storage (e.g., peak power and regenerative braking) (He and Zhang, 2022).

Electrochemical storage systems are flexible and are widely used in energy systems, mainly in sectors needing high efficiency, modularity, and scalability.

1.3.3 Electrical storage systems

Electrical ESS stores energy in electric and magnetic fields and is characterized by its speed and efficiency. ESS is ideal for higher power-consuming applications and high cycling rates. The energy is stored in conventional capacitors in an electric field between two parallel plates separated by an insulator. Due to their low energy density, conventional capacitors can deliver energy very quickly and are used in applications that require power conditioning and in electrical circuits. Supercapacitors are a type of capacitor that lies between conventional capacitors and batteries. It has a higher energy density than conventional capacitors but still have a high-power density. Their fast charging and discharging with high cycle life allow them to be used in high-cycle applications. SMES devices are efficient (more than 95%) due to the zero resistance of superconductors. They also respond quickly and so are well-suited for power quality and grid stability. However, their requirement for cooling and expensive materials limits their application. Electrical ESSs are used for niche applications that require high-power quality, fast response, and high efficiency but not for large-scale energy storage (Adetokun et al., 2022).

1.3.4 Thermal storage systems

Sensible heat storage uses a temperature rise of a medium such as molten salts, water, or solids. The specific heat capacity and temperature rise are used for energy storage. This is a simple and low-cost method, and is widely used in district heating and concentrated solar power. Latent heat storage makes use of the PCMs to store energy when phase changes occur (melting and freezing). These storage methods have a higher energy density than sensible heat storage and can offer an almost isothermal phase change, which is useful for thermal management. Thermochemical storage involves energy storage and release by reversible chemical reactions. They are energy efficient with high energy density and can potentially provide long-term energy storage with minimal losses. But they are still in the development stage and face material stability and system design problems. Thermal storage systems are critical for integration of RE (e.g., solar thermal collectors) and energy efficiency in industrial and buildings (Farullaet al., 2020).

1.3.5 Chemical storage systems

Chemical ESS stores energy in the form of chemical bonds, enabling long-term and large capacity energy storage. These systems are crucial for seasonal storage and decarbonizing challenging-to-electrify industries. One type of chemical energy storage is hydrogen energy storage. Hydrogen is created using excess electricity to power electrolysis, stored and then converted back to electrical energy through fuel cells or combustion. Hydrogen has a high specific energy and is storable, so it is suitable for large-scale and long-term storage. But there are challenges of storage, transport, and conversion efficiency (McGillicuddyet al., 2020).

Chemical storage can also be in the form of synthetic fuels, such as methanol and ammonia. These fuels are made from RE and can be used in existing transport and industrial applications. These have advantages in terms of energy density and compatibility but have complex production processes and conversion losses. Chemical ESS provides an opportunity to integrate RE into challenging sectors such as aviation, shipping, and industry. They also enable the development of energy systems that simultaneously generate electricity, heat, and fuel (Farullaet al., 2020).

In summary, the diversity of ESS reflects the diversity of energy system requirements. Mechanical systems are widely used in large-scale applications, electrochemical systems have flexibility and scalability, electrical systems have a rapid response, thermal systems enhance energy efficiency, and chemical systems allow for long-term and multisectoral energy storage. Knowledge of these technologies is essential to building integrated and sustainable energy systems to meet future energy challenges.

1.4 Applications and integration

ESS are indispensable in most of the applications in modern energy systems. ESS plays a vital role in preserving reliability, efficiency, and flexibility in the shift to greater penetration of RE, decentralization, and electrification. These systems not only store energy but also manage energy, enabling intelligent energy exchanges between generation, transmission, distribution, and loads. Their use in power systems allows both bulk power system and distributed system designs and is key to the transition to sustainable energy systems. ESS is multifunctional. Such applications include energy arbitrage, peak shaving, frequency regulation, voltage support, back-up power, and intermittent energy source integration. The advantages of ESS are not only related to energy storage but also to their responsiveness and flexibility to system requirements. The following subsections provide some of the important uses of ESS.

1.4.1 Renewable energy integration

Probably the most critical application of ESS is to help the integration of intermittent RES, including wind and solar. RE generation is intermittent and can lead to problems in energy supply. Energy storage solves these problems by decoupling energy and time, enabling temporal matching of supply and demand. Energy can be captured when the RE supply is high, such as during the day (solar) or when the wind is blowing. This stored energy can then be used when there is low generation or high demand, thus flattening the supply curve. This is also called energy shifting or load leveling and is beneficial for increasing the efficiency of RE use and reducing the reliance on fossil fuel back-up systems (Hussain et al., 2025).

Energy storage is also important in the case of short-term variability of RE. Variability of solar energy (due to cloud cover) and wind energy can cause fluctuations in the power output and affect grid stability. Energy storage (with high power density and fast response) can quickly adjust to these variations to provide a constant power output. Also, storage systems increase the penetration of renewables by avoiding issues of grid congestion and curtailment. Local storage can prevent expensive transmission system upgrades and improve efficiency. They also help integrate local RE generation, such as rooftop PV systems, by providing local balancing and back-up. Further to this, in off-grid systems, energy storage is essential to make stand-alone renewable energy systems (RESs) feasible. Hybrid solar-wind-storage systems can provide a sustainable and uninterrupted energy supply in off-grid systems, enhancing energy access (Wang et al., 2025).

1.4.2 Grid stability and load management

Power system stability is crucial, and energy storage plays a key role in this. Generation and load balancing are vital for maintaining the frequency and voltage of the system. If this is not achieved, the system can be unstable, equipment can be damaged, or even blackouts can occur. Frequency regulation is the adjustment of the system's frequency within a specified range to counter short-term imbalances. Batteries and flywheels have millisecond response times and are well-suited for this task. A second major application is voltage regulation, which is becoming increasingly important in distribution systems with increasing penetrations of distributed energy resources (Li and Deusen, 2025).

ESS can also be used for load management. Peaking power plants (which are expensive and inefficient) may need to be used during peak periods. Energy storage can be used to store energy when it is cheap and discharge it when it is expensive, facilitating peak shaving, thus reducing the cost of energy generation and relieving pressure on the grid. Energy storage also enables demand response, in which consumers adjust their demand based on price or supply. Energy storage can help achieve this by automatically storing energy in lower price conditions and releasing it in higher price settings, increasing economic efficiency (Ramkumaret al., 2025).

ESS also enhances grid reliability by providing back-up power during blackouts and enabling a black start. Black start refers to starting a power station without a power supply, which is needed to restore power after blackouts. ESS can supply power for this, speeding up recovery.

1.4.3 Distributed systems and electric mobility

Electric mobility is a major transition in the energy and transport sectors, and ESSs are key to this. Batteries are the primary power source for EVs, enabling zero-emission driving and reducing GHGs emissions from transport. Batteries also enable better range, efficiency, and affordability for EVs, which has led to their rapid adoption around the world. Batteries also assist in EV charging. Fast-charging stations can be battery-intensive. ESS, combined with charging stations, can be used to smooth out these peaks and valleys and help to stabilize the grid. This helps to balance the grid and enables charging stations to be deployed in areas with limited grid capacity (Jose and Shrivastava, 2024).

V2G is a new application of EVs as storage. In this case, EV batteries may be employed to supply energy to the grid when it is in need and to store energy when the grid has excess capacity. This energy bidirectional flow enhances grid flexibility and provides opportunities for energy management and revenue opportunities. Energy storage is also important for distributed energy systems, such as microgrids and prosumer networks. These systems allow consumers to produce and manage energy, often using rooftop photovoltaic (PV) systems and storage technologies. Energy storage enables these prosumers to store excess energy, reduce grid dependency and enhance self-reliance (Sovacool and Hirsh, 2009).

Energy storage is used in microgrids, which can operate independently or in grid-connected mode, to supply power. Energy storage provides load balancing, frequency regulation, and back-up power, enabling seamless grid-connected and islanded operation. This is critical for load operations such as hospitals, the military, and data centers. Finally, the deployment of ESS in distributed systems allows for the development of smart grids, which are able to manage energy using advanced communication and control technologies. ESSs are crucial for these, as they enable real-time energy management, improve efficiency, and allow the adoption of RES. In summary, ESS plays a key role in various applications in energy systems. From RE integration, grid stability, electric transportation, to distributed energy systems, ESS provides the flexibility and resilience required for the energy transition. Further development and implementation will be essential for smart and sustainable energy systems (Obakhume and Opatola, 2025).

1.5 Challenges, modeling, and future perspectives

ESSs are now increasingly playing a key role in modern energy systems, particularly in RE integration, electrification, and decarbonization. Although major advances have been achieved and increasing installation in a range of applications, some challenges are still present. The challenges include performance evaluation, model uncertainties, material sustainability, economic sustainability, and integration. Recent research indicates that to solve these issues, a holistic and interdisciplinary approach is required, taking into account innovations in materials, systems engineering, data, and policy modeling. Meanwhile, technology development suggests a paradigm shift to smart, hybrid, and long-duration ESS, which will be the future of energy systems.

1.5.1 Performance assessment and models

Performance assessment and modeling are crucial to design, optimize, and integrate EST. As modeling methods have evolved, from empirical to complex multiphysics and empirical models. These approaches can be expanded into empirical models, physics-based models, hybrid or data-driven models, with different levels of accuracy and efficiency of the models (Li and Deusen, 2025). Simpler empirical and equivalent circuit models (ECMs) are commonly employed owing to their efficiency and simplicity. These models represent ESS using electrical components such as resistors, capacitors, and current or voltage sources. However, they are limited in terms of providing detailed information about the processes within the system, degradation, and thermal issues (Kang et al., 2024).

Physics-based models provide a greater level of detail for ESS by taking into account the fundamentals of electrochemical reactions, mass transport, and heat transfer. They can include fine details in the system, like ion diffusion, reaction rate, or heat generation. Consequently, they are invaluable for predicting the performance, safety, and design. But their computational demands can be a barrier to real-time and large-scale use. Data-driven modeling is on the rise, particularly involving machine learning and artificial intelligence (Ali et al., 2025; Mohammed et al., 2026). These models learn from big data and predict system behavior. These are used for state estimation, fault detection, prognostics, and efficiency optimization (Tian et al., 2025).

Lastly, the concept of digital twins—digital replicas of physical ESS—is also emerging. Digital twins enable real-time monitoring, simulation, and optimization by integrating real operating data and models. This approach allows for predictive maintenance, enhanced reliability, and real-time decision-making under various scenarios. Whole system modeling does not only include ESS but also grid interconnections, market dynamics, and multienergy systems. Energy management and planning techniques such as mixed-integer linear programming, dynamic programming, and stochastic optimization are typically employed. These methods consider variability in RE generation, demand, and market prices, enabling adaptable and robust operations (Ustolin et al., 2020). But a trade-off between model accuracy and speed is needed. In addition, the lack of standard modeling paradigms and verification/validation practices makes it difficult to compare and reproduce results. The next frontier of energy storage research is to develop scalable, interoperable, and multiscale modeling platforms to appropriately represent integrated energy systems.

1.5.2 Technical and economic issues

There exist several technical and economic challenges to be solved to enable large-scale deployment of EST, which may provide statistically and economically viable systems. From a technical perspective, the availability of materials, degradation, safety, and integration issues are key. Sustainability and availability of materials are key issues. Lithium, cobalt, and rare earths are used in many new EST. The extraction and processing of these materials present environmental, resource availability, and geopolitical challenges. And the recycling and life cycle of storage systems are not fully developed, adding to environmental and economic risks (Njemaet al., 2024).

Life cycle and degradation are important factors that impact reliability and costs. Storage systems suffer from capacity fade, increased internal resistance, and efficiency degradation due to chemical, thermal, and mechanical degradation processes. These processes are influenced by temperature, charge/discharge rates, and depth of discharge. Degradation modeling and mitigation strategies are required to extend the lifetime and reduce the levelized cost of the energy system (LCOES) (Boa Morteet al., 2023).

Safety is an important consideration, particularly for electrochemical systems. Concerns include thermal runaway, fire, and leakage of the electrolyte, which demands advanced battery management and safety systems. Safely operating in various dynamic environments is essential to build trust and confidence (Lyuet al., 2020).

Energy system integration is also complex. The dynamic nature of ESS and variable nature of renewables pose challenges for grid management. Hybrid storage systems employing multiple technologies add to the complexity of system design, control, and optimization (Mishra et al., 2022).

There are also economic concerns despite the substantial reduction in the cost of EST; high capital costs remain an impediment. Apart from capital cost, the cost of operation and maintenance must be considered. The regulatory and market environment and revenue streams influence the economic feasibility of the storage system (Liu and Zhang, 2025).

Part of the problem is the lack of integrated market structures that value the complexities of services offered by ESS. These services include energy arbitrage, frequency regulation, capacity support, and grid regulation. These are not always valued in the market appropriately, limiting investment. This is coupled with regulatory uncertainty and inconsistencies.

Technology readiness and scalability. Some technologies (e.g., pumped hydro storage and lithium-ion batteries) are commercially mature, while others are in research or pilot stage. Moving from lab-scale to large-scale deployment of R&D can be costly and requires funding, infrastructure and policy mechanisms.

1.5.3 Potential developments

There are many disruptive developments for the future of ESS to overcome the challenges and to create new uses. A prime area of interest is the development of next-generation EST based on new materials. Scientists are working on metal-air batteries, solid-state batteries, and new flow batteries to provide greater energy density, safety, and durability. And there is a strong interest in hybrid ESS. These technologies combine different EST, such as batteries and supercapacitors, to have the benefits of both, such as higher power as well as energy density. This enables better performance for a diverse range of applications, particularly RE storage and energy management (He and Zhang, 2022).

Digitalization is increasingly crucial to the development of EST. AI, big data, and IoT technologies are used to enable advanced monitoring, maintenance, and optimization. Intelligent energy management systems can automatically adjust system parameters to improve efficiency, cost, and longevity (Prasshanth et al., 2024).

ESSs, which can store energy for longer periods, are also gaining attention to overcome the time mismatch between RE generation and use, long-duration (hours to months) ESS, thermal energy storage, and chemical energy storage (e.g., hydrogen).

We are also seeing an increased focus on sector coupling, which uses EST to link the electricity sector with the heating, cooling, and transport sectors. Power-to-X technologies will be applied to convert electricity into chemical fuels, creating new energy storage and utilization opportunities. This enhances flexibility and makes decarbonization of hard-to-abate transport sectors possible (Burreet al., 2020).

EST is increasingly considering environmental sustainability and the circular economy. The research focuses on using sustainable materials, improving recyclability, and reducing the environmental footprint of energy storage solutions. Life cycle assessment tools are being developed to assess the overall impact. Policy and regulations are also crucial for energy storage. Institutions and policies play a vital role in offering subsidies, reforms in the market, and research funding to support the adoption of energy storage. Bylaws, codes, and regulations are crucial in reducing investment risks and promoting innovation.

In conclusion, despite the present challenges of performance, cost, and system integration, the ongoing developments in modeling, materials, and system design are resulting in rapid progress. The combination with digital smart grids, hybrid technologies, and thermal storage is expected to bring about a transition in the RES toward a flexible, sustainable, and efficient energy future.

1.6 Conclusion

Energy storage systems are employed in the transformation of energy systems toward sustainability, reliability, and efficiency. The discussion in this chapter highlights that energy storage is no longer an auxiliary technology but rather an essential element to enable high RE integration, grid stability, and electrification of multiple sectors. The diversity of the EST, such as electrochemical, mechanical, chemical, electrical, and thermal storage, demonstrates that there is no "one size fits all"solution to the application needs, hence, the need for context-specific and holistic solutions. The conversations about the fundamental principles and performance metrics point out the importance of understanding the trade-offs between energy/power density, efficiency, and life-cycle characteristics in the selection of storage technologies. As well, the evolution of storage modeling and performance evaluation suggests a clear trend toward hybrid and data-driven techniques, with which storage systems can be predicted, optimized, and controlled in real time. Despite these advancements, there exist many technical and economic challenges, such as the sustainability of the materials, degradation of the system, safety concerns, and higher costs. These challenges can be addressed by a collective approach of research, policy, and industrial development. The emerging trends, such as digitalization, hybrid ESS, long-duration storage technologies, and sector coupling, are good opportunities to overcome the current challenges.

AI disclosure

During the preparation of this work the author(s) used Grammerly in order to perform grammar refinement, clarity enhancement, and language editing. After using this tool/service, the author(s) reviewed and edited the content as needed and take(s) full responsibility for the content of the publication.

References

Adetokun, B.B., Oghorada, O., Abubakar, S.J.a., 2022. Superconducting magnetic energy storage systems: prospects and challenges for renewable energy applications. J. Energy Storage 55. https://doi.org/10.1016/j.est.2022.105663.

Ali, Q., Ali, U.F., Asghar, T., Khan, T.A., Fatima, N., 2025. A hybrid machine learning and SCAPS 1D expedition for optimization of lead-free $BeSiP_2$ perovskite solar cells. Maejo Int. J. Energy Environ. Commun. 7 (3), 1–13. https://doi.org/10.54279/mijeec.v7i3.261848.

Balali, Y., Stegen, S., 2021. Review of energy storage systems for vehicles based on technology, environmental impacts, and costs. Renew. Sustain. Energy Rev. 135, 110185. https://doi.org/10.1016/j.rser.2020.110185.

Boa Morte, I.B., Araújo, O.d.Q.F., Morgado, C.R.V., de Medeiros, J.L., 2023. Electrification and decarbonization: a critical review of interconnected sectors, policies, and sustainable development goals. Energy Storage Saving 2 (4), 615–630. https://doi.org/10.1016/j.enss.2023.08.004.

Burre, J., Bongartz, D., Brée, L., Roh, K., Mitsos, A., 2020. Power-to-X: between electricity storage, e-production, and demand side management. Chem.-Ing.-Tech. 92 (1–2), 74–84. https://doi.org/10.1002/cite.201900102.

Chen, H., Cong, T.N., Yang, W., Tan, C., Li, Y., Ding, Y., 2009. Progress in electrical energy storage system: a critical review. Prog. Nat. Sci. 19 (3), 291–312. https://doi.org/10.1016/j.pnsc.2008.07.014.

Chougale, A., Shelke, H.D., Prasad, B., Jadkar, S.R., Naik, N., Pathan, H.M., Shinde, D.R., 2023. Simulation of p-CdTe and n-TiO2 heterojunction solar cell efficiency. J. Comput. Mech. Manag. 2 (1), 35–42. https://doi.org/10.57159/gadl.jcmm.2.1.23036.

Eftekhari, A., 2017. Lithium-ion batteries with high rate capabilities. ACS Sustain. Chem. Eng. 5 (4), 2799–2816. https://doi.org/10.1021/acssuschemeng.7b00046.

Farulla, G.A., Cellura, M., Guarino, F., Ferraro, M., 2020. A review of thermochemical energy storage systems for power grid support. Appl. Sci. 10 (9). https://doi.org/10.3390/app10093142.

Haas, J., Prieto-Miranda, L., Ghorbani, N., Breyer, C., 2022. Revisiting the potential of pumped-hydro energy storage: a method to detect economically attractive sites. Renew. Energy 181, 182–193. https://doi.org/10.1016/j.renene.2021.09.009.

He, X., Zhang, X., 2022. A comprehensive review of supercapacitors: properties, electrodes, electrolytes and thermal management systems based on phase change materials. J. Energy Storage 56, 106023. https://doi.org/10.1016/j.est.2022.106023.

Hussain, A., Huseynova, A., Hakimova, Y., Nassani, A.A., Bo, B., 2025. Energy efficiency and emission flexibility: management and economic insights for renewable energy integration. Energy Strategy Rev. 57. https://doi.org/10.1016/j.esr.2024.101631.

Jose, A., Shrivastava, S., 2024. Evolution of electrical vehicles, battery state estimation, and future research directions: a critical review. IEEE Access 12, 158627–158646. https://doi.org/10.1109/access.2024.3482698.

Kang, D.H., An, J.H., Lee, C.J., 2024. Numerical modeling and optimization of thermal insulation for liquid hydrogen storage tanks. Energy 291. https://doi.org/10.1016/j.energy.2023.130143.

Karden, E., Ploumen, S., Fricke, B., Miller, T., Snyder, K., 2007. Energy storage devices for future hybrid electric vehicles. J. Power Sources 168 (1), 2–11. https://doi.org/10.1016/j.jpowsour.2006.10.090.

Kumar Jha, A., Darlami, H.B., Bhattarai, N., Karn, S., Neupane, G., 2025. Projecting energy demand and GHGreduction with electric vehicle adoption in Nepal. Appl. Eng. Lett.J. Eng. Appl. Sci. 10 (3), 160–170. https://doi.org/10.46793/aeletters.2025.10.3.4.

Lakshmi, K.C.S., Vedhanarayanan, B., 2023. High-performance supercapacitors: a comprehensive review on paradigm shift of conventional energy storage devices. Batteries 9 (4). https://doi.org/10.3390/batteries9040202.

Li, Z., Deusen, D., 2025. Role of energy storage technologies in enhancing grid stability and reducing fossil fuel dependency. Int. J. Hydrogen Energy 102, 1055–1074. https://doi.org/10.1016/j.ijhydene.2024.12.489.

Liu, Y., Zhang, Y., 2025. Economic and environmental assessment of different energy storage methods for hybrid energy systems. Sci. Rep. 15 (1). https://doi.org/10.1038/s41598-025-09732-2.

Lyu, P., Liu, X., Qu, J., Zhao, J., Huo, Y., Qu, Z., Rao, Z., 2020. Recent advances of thermal safety of lithium ion battery for energy storage. Energy Storage Mater. 31, 195–220. https://doi.org/10.1016/j.ensm.2020.06.042.

Matos, M., Almeida, J., Gonçalves, P., Baldo, F., Braz, F.J., Bartolomeu, P.C., 2024. A machine learning-based electricity consumption forecast and management system for renewable energy communities. Energies 17 (3). https://doi.org/10.3390/en17030630.

May, G.J., Davidson, A., Monahov, B., 2018. Lead batteries for utility energy storage: a review. J. Energy Storage 15, 145–157. https://doi.org/10.1016/j.est.2017.11.008.

McGillicuddy, R.D., Thapa, S., Wenny, M.B., Gonzalez, M.I., Mason, J.A., 2020. Metal-organic phase-change materials for thermal energy storage. J. Am. Chem. Soc. 142 (45), 19170–19180. https://doi.org/10.1021/jacs.0c08777.

Mishra, S., Saini, G., Saha, S., Chauhan, A., Kumar, A., Maity, S., 2022. A survey on multi-criterion decision parameters, integration layout, storage technologies, sizing methodologies and control strategies for integrated renewable energy system. Sustain. Energy Technol. Assess. 52, 102246. https://doi.org/10.1016/j.seta.2022.102246.

Mohammed, A., Singh, R.S., Aslam, S., 2026. Humidity-aware hybrid transformer-LSTM framework for IoT-enabled photovoltaic power prediction. J. Comput. Mech. Manag. 5 (2), 25–38. https://doi.org/10.57159/jcmm.5.2.25394.

Njema, G.G., Ouma, R.B.O., Kibet, J.K., Ponnusami, V., 2024. A review on the recent advances in battery development and energy storage technologies. J. Renew. Energy 2024, 1–35. https://doi.org/10.1155/2024/2329261.

NREL, 2024. Industrial Energy Storage Review Report.

Obakhume, K.A., Opatola, F.F., 2025. Enhancing renewable energy integration through energy storage and smart grid innovations: a systematic review. J. Emerg. Sci. Eng. 3 (2). https://doi.org/10.61435/jese.2025.e38.

Ovejas, V.J., Cuadras, A., 2019. Effects of cycling on lithium-ion battery hysteresis and overvoltage. Sci. Rep. 9 (1). https://doi.org/10.1038/s41598-019-51474-5.

Prasshanth, C.V., Lakshminarayanan, A.K., Ramasubramanian, B., Ramakrishna, S., 2024. Progress of machine learning in materials design for Li-Ion battery. Next Mater. 2. https://doi.org/10.1016/j.nxmate.2024.100145.

Qiu, Y., 2023. Application trend analysis of compressed air energy storage and gas storage . Highlights in science. Eng. Technol. 33, 14–19. https://doi.org/10.54097/hset.v33i.5238.

Raman, N.S., Barooah, P., 2020. On the round-trip efficiency of an HVAC-Based virtual battery. IEEE Trans. Smart Grid 11 (1), 403–410. https://doi.org/10.1109/TSG.2019.2923588.

Ramaraj, R., Murugesan, S., Unpaprom, Y., 2025. Recent advances in microalgae-based biohydrogen production: a review of innovative strategies, pretreatment, genetic engineering, and bioelectrochemical technologies. Maejo Int. J. Energy Environ. Commun. 7 (1), 42–56. https://doi.org/10.54279/mijeec.v7i1.262587.

Ramkumar, M.S., Subramani, J., Sivaramkrishnan, M., Munimathan, A., Michael, G.K.O., Alam, M.M., 2025. Optimal energy management for multi-energy microgrids using hybrid solutions to address renewable energy source uncertainty. Sci. Rep. 15 (1). https://doi.org/10.1038/s41598-025-90062-8.

Ren, Z., Li, H., Yan, W., Lv, W., Zhang, G., Lv, L., Sun, L., Sun, Z., Gao, W., 2023. Comprehensive evaluation on production and recycling of lithium-ion batteries: a critical review. Renew. Sustain. Energy Rev. 185, 113585. https://doi.org/10.1016/j.rser.2023.113585.

Senthilkumar, N., Yuvaperiyasamy, M., Deepanraj, B., Kavitha, S., 2024. Performance analysis of pyramid solar still with natural banana fibers and Kanche marbles. Interactions 245, 148. https://doi.org/10.1007/s10751-024-01999-7.

Serarslan, B., 2025. A review and case study about the impact of temperature and ageing on electric vehicle batteries. Int. J. Autom. Eng. Technol 14 (4), 229–244. https://doi.org/10.18245/ijaet.1519648.

Shafie, S.M., Nu'man, A.H., 2024. Energy storage integration for renewable energy supply in Malaysia: a review. Int. J. Energy Econ.Policy 14 (5), 242–250. https://doi.org/10.32479/ijeep.16242.

Sovacool, B.K., Hirsh, R.F., 2009. Beyond batteries: an examination of the benefits and barriers to plug-in hybrid electric vehicles (PHEVs) and a vehicle-to-grid (V2G) transition. Energy Policy 37 (3), 1095–1103. https://doi.org/10.1016/j.enpol.2008.10.005.

Srinivasarao, P., Mohamed, M.R., 2024. A Comprehensive Review of Electric Vehicle Energy Management Techniques, Optimizations, and Controllers. … Communication.

Thirumalvalavan, S., Thanikasalam, A., Senthilkumar, N., Sabar, K., Yuvaperiyasamy, M., 2025. Investigation of the emission and performance characteristics of a direct injection diesel engine fuelled with dual biodiesel blends. Adv. Eng. Lett. 4 (3), 105–117. https://doi.org/10.46793/adeletters.2025.4.3.1.

Tian, X., Huang, Z., Gong, Y., Li, H., Kong, Y., 2025. Physics-based data-driven proxy model for geothermal energy storage system. J. Energy Storage 132, 117619. https://doi.org/10.1016/j.est.2025.117619.

Ustolin, F., Lamb, J.J., Burheim, O.S., Pollet, B.G., 2020. Energy and safety of hydrogen storage. Hydr. Biomass Bioenergy 133–153. https://doi.org/10.1016/B978-0-08-102629-8.00008-6.

Wang, H., Zhao, W., 2023. Application of hydrogel for energy storage and conversion. Next Mater. 1 (4), 100049. https://doi.org/10.1016/j.nxmate.2023.100049.

Wang, C., Ma, Y., Xie, J., Ouyang, Q., 2025. Multi-objective energy dispatch with deep reinforcement learning for wind–solar–thermal–storage hybrid systems. J. Energy Storage 105, 114635. https://doi.org/10.1016/j.est.2024.114635.

Wu, T., He, J., Zou, J., Yao, Y., Yan, X., Yang, A., 2024. Analysis and detection of charge and discharge characteristics of lithium battery based on multi-sensor fusion. Electr. Eng. 106 (3), 2627–2636. https://doi.org/10.1007/s00202-023-02083-z.

Xu, K., Guo, Y., Lei, G., Zhu, J., 2023. A review of flywheel energy storage system technologies. Energies 16 (18), 6462. https://doi.org/10.3390/en16186462.

Zhang, X., Zhu, J., Sahraei, E., 2017. Degradation of battery separators under charge–discharge cycles. RSCAdv. 7 (88), 56099–56107. https://doi.org/10.1039/c7ra11585g.

Fundamentals of energy storage

2

Onur Karaman, Ali Ihsan Komur and Ceren Karaman
Akdeniz University, Antalya, Turkey

2.1 Introduction

Energy storage has emerged as a pivotal and unavoidable supplementary of the modern energy system, playing a critical role in its functioning and evolution (Kalair et al., 2021; Qin et al., 2025). This concept the bridge between flexible supply and dynamic request, among electrons generated and energy services delivered (Kömür et al., 2025). As societies progress on electrification of transportation, digitizing industries, and integration of renewable energies, storage is becoming a foundational capability instead of a peripheral accessory (Horzela-Miś and Semrau, 2025). In addition, it fosters grid reliability, improves entity utilization, facilitates the refinement of novel market designs, and paves the way for substantial decarbonization (Qin et al., 2025; Horzela-Miś and Semrau, 2025). Therefore, understanding how storage works-physically, chemically, and economically-is essential for researchers, engineers, planners, policymakers, and entrepreneurs alike.

Energy storage technologies may be broadly classified according to the physical or chemical form in which energy is stored, as schematically summarized in Fig. 2.1. The fundamental principle of energy storage includes converting energy from a primary form into a storable medium, retaining it with minimal energy loss, and reconverting it into useful work or electricity when required (Mitali et al., 2022; Amir et al., 2023). This simple idea assists a considerably diverse range of technologies. For instance, electrochemical systems, including lithium-ion batteries, exhibit quick response and modular implementation, commonplace in practices from smartphones to utility-scale energy storage facilities (Rouholamini et al., 2022; Singh et al., 2025). Mechanical storage systems, such as pumped hydro, compressed air, liquid air, and flywheels, use mass and motion to store large quantities of energy or provide rapid bursts of power (Elalfy et al., 2024; Mahadevan et al., 2025). Thermal storage, both sensible and latent, buffers heat and cold to decouple generation from utilization in buildings and industry, while emerging thermochemical process promise higher energy densities and longer durations (Bao and Ma, 2022). On the other hand, chemical carriers including hydrogen, ammonia, and synthetic fuels broaden storage from hours to seasons and across sectors, connecting electricity with transport, heavy industry, and heating (Müller et al., 2024; Lan et al., 2024).

Each storage path is characterized by a balanced set of key performance characteristics, and power capacity defines instantaneous output (Mahadevan et al., 2025). Additionally, energy capacity identifies the discharge time (Mahadevan et al., 2025). The concept of round-trip efficiency take possession of conversion losses,

Energy Storage Technologies. https://doi.org/10.1016/B978-0-443-45858-3.00005-X

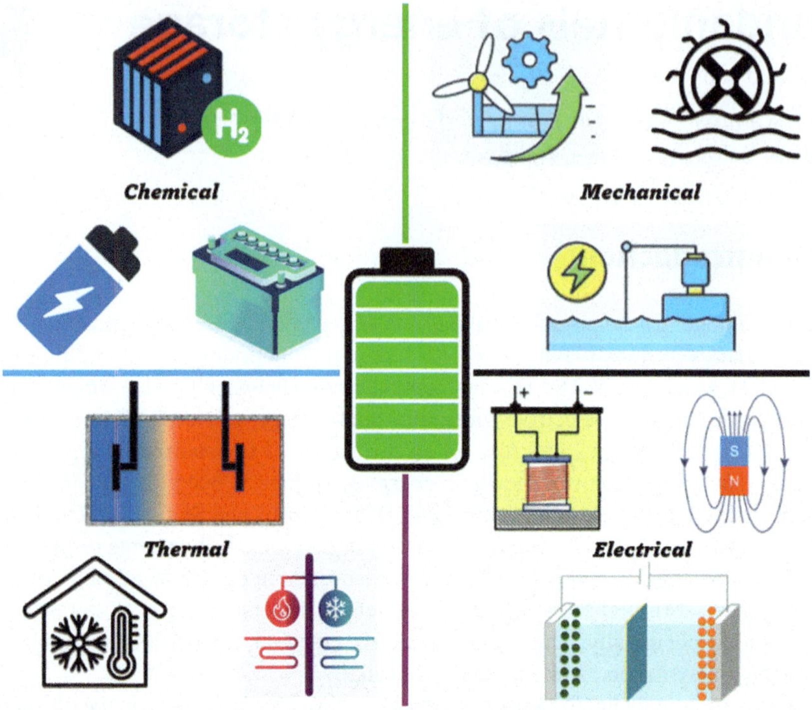

Figure 2.1 A visual overview of energy storage technologies. Schematic classification of energy storage systems based on the physical and chemical form in which energy is stored. The diagram highlights the main categories of chemical, mechanical, electrical, and thermal energy storage mechanisms and illustrates their conceptual relationships.

whereas response time, cycle life, and self-discharge affect operational value and require maintenance (Cesaro et al., 2020). Furthermore, security attributes, environmental footprints, and supply chain limitations shape practicality and social license. Eventually, costs—capex, operational expenditure, and degradation—must be evaluated entirely by the lens of applications and revenue streams, rather than in isolation. Because technology designed for millisecond response in frequency regulation is likely to differ considerably from system optimized for days-long power transitions throughout wind calms or winter peaks (Hassan et al., 2023).

The factors influencing the adoption of energy storage adoption are varied and span the supply, demand, grid, and market dimensions. On the supply side, changeable renewable energy introduces temporal mismatches, which storage can smooth, turning excess midday solar into evening capacity or storing wind formed overnight for morning ramps (Enasel and Dumitrascu, 2025). From a demand-side perspective, electrification adds novel load profiles—from fast-charging depots to industrial heat pumping systems—that benefit from storage for demand charge management, resilience, and power quality behind the meter (Boampong and Brown, 2020). At the grid standard, storage has the capacity to defer transmission and distribution boosts, ensure services analogous to inertia, and improve black-start capabilities (Zerrahn et al., 2018). In markets, it is able to arbitrate time-varying prices, stacks ancillary services, and supports capacity adequacy, maintain regulations recognize and compensate its multifunctionality (Zerrahn et al., 2018). In this context, historically, energy storage technologies have co-evolved in tandem with energy production methods. For instance, early forms such as pumped hydroelectric storage and compressed air energy storage (CAES) were improved to balance large-scale electrical grids. However, there have been rapid developments in electrochemical storage systems (particularly batteries and supercapacitors), in recent years, which have had a significant impact on portable electronic devices, electric vehicles, and distributed energy systems. Consequently, the fundamentals of energy storage technologies are experiencing a process of continuous renewal and diversity on a daily basis.

In view of the valuable information provided, this chapter aims to provide a comprehensive overview of the fundamental principles underlying energy storage. In this study, investigates electrochemical, mechanical, thermal and chemical technology classes, emphasizing their principles, performance metrics and latest trends. Then, starting with the physics and chemistry of storage media, energy density, power density and thermodynamic limits are clarified. Subsequently, methodologies are presented to evaluate techno-economics such as storage cost balancing, degradation-aware distribution, and value stacking. By the end of this chapter, readers ought to be able to distinguish between major storage technologies and their operating principles and interpret critical performance metrics and their benefits. Ultimately, the fundamentals of energy storage stand for not only scientific discipline but also a strategic catalyst for global energy sustainability. Researchers, engineers, and policymakers who master foundational concepts will be better able to navigate the complex landscape of such emerging energy technologies and thus contribute to a more resilient and low-carbon energy future.

2.2 Thermodynamic principles of energy storage

Thermodynamics is the branch of physical science that quantifies how energy is transformed, conserved, and degraded within materials and systems. It establishes the universal laws governing heat, work, chemical reactions, phase transitions, and energy conversion processes, thereby providing the theoretical foundation for analyzing all forms of energy storage (Gedde, 2020). In its broadest sense, thermodynamics characterizes the relationships among energy, entropy, temperature, pressure, and chemical potential. This enables prediction of which transformations are feasible, how much energy can be stored or recovered, and what portion of that energy is inevitably lost to irreversibility.

Energy storage processes whether rooted in electrochemical transformations, mechanical work, thermal enthalpy, or chemical reactions are fundamentally governed by the laws of thermodynamics. Because every storage technology involves charging, discharging, reconfiguration of matter, or heat transfer, thermodynamics becomes the primary framework.

Thermodynamics is traditionally structured around four principal laws including the Zeroth, First, Second, and Third Laws (Xu, 2026). These laws collectively define temperature as a measurable property, energy conservation as a universal constraint, entropy as a measure of irreversibility, and absolute zero as a limiting thermodynamic state (Xu, 2026). The Zeroth Law of Thermodynamics states that if two systems are each in thermal equilibrium with a third system, they are also in equilibrium with one another (Zohuri, 2018) (Fig. 2.2). Although concise, this principle provides the formal basis for defining temperature as a thermodynamic coordinate and underpins all heat-transfer and thermal-management considerations in energy storage (Honig, 2007; Zohuri, 2018). Its implications extend to the design of thermal energy storage media, the assessment of temperature gradients during battery cycling, and the evaluation of thermal stability across mechanical and chemical storage systems (Zohuri, 2018).

These thermodynamic laws govern a vast range of energy storage phenomena from redox reactions in batteries and ion mobility in electrolytes, to gas compression in mechanical storage, heat accumulation in molten salts, and bond formation in hydrogen carriers (AlShafi and Bicer, 2021; Ding, 2021). Thermodynamics describes the internal energy changes associated with charging, the entropy generation that reduces round-trip efficiency, the Gibbs free-energy changes that determine battery voltages, the chemical potential gradients that drive mass transport, and the exergy destruction that limits overall work recovery (Lijesh and Khonsari, 2024; Li et al., 2025).

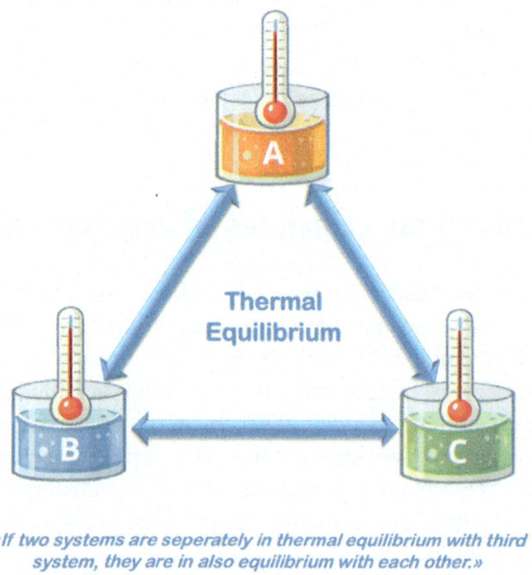

Figure 2.2 Illustration of the zeroth law of thermodynamics demonstrating thermal equilibrium among three systems. If two systems are each in thermal equilibrium with a third system, they are also in thermal equilibrium with one another, enabling the definition of temperature as a state variable.

In practical applications, thermodynamics serves multiple critical functions in energy storage science and engineering. It establishes the theoretical voltage and capacity limits of electrochemical cells, guides the design of materials with favorable reaction enthalpies or entropy profiles, and determines the feasibility of thermochemical heat-storage reactions (Ding, 2021; Hadidi, 2024). It also governs the efficiency boundaries of heat engines integrated with thermal storage, dictates how much work can be recovered from compressed air or elevated fluids, and identifies the conditions under which phase changes or chemical reactions will proceed reversibly or degrade system components (Ding, 2021; Hadidi, 2024). As storage technologies scale to support renewable integration, electric mobility, industrial heat, and grid stability, thermodynamic analysis becomes essential for identifying optimal operating strategies, minimizing losses, and ensuring safe and durable operation.

The following subsections translate these foundational principles into the context of modern energy storage systems, beginning with the First Law of Thermodynamics and its implications for energy conservation, followed by discussions on entropy generation, free-energy landscapes, chemical potential, and exergy. All of these shape the ultimate performance and feasibility of storage technologies.

2.2.1 The first law of thermodynamics and energy conservation

A rigorous analysis of energy storage begins with the general conservation of energy, which applies to all physical systems regardless of their scale or working principle. In its most comprehensive form, the energy balance for a control volume accounting for heat transfer, work interaction, and mass transport is expressed as in Eq. (2.1):

$$\frac{dE}{dt} = \dot{Q} - \dot{W} + \sum_{\text{in}} \dot{m}_{\text{in}} \left(h + \frac{V^2}{2} + gz \right) - \sum_{\text{out}} \dot{m}_{\text{out}} \left(h + \frac{V^2}{2} + gz \right) \tag{2.1}$$

where E represents total stored energy (internal + kinetic + potential), $\dot{Q}$ is heat transfer rate, $\dot{W}$ is the work, $\dot{m}$ stands for mass flow, h is enthalpy, V is velocity, g is the gravitational acceleration, and z is elevation.

This formulation includes contributions from internal, kinetic, and potential energies, as well as the enthalpy of streams crossing the system boundary (Zohuri, 2018; Iwase, 2014) Although essential for thermodynamic systems involving mass flow such as combustion chambers, turbines, or open heat engines, this full expression simplifies significantly when applied to most modern energy storage devices.

Electrochemical cells, supercapacitors, flywheels, thermal energy reservoirs, and compressed-air chambers commonly operate as closed systems during charge and discharge cycles (Matsui et al., 2022; Koohi-Fayegh and Rosen, 2020). This means mass does not cross their boundaries, volume remains essentially fixed, and changes in kinetic or potential energy are negligible compared to internal energy transformations (Matsui et al., 2022; Koohi-Fayegh and Rosen, 2020). Also, storage cycles are *quasi-steady* in terms of mass and volume, even though electrical conditions change over time (Sánchez and Williams, 2014; Jung et al., 2015). Under these constraints, the balance reduces to the classical closed-system form of the First Law expressed by Eq. (2.2) (Fig. 2.3)

$$\Delta U = Q - W \tag{2.2}$$

here, U represents the internal energy that is the total microscopic energy stored within a material or system. Q is heat transfer, and W stands for work interactions between the system and surroundings.

This simplified formulation of the energy conversion law is particularly relevant for energy storage because it distinguishes how energy is introduced, retained, and later recovered. During charging, electrical, mechanical, or thermal work is applied to the system, increasing its internal energy through redox transformations, pressure accumulation, thermal enthalpy increase, or rotational kinetic storage (Njema et al., 2024). Inefficiencies such as Joule heating in conductors, viscous dissipation in compressors, or heat leakage in thermal reservoirs appear as Q, which must be managed to maintain system stability and component longevity (Kandhasamy et al., 2024).

During discharge, the internal energy decreases as useful electrical or mechanical work is delivered to the external environment (Njema et al., 2024). The First Law provides the quantitative boundary for this conversion but does not specify the fraction of stored energy that is recoverable. This limitation is imposed by the Second Law, which governs entropy generation and irreversibility. Nonetheless, the First Law remains essential for determining the magnitude of internal energy changes associated with charge-discharge cycles and for constructing predictive models of system behavior under various loading conditions.

In electrochemical energy storage systems, the First Law governs the balance between electrical work performed on the cell during charging and the resulting increase

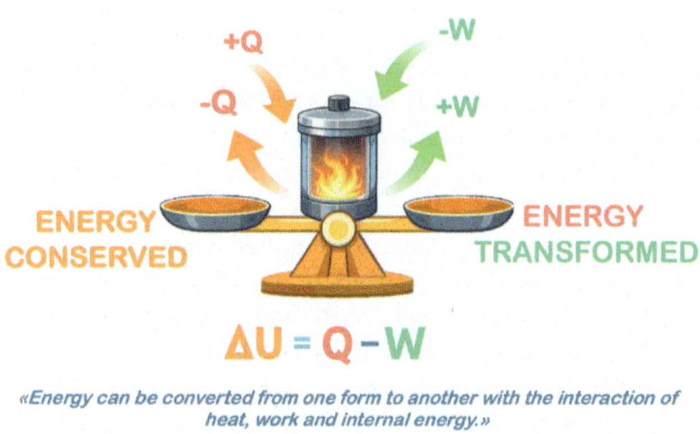

Figure 2.3 Schematic representation of the first law of thermodynamics for closed systems. Conceptual illustration of the First Law of Thermodynamics, depicting energy conservation within a system through the balance of heat exchange (Q) and work interactions (W). The diagram emphasizes that changes in internal energy (ΔU) arise from energy transferred as heat or work, highlighting energy transformation rather than creation or loss.

in chemical potential of active materials (Maier, 2013; Osara and Bryant, 2021). Electrical work input contributes partly to driving redox transformations (the desired storage pathway) and partly to parasitic losses such as ohmic heating, activation overpotentials, and mass-transport limitations (Barrera et al., 2024). These losses manifest as heat generation Q, which must be managed through thermal strategies to preserve material stability. Conversely, during discharge, the decrease in internal energy corresponds to electrical work delivered to an external circuit, again reduced by irreversible losses (Barrera et al., 2024). The First Law therefore provides a quantitative framework for distinguishing between reversible energy storage which is associated with changes in chemical free energy and irreversible dissipation, which diminishes round-trip efficiency.

In mechanical storage, internal energy manifests as gravitational, elastic, or rotational potential, while in thermal systems, it appears as changes in enthalpy or latent heat associated with temperature variations or phase transitions (Steinmann, 2022). Across all classes, Eq. (2.1) provides a unifying framework for energy accounting, enabling comparison of storage methods that operate through fundamentally different physical mechanisms. As an example of mechanical energy storage, in pumped-hydro energy storage (PHES), the applied electrical work during charging is used to elevate water to a higher reservoir. The stored potential energy is given by (Kandhasamy et al., 2024):

$$W_{\text{stored}} = mgz \tag{2.3}$$

Where m is the mass transfer of water, g is the gravitational acceleration, and z is the elevation difference between lower and upper reservoirs.

This potential energy is later converted back to electricity through turbine-generator assemblies. In real systems, only a fraction of the theoretical potential energy is recoverable due to hydraulic friction, turbulent flow losses, penstock resistance, and generator inefficiencies (Kandhasamy et al., 2024). These losses manifest as heat generation and energy degradation within the First Law framework.

A similar thermodynamic balance governs CAES. Here, mechanical energy is stored by compressing air in an underground cavern or high-pressure vessel. The theoretical work required for gas compression is expressed as:

$$W_{\text{stored}} = \int_{V_1}^{V_2} P dV \tag{2.4}$$

where P is the instantaneous gas pressure, V is system volüme, and V_1 and V_2 denote the initial and final volumes.

In practice, compression rarely follows an ideal isothermal or adiabatic path. Real compressors generate significant heat due to finite compression rates, friction, and aerodynamic losses. If this heat is not recovered (as in diabatic CAES), it represents an irreversible loss that lowers the recoverable energy. Advanced configurations, such as adiabatic CAES (A-CAES), attempt to capture and store the compression heat for reuse during discharge, improving thermodynamic performance (Dupin and Teixeira, 2025).

Thermal energy storage (TES) systems also adhere strictly to the First Law. When heat is supplied to a storage medium, its internal energy or enthalpy increases according to Eq. (2.5). This relation defines sensible heat storage, where energy is stored through temperature changes of liquids, solids, or molten salts.

$$\Delta H = \int_{T_1}^{T_2} C_p(T)\, \mathrm{d}T \tag{2.5}$$

where ΔH is the change in enthalpy (J), $C_p(T)$ is the temperature-dependent heat capacity at constant pressure $\left(\mathrm{J\ kg^{-1}\ K^{-1}}\right)$, T_1 and T_2 are the initial and final temperatures (K).

For latent heat storage, energy is stored through phase change (Eq. 2.6). Latent heat storage offers higher energy densities and near-isothermal behavior, beneficial for applications requiring stable output temperatures. However, phase segregation, subcooling, and thermal degradation introduce additional thermodynamic irreversibilities that must be carefully managed.

$$\Delta H_{\mathrm{latent}} = mL \tag{2.6}$$

where m is the mass of the phase-change material (kg), L is the latent heat of fusion, vapourization, or solid-solid transition $\left(\mathrm{J\ kg^{-1}}\right)$.

Although the First Law governs the amount of heat or work transferred into thermal energy storage media, it does not dictate the fraction of stored thermal energy that can be converted back into mechanical or electrical power (Saha and Hassan Rupam, 2023; Tripathi et al., 2023). That limitation emerges from the Second Law, which imposes temperature-dependent efficiency ceilings on heat engines (e.g., Carnot efficiency) and governs entropy generation due to heat loss, conduction through temperature gradients, and material irreversibilities (Barbir, 2012). According to the Second law, heat transfer occurs spontaneously from a hot body to a cold body, and it is impossible to convert heat completely into work (Fig. 2.4). Consequently, the total entropy of a system and its surroundings never decreases. Nevertheless, the First Law remains essential for modeling TES charge-discharge cycles, sizing storage materials, estimating auxiliary heating requirements, and quantifying thermal leakage over long durations (Jouhara et al., 2020).

Across all storage modalities, the First Law establishes the fundamental energy accounting framework *"energy can be stored, transformed, and released, but never created or destroyed."* The proportion of energy that can be returned as useful work depends not on energy conservation itself but on system-specific properties, design choices, and the irreversible losses governed by the Second Law. Consequently, the First Law provides the quantitative baseline upon which advanced analyses of entropy production, exergy destruction, reaction equilibria, and kinetic

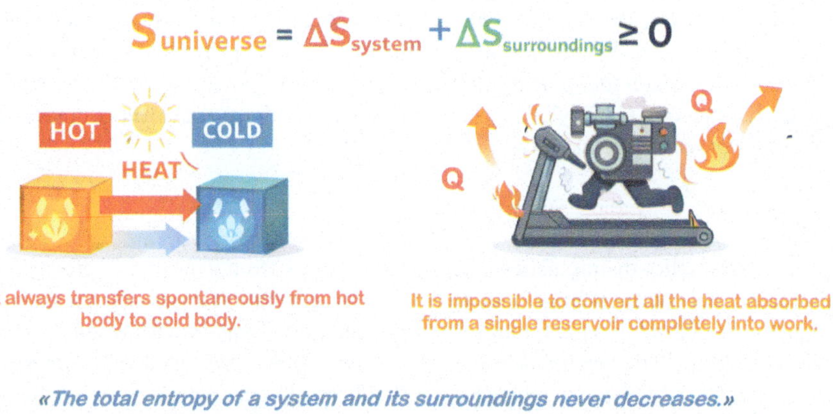

Figure 2.4 Schematic illustration of the second law of thermodynamics. Conceptual illustration of the Second Law of Thermodynamics, emphasizing entropy generation and irreversibility. The figure highlights the spontaneous transfer of heat from hot to cold reservoirs and the fundamental limitation that heat absorbed from a single reservoir cannot be completely converted into work, consistent with the requirement that total entropy production of the universe is nonnegative.

limitations must be constructed. Without this foundational understanding, neither efficiency evaluation nor predictive modeling of modern energy storage technologies can be rigorously undertaken.

2.2.2 The second law of thermodynamics: Entropy generation and irreversibility

While the First Law establishes the quantitative balance of energy within a storage system, it alone cannot explain why only a portion of the stored energy is recoverable as useful work. This limitation is addressed by the Second Law of Thermodynamics, which introduces "*entropy*" (S; kJ) as the governing variable that determines the directionality, feasibility, and quality of energy transformations.

In energy storage systems irreversibilities arise whenever processes deviate from ideal equilibrium conditions, leading to entropy production and a permanent degradation of the system's work potential. These irreversibilities originate from diverse mechanisms, including finite-rate heat transfer, ionic and molecular diffusion, viscous and frictional dissipation, charge-transfer kinetics, phase transitions, microstructural rearrangements, and parasitic side reactions (Horlock, 2003). Because modern storage technologies operate far from equilibrium during charge and discharge, entropy generation becomes an intrinsic and unavoidable consequence of practical operation (Huo et al., 2023; Qu et al., 2023). Understanding how entropy is created, transported, and accumulated is therefore essential for quantifying round-trip efficiency, predicting thermal behavior, diagnosing degradation pathways, and establishing the fundamental thermodynamic limits of storage performance.

The Second Law provides the conceptual and mathematical framework for analyzing how useful work is destroyed and how system design can suppress irreversibilities to approach theoretical efficiency limits. In its general form, the Second Law states that the total entropy of a system and its surroundings can never decrease during an actual process, expressed as

$$\Delta S_{\text{total}} = \Delta S_{\text{system}} + \Delta S_{\text{surroundings}} \geq 0 \tag{2.7}$$

This inequality distinguishes reversible processes from irreversible processes that dominate real energy storage operation. Entropy generation arises when a system deviates from thermodynamic equilibrium or when gradients in temperature, pressure, concentration, or electrical potential drive finite-rate transport phenomena (Struchtrup, 2020). Energy storage devices inherently operate under such nonequilibrium conditions, and therefore entropy production is unavoidable regardless of the storage modality (Struchtrup, 2020).

The sources of entropy generation within energy storage systems are multifaceted. Heat transfer across finite temperature differences produces irreversibility whenever thermal energy flows from hot to cold reservoirs, as quantified by Eq. (2.8).

$$\Delta S_{\text{gen}} = Q\left(\frac{1}{T_{\text{c}}} - \frac{1}{T_{\text{h}}}\right) \tag{2.8}$$

where Q is the heat transferred and T_{h} and T_{c} are the temperatures of the hot and cold reservoirs, respectively. This effect is particularly relevant in thermal energy storage systems, battery pack thermal management, and compressed-air systems in which temperature gradients are significant. Mechanical storage systems experience entropy generation through viscous dissipation, fluid friction, and turbulence, all of which degrade useful mechanical work into heat (Yaseen et al., 2025). In electrochemical systems, entropy is produced through Joule heating, charge-transfer overpotentials, concentration polarization, and mass transport limitations. The irreversible passage of current through resistive pathways contributes to entropy generation according to Eq. (2.9) (Long et al., 2026; Randriamahazaka, 2010).

$$\Delta S_{\text{gen,IR}} = \frac{I^2 R}{T} \tag{2.9}$$

where I is current, R is internal resistance, and T is temperature. Charge-transfer processes at electrodes further contribute to irreversibility. Activation overpotentials represent kinetic barriers that require additional driving force. And additional energy input. This is subsequently dissipated as heat rather than stored or delivered as useful work. Mass transport processes also generate entropy whenever concentration gradients exist, with the associated entropy production often expressed through flux-force relationships such as Eq. (2.10) (Ramírez-Minguela et al., 2018).

$$\Delta S_{\text{gen,diffusion}} \propto \int J_i \cdot \nabla \mu_i \, \mathrm{d}V \tag{2.10}$$

where J_i is the mass flux of species i and μ_i is its chemical potential. Diffusion, intercalation, desorption, and phase transformations in electrodes and electrolytes all involve such gradients and contribute to irreversible losses.

These various mechanisms collectively reduce the amount of useful work that can be recovered from a storage system. If W_{rev} denotes the maximum theoretical work obtainable in a reversible process, then the actual work output is diminished by the entropy generated during operation, captured by the relationship Eq. (2.11).

$$W_{actual} = W_{rev} - T_0\,\Delta S_{gen} \tag{2.11}$$

where T_0 is the ambient temperature. This expression highlights the direct thermodynamic penalty imposed by entropy generation. Every increment of irreversibility reduces the system's available work by an amount proportional to the environmental temperature. In thermal energy storage combined with power generation, the Second Law imposes an additional limitation through the Carnot efficiency (η_{Carnot}),

$$\eta_{Carnot} = 1 - \frac{T_{cold}}{T_{hot}} \tag{2.12}$$

which represents the theoretical maximum fraction of heat that can be converted to work. Any temperature degradation during storage or discharge lowers this limit, and entropy generation further restricts practical efficiencies (Long et al., 2026). In electrochemical systems, the Second Law explains why charge and discharge voltages differ from the equilibrium potential predicted by thermodynamics; overpotentials caused by kinetic and transport processes represent irreversibility, and the area between charge and discharge curves in a voltage-capacity profile corresponds to energy lost through entropy generation (Long et al., 2026).

Understanding entropy production is therefore essential for analyzing and improving the round-trip efficiency, thermal behavior, and long-term durability of energy storage systems. Reducing entropy generation directly enhances performance by preserving the useful work potential. The Second Law thus not only delineates the theoretical limits of energy recovery but also provides a framework for diagnosing inefficiencies, guiding materials development, and informing operational strategies. In combination with the First Law, it establishes a complete thermodynamic foundation for the study and advancement of modern energy storage technologies. Moreover, while entropy generation describes the presence and origin of irreversibility, its quantitative impact on recoverable work is more rigorously captured through the concept of exergy, discussed in the following Section 2.2.5.

2.2.3 Gibbs free energy and the limits of electrochemical storage

Electrochemical energy storage is fundamentally governed by the thermodynamics of chemical transformations occurring at the electrode-electrolyte interfaces (Huang, 2024). The central quantity controlling these transformations is the Gibbs free energy

(G; kJ), which quantifies the maximum nonexpansion work obtainable from a system at constant temperature and pressure (Huang, 2024). In the context of electrochemical cells, Gibbs free energy determines both the feasibility and the theoretical voltage of the redox reactions that serve as the basis for charge and discharge. When an electrochemical cell operates reversibly, the change in Gibbs free energy ΔG is directly related to the electrical work extracted through the transfer of electrons, expressed by Eq. (2.13) (Liu et al., 2016).

$$\Delta G = -nFE_{\mathrm{rev}} \tag{2.13}$$

where n is the number of moles of electrons transferred, F is Faraday's constant $\left(96485 \text{ C mol}^{-1}\right)$, and E_{rev} is the reversible cell potential (V). Eq. (2.13) establishes the fundamental thermodynamic limit of any electrochemical energy storage device. The maximum electrical energy obtainable per mole of reactants is constrained entirely by the free-energy difference between the initial and final chemical states.

Under equilibrium conditions, the reversible cell voltage depends not only on the intrinsic standard-state free energy but also on the activities of reactants and products (Liu et al., 2016). The Nernst equation expresses this dependence as following Eq. (2.14):

$$E = E^{\circ} - \frac{RT}{nF}\ln\left(\frac{a_{\mathrm{products}}}{a_{\mathrm{reactants}}}\right) \tag{2.14}$$

where R is the universal gas constant (8314.3 J/mol K), T is temperature (K), and $a_i$denotes the activity of species i. This relationship demonstrates that the chemical potential and thus the cell voltage varies continuously with composition, concentration, and phase state of the active materials. Intercalation electrodes, conversion materials, and metal-electrolyte interfaces each exhibit distinct voltage-composition profiles that reflect their underlying thermodynamic landscapes (Liu et al., 2016). The slope and curvature of these profiles are governed by the derivative of the Gibbs free energy with respect to composition, directly linking material thermodynamics to practical voltage characteristics.

Temperature also exerts a significant influence on electrochemical free energy. The Gibbs-Helmholtz equation provides the temperature dependence of the equilibrium potential:

$$\left(\frac{\partial E_{\mathrm{rev}}}{\partial T}\right)_P = \frac{\Delta S}{nF} \tag{2.15}$$

where ΔS is the entropy change of the cell reaction. This expression indicates that processes accompanied by positive reaction entropy exhibit increasing voltage with temperature, whereas reactions with negative entropy display the opposite trend. The sign and magnitude of ΔS therefore determine thermal sensitivity, influencing both operational stability and thermal management strategies. Electrochemical systems that undergo structural rearrangement, phase transitions, or solvation changes typically exhibit large entropy variations, thereby imposing thermodynamic constraints on temperature-dependent performance.

Although the reversible free-energy change sets the theoretical upper bound on energy storage, real systems seldom operate at this reversible limit (Haynes, 2001). Overpotentials arising from activation barriers, ohmic resistance, and diffusion gradients cause the operating voltage to deviate from E_{rev}, producing entropy and reducing the work extractable during discharge. The First and Second Laws together describe the consequences that internal energy must increase during charging. However, entropy generation during both charge and discharge diminishes the portion of stored free energy convertible to useful work. The actual electrical work delivered W_{actual} is therefore

$$W_{actual} = -\Delta G - T_0 \Delta S_{gen} \tag{2.16}$$

where T_0 is the environmental reference temperature and ΔS_{gen} is entropy generated through irreversibilities. Eq. (2.16) formalizes the thermodynamic penalty imposed by finite reaction rates, resistive pathways, and mass-transport limitations, all of which degrade exergy and impose practical limits on the efficiency and power capability of electrochemical storage.

Gibbs free energy also governs the thermodynamic stability of electrode-electrolyte interfaces (Wang et al., 2025). Electrolyte decomposition, surface film formation (such as the solid-electrolyte interphase), and gas evolution occur when electrode potentials exceed the stability window defined by the electrolyte's free-energy limits (Wang et al., 2025). These reactions lower system efficiency by consuming electrolyte constituents and irreversibly modifying the interfacial environment, further distancing practical operation from ideal thermodynamic behavior. Thus, free-energy considerations are central not only to defining equilibrium voltages but also to constraining safe and stable operating conditions.

In summary, Gibbs free energy provides the thermodynamic foundation for understanding the maximum energy a cell can store or deliver, the voltage profile across composition and temperature, and the intrinsic stability of electrochemical interfaces. While kinetics and transport phenomena ultimately determine operational performance, it is the free-energy landscape that establishes the theoretical ceiling of electrochemical storage technologies. Recognizing these limits is essential for evaluating new electrode materials, predicting cell behavior under varying thermodynamic conditions, and interpreting mechanistic phenomena discussed in subsequent sections.

2.2.4 Chemical potential and energy storage systems

Chemical potential is one of the most fundamental thermodynamic quantities governing the behavior of energy storage materials. It represents the change in a system's Gibbs free energy with respect to the addition of an infinitesimal amount of a chemical species, holding temperature, pressure, and all other species constant. Formally, for species i, the chemical potential ($\mu_i; J/mol$) is defined as

$$\mu_i = \left(\frac{\partial G}{\partial n_i}\right)_{T,P,n_{j\neq i}} \tag{2.17}$$

where G is the Gibbs free energy and n_i is the number of moles of the species. This definition captures the intrinsic thermodynamic driving force for mass transport, phase transformations, and redox reactions. In the context of energy storage systems, the chemical potential dictates how species distribute themselves within the system and how energy is stored, transferred, or released through compositional changes.

In electrochemical energy storage materials, the chemical potential determines the equilibrium voltage and the direction of ionic migration (Dulgerbaki et al., 2016; Boettcher et al., 2021). Differences in chemical potential across electrodes create an electromotive force, and the measurable cell voltage is directly proportional to the difference in the electrochemical potentials of electrons in the two electrodes (Boettcher et al., 2021). When ions intercalate into or deintercalated from host structures, the chemical potential of the intercalant varies with concentration, occupancy of crystallographic sites, and interactions among inserted species (Kömür et al., 2025; Liu et al., 2016; Boettcher et al., 2021). These variations manifest as composition-dependent voltage plateaus or slopes in electrochemical profiles. A steep change in chemical potential may correspond to a phase transition, while a smooth variation reflects solid-solution behavior (Liu et al., 2016; Boettcher et al., 2021). Such thermodynamic insights are vital for evaluating energy density, predicting voltage hysteresis, and understanding the intrinsic limitations of active materials.

Chemical potential also governs diffusion and mass transport within energy storage media. Fick's diffusion, ion migration, and concentration polarization all arise from gradients in chemical potential. The flux of species i, J_i, the molar rate at which species $i$moves per unit area, is driven by gradients in chemical potential and expressed as (Boettcher et al., 2021; Ferguson, 2012):

$$J_i = -\frac{D_i}{RT} c_i \nabla \mu_i \tag{2.18}$$

where D_i is diffusivity of species i, and c_i its concentration. This equation highlights that mass transport is fundamentally driven by spatial variations in chemical potential rather than concentration alone. As a result, diffusion slows dramatically when chemical potential gradients diminish, impacting power capability and contributing to polarization losses. In storage systems such as lithium-ion batteries, metal hydrides, solid electrolytes, and redox-flow systems where ion motion or mass redistribution is essential, the chemical potential landscape governs both kinetics and thermodynamic limits.

In thermal energy storage systems, chemical potential plays a critical role in materials undergoing phase transitions or reversible thermochemical reactions. Latent heat storage relies on the coexistence of phases with distinct chemical potentials, and the equilibrium temperature for melting or solidification is determined by conditions that equalize their chemical potentials (Salunkhe and Jaya Krishna, 2017). Similarly, in thermochemical storage systems employing reversible decomposition or hydration reactions, the chemical potential of reactants and products controls the equilibrium extent of reaction and therefore the temperature and pressure conditions under which energy can be absorbed or released (Salunkhe and Jaya Krishna, 2017). For such systems, the Gibbs free energy change and thus the chemical potentials govern not only energy density but also practical operating windows and stability against undesired side reactions.

Mechanical-chemical hybrid systems, including hydrogen storage in metal hydrides or adsorption-based storage media, depend critically on the chemical potential of gas-phase species (Bae et al., 2017). The driving force for hydrogen absorption into a solid is the gradient between the chemical potential of hydrogen in the gas phase and that in the solid host (Bae et al., 2017). The pressure-composition-temperature (PCT) behavior observed in these systems reflects equilibrium conditions where chemical potentials are balanced (Xiao et al., 2023). Deviations from equilibrium introduce hysteresis and entropy production, limiting round-trip efficiency and dictating material design strategies.

The chemical potential also provides a unifying thermodynamic interpretation of degradation mechanisms in storage media. Processes such as phase segregation, electrolyte decomposition, morphological instability, and interfacial film formation occur when local chemical potentials exceed stability thresholds (Fang et al., 2025; Jeon et al., 2025). For example, in electrochemical systems, if the chemical potential of electrons at the electrode surface lies outside the electrochemical stability window of the electrolyte, reduction or oxidation reactions spontaneously initiate, consuming electrolyte constituents and forming passivation layers (Fang et al., 2025; Jeon et al., 2025). These reactions lower system efficiency and reduce the amount of free energy recoverable in subsequent cycles.

Overall, chemical potential serves as the thermodynamic foundation linking material composition, equilibrium properties, mass transport, interfacial stability, and energy storage capability. It defines the direction and magnitude of thermodynamic driving forces, establishes equilibrium conditions, and constrains reversible performance. Understanding chemical potential landscapes within energy storage media is therefore essential for predicting system behavior, optimizing materials for high energy and power, and guiding the development of stable, efficient, and long-lived energy storage technologies.

2.2.5 Exergy and maximum recoverable work

While internal energy and Gibbs free energy quantify the total and chemically available energy within a system, only a portion of this energy can be converted into useful work under real conditions (Kamran, 2021; Bianciardi and Ulgiati, 2004). The concept of "***exergy***" which is often referred to as available energy or maximum recoverable work captures this distinction by measuring the theoretical upper limit of work that can be extracted as a system comes to equilibrium with its surroundings. Exergy is therefore a property of both the system and the environment, and it decreases irreversibly as entropy is generated during charging, discharging, heat transfer, or material transformations. For a system exchanging heat and work with an environment characterized by temperature T_0 and pressure P_0, the reversible (maximum) work obtainable is given by the change in the system's exergy:

$$W_{rev} = \Delta U + P_0 \Delta V - T_0 \Delta S \tag{2.19}$$

where ΔU, ΔV, and ΔS represent changes in internal energy, volume, and entropy, respectively. The term $T_0 \Delta S$ quantifies the portion of energy that is unavailable for conversion to work due to entropy production, thus linking exergy directly to irreversibility. This formulation underscores that even when the First Law permits significant energy release, the Second Law may restrict the fraction recoverable as useful work.

In electrochemical systems, exergy is closely related to the Gibbs free energy, which defines the maximum nonexpansion work obtainable from a chemical reaction at constant temperature and pressure (Agbaje and Akkaya, 2023). Because redox reactions in batteries occur nearly isothermally and without significant volume change, the reversible electrical work is approximately equal to the decrease in Gibbs free energy (Agbaje and Akkaya, 2023). However, the actual electrical energy delivered to an external load is reduced by kinetic overpotentials, resistive losses, and mass-transport limitations (Zhang et al., 2024). These irreversible processes generate entropy and destroy exergy, resulting in a discharge voltage lower than the equilibrium voltage predicted by thermodynamics. The exergy destroyed during a process can be quantified as

$$X_{dest} = T_0 \Delta S_{gen} \tag{2.20}$$

where X_{dest} is exergy destruction and $\Delta S_{gen} \geq 0$ is the entropy generated by internal irreversibilities. This expression highlights the direct and unavoidable link between entropy generation and loss of usable energy. Each source of irreversibility discussed previously (such as finite-rate heat transfer, activation kinetics, viscous dissipation, Joule heating, and diffusion gradients) reduces exergy and therefore lowers round-trip efficiency.

In thermal energy storage, exergy analysis is equally important. While enthalpy change quantifies the total stored thermal energy, only a fraction of this energy is convertible into work through a heat engine, and this fraction is bounded by the Carnot limit (Rosen and Dincer, 2003). A high-temperature reservoir possesses greater exergy because it can drive a larger temperature difference relative to the environment. As heat leaks or as the system cools toward environmental temperature T_0, exergy is lost even if total energy remains nearly constant. Thus, thermal systems may retain energy but lose usefulness due to thermodynamic degradation, not simply heat loss, but the loss of temperature potential that defines work-producing capability. Exergy therefore provides a more meaningful measure of performance than energy alone, especially in high-temperature and thermochemical storage systems.

Mechanical storage systems, such as pumped hydro and compressed air systems, also exhibit exergy destruction through frictional losses, turbulence, aerodynamic drag, and nonideal compression and expansion processes (Yang et al., 2024a). Even in the absence of chemical transformations, mechanical processes generate entropy because practical fluid motion is inherently irreversible. As a result, the recoverable gravitational or pressure energy during discharge is always less than the energy supplied during charging, reinforcing the idea that exergy, not energy, defines the true upper bound on efficiency.

Exergy analysis offers a powerful framework for comparing disparate energy storage technologies on a common thermodynamic basis. While energy quantifies capacity, exergy quantifies *quality*. Exergy reveals how operating temperature, reaction pathways, material selection, and system architecture influence the ability to recover stored energy (Rosen and Dincer, 2003). It also serves as a diagnostic tool: identifying dominant sources of irreversibility, guiding optimization of thermal management strategies, informing selection of high-exergy storage materials, and enabling rigorous techno-economic assessments (Rosen and Dincer, 2003; Meng et al., 2024). Because every Joule of destroyed exergy corresponds to lost work potential, exergy analysis directly connects thermodynamic inefficiencies to performance degradation and economic cost.

In summary, exergy provides a unified thermodynamic metric for evaluating the maximum useful work obtainable from energy storage systems. It incorporates both the First and Second Laws, capturing the combined influence of internal energy, entropy, and environmental conditions. As storage technologies evolve toward higher efficiency, higher temperature operation, and more aggressive cycling demands, exergy will remain central to understanding fundamental limits, diagnosing losses, and optimizing the design and operation of advanced energy storage systems.

2.2.6 Thermodynamic stability and degradation pathways

The long-term performance and reliability of energy storage systems are ultimately constrained by the thermodynamic stability of their constituent materials and interfaces (Kiyabu et al., 2026). Stability determines whether a material can persist in a given chemical, electrochemical, thermal, or mechanical environment without undergoing spontaneous transformation into a lower free-energy state (Kömür et al., 2025; Dulgerbaki et al., 2016). From a thermodynamic perspective, degradation occurs when the system's chemical potentials, free energies, or interfacial equilibria lie outside the stability domain defined by the surrounding conditions, thereby activating irreversible pathways that consume exergy, generate entropy, and progressively reduce performance.

For any material or interface, stability is dictated by the minimization of Gibbs free energy. A phase is stable when its free energy is lower than all competing phases; it is metastable when higher in free energy but kinetically protected by large activation barriers (Huang et al., 2019). Electrodes, electrolytes, and interphases in energy storage devices often operate in metastable regimes, where repeated cycling, elevated temperature, or applied potentials gradually lower activation barriers and accelerate transitions toward equilibrium states (Huang et al., 2019). When an electrode is driven outside its thermodynamic stability window, decomposition, dissolution, gas evolution, or structural collapse may occur, producing entropy and shifting the material to a more thermodynamically favorable configuration. These processes reduce the amount of recoverable work and alter the free-energy landscape that governs subsequent operation.

Interfacial stability plays a particularly critical role in electrochemical energy storage. The electrode-electrolyte interface is thermodynamically defined by the alignment of chemical potentials for electrons and ions. If the electrode potential exceeds the decomposition limits of the electrolyte, reduction or oxidation reactions become spontaneous, forming new interfacial phases that may be more stable than the original components (Mohandas et al., 2024). While some passivation layers, such as the solid-electrolyte interphase (SEI), can confer kinetic protection, their formation is inherently irreversible and is accompanied by entropy production and loss of useful free energy (Mohandas et al., 2024). Continued growth or instability of these layers shifts interfacial chemical potentials, increases transport resistance, and ultimately degrades electrochemical performance. The same principles govern the degradation of cathode surfaces, where high oxidation potentials can drive electrolyte oxidation, transition-metal dissolution, and surface reconstruction processes.

Thermal and thermochemical storage systems are likewise governed by stability considerations. Reversible phase transitions require well-defined chemical potentials and minimal accumulation of irreversible structural changes. However, repeated cycling across phase boundaries may induce segregation, coarsening, or incomplete reversibility of the phase transition, driving the system toward lower-energy configurations that do not efficiently store or release heat. In thermochemical storage, side reactions, irreversible decomposition, and shifts in equilibrium constants reduce exergy and diminish the extent of reversible reaction over time (Zou et al., 2024; Yang et al., 2024b). These degradative processes result from the system seeking its thermodynamic equilibrium in the presence of repeated perturbations.

Mechanical and mechanical-chemical hybrid systems exhibit analogous stability constraints. Metal hydrides used for hydrogen storage, for example, may undergo disproportionation reactions, surface oxidation, or loss of crystallographic reversibility when subjected to repeated absorption-desorption cycles (Wang et al., 2024). These transformations reflect the system's migration toward a lower free-energy configuration and are driven by gradients in chemical potential between the stored species and the host lattice (Wang et al., 2024). In compressed-air and pumped-hydro systems, thermodynamic stability manifests in different forms: mechanical fatigue, fluid-material interactions, and thermal gradients generate irreversible entropy-producing pathways that erode system efficiency and long-term durability.

Thermodynamic degradation is inherently linked to entropy generation. Each irreversible pathway produces entropy and reduces the exergy available for future cycles.

Over time, small increments of irreversibility accumulate, diminishing the maximum recoverable work and redefining the effective free-energy landscape of the system. This cumulative degradation reflects the system's progressive migration toward equilibrium with its environment. Materials and interfaces may evolve through changes in composition, microstructure, defect concentration, or morphology, each corresponding to a reduction in free energy facilitated by entropy production (Wang et al., 2024).

Understanding thermodynamic stability therefore requires mapping not only equilibrium free-energy surfaces but also the pathways leading toward degradation. The driving forces for these transformations can be expressed in terms of gradients in chemical potential, reaction affinities, and environmental perturbations such as temperature, pressure, or applied potential. Mitigating degradation involves minimizing entropy generation, stabilizing metastable states through kinetic barriers, and designing storage media and interfaces that maintain their thermodynamic integrity under operating conditions. This perspective provides a foundation for predicting long-term performance, evaluating material selection, and informing the design of next-generation energy storage technologies capable of sustaining high efficiency and durability over extended cycling.

2.3 Electrochemical mechanisms

The thermodynamic framework developed in Section 2.2 establishes the fundamental limits governing energy storage systems, defining how energy is conserved, degraded, and constrained by free-energy landscapes, entropy generation, chemical potential gradients, exergy destruction, and stability boundaries. While these principles determine what is theoretically possible, they do not describe how energy is stored and released in practice. Real energy storage devices operate through specific physicochemical processes that translate thermodynamic driving forces into observable electrical behavior. In electrochemical systems, this translation occurs through coupled charge-transfer reactions, ion transport, and interfacial phenomena that enable reversible energy storage within finite time scales and material architectures (Hammes-Schiffer and Stuchebrukhov, 2010; Pascoe Bilibana, 2023). Therefore, this section shifts focus from thermodynamic limits to the electrochemical mechanisms that govern how these limits are approached or departed from during operation, providing a mechanistic foundation for understanding performance, efficiency, and degradation in practical electrochemical energy storage technologies.

2.3.1 Charge transfer and redox reactions

Electrochemical energy storage is fundamentally enabled by charge-transfer reactions that convert chemical potential differences into electrical current through coupled electron and ion motion. These reactions occur at interfaces where electronic conductors and ionic conductors coexist, allowing redox processes to proceed while maintaining macroscopic electroneutrality (Zhang et al., 2025). During operation, oxidation and reduction reactions at opposing electrodes are spatially separated yet electrically coupled through the external circuit, giving rise to a measurable cell voltage and current flow (Zhang et al., 2025).

At the molecular level, a charge-transfer reaction involves the redistribution of electronic density between reactant species and the electrode surface. A general redox process may be expressed as;

$$\text{Oxidation} + ne^{-} \rightleftharpoons \text{Reduction} \tag{2.21}$$

where the forward and reverse reactions correspond to reduction and oxidation, respectively. The extent to which this reaction proceeds is governed by the energetic alignment between the electronic states of the electrode and the redox-active species in the electrolyte or solid phase. Electron transfer requires sufficient orbital overlap and favorable energetic alignment, while simultaneously accommodating changes in local coordination, solvation structure, and lattice configuration (Kellett et al., 2018).

Charge-transfer reactions are inherently activated processes. Even when thermodynamically favorable, electron transfer across an interface requires overcoming an energy barrier associated with electronic reorganization and nuclear rearrangement of the reacting species (Mentel et al., 2018; Cao et al., 2025). This barrier gives rise to finite reaction rates and necessitates a deviation of the electrode potential from its equilibrium value in order to sustain measurable current densities (Mentel et al., 2018; Cao et al., 2025). The magnitude of this deviation reflects the intrinsic kinetics of the redox couple and the physicochemical properties of the interface, including electronic conductivity, surface chemistry, and local dielectric environment (Gupta et al., 2025).

In practical energy storage systems, charge-transfer reactions are rarely elementary. Instead, they often involve multistep pathways, intermediate states, or coupled ion–electron transfer processes. In solid-state electrodes, electron transfer is frequently synchronized with the insertion or extraction of charge-compensating ions, requiring coordinated transport across phase boundaries (Alsaçet al., 2025). In systems involving conversion reactions or metal deposition, charge transfer is accompanied by bond rupture, nucleation, and growth phenomena that substantially increase energetic barriers and introduce spatial heterogeneity in reaction rate (Carrasco, 2024; Alsaç et al., 2025).

The spatial localization of charge-transfer reactions further constrains electrochemical performance. Because electrons and ions are transported through different phases, charge transfer is confined to regions where both carriers are simultaneously accessible (Carrasco, 2024; Alsaç et al., 2025). The density and distribution of these electrochemically active sites directly influence achievable current densities and uniformity of reaction (Carrasco, 2024; Alsaç et al., 2025). As a result, the effective surface area, electronic structure, and morphological stability of electrodes exert a strong control over charge-transfer behavior, even when bulk thermodynamic properties remain unchanged.

Charge-transfer processes also constitute a primary source of irreversibility in electrochemical energy storage. The energy expended to overcome activation barriers is dissipated as heat, contributing to entropy generation and reducing the fraction of stored chemical energy that can be recovered as useful electrical work Under high-rate operation or extreme conditions, these losses intensify, leading to voltage hysteresis, reduced efficiency, and accelerated degradation (Carrasco, 2024; Alsaç et al., 2025). Consequently, charge-transfer kinetics form the critical mechanistic link between fundamental energetic constraints and real-world performance limitations in electrochemical energy storage devices.

2.3.2 Ion transport in electrolytes and solids

Ion transport constitutes a central process in electrochemical energy storage systems, as it governs the redistribution of charge carriers required to sustain redox reactions throughout the device. While interfacial charge-transfer reactions enable the conversion between chemical and electrical energy, ion transport ensures global charge neutrality by facilitating the movement of ionic species within electrolytes and solid phases (Carrasco, 2024). Consequently, the efficiency, rate capability, and polarization behavior of electrochemical devices are strongly dictated by the mechanisms and limitations of ionic motion across multiple length scales.

In liquid electrolytes and ion-conducting media, ionic transport is commonly described by the Nernst-Planck equation (Eq. 2.22), which expresses the molar flux of an ionic species as the superposition of diffusion driven by concentration gradients, convection associated with bulk fluid motion, and migration under an applied electric field (Li et al., 2013),

$$J_i = -D_i \nabla c_i + cV + \frac{z_i e D_i}{k_B T} c_i \mathrm{E} \tag{2.22}$$

where J_i is the ionic flux, D_i is the diffusion coefficient, c_i is the concentration, V is the bulk velocity of electrolyte, z_i is the ionic charge number, e is the elementary charge, k_B is the Boltzmann constant, T is the absolute temperature, and E represents electric field. The three terms represent diffusive, convective, and migrative contributions, respectively. In most electrochemical energy storage systems—particularly in quiescent liquid electrolytes and solid-state configurations—the convective term is negligible, and ionic transport is dominated by diffusion and electric-field-driven migration. At the device scale, this coupled ionic motion through the electrolyte and separator enables continuous charge compensation between spatially separated electrodes during operation, as schematically illustrated in Fig. 2.5A (Cappabianca et al., 2023).

In concentrated electrolytes and solid-state systems, ion transport is more accurately described in terms of chemical potential gradients rather than concentration alone. Interactions among ions, solvation effects, and nonideal behavior alter the effective driving force for transport, leading to deviations from dilute-solution approximations (Tournassat et al., 2020; Boldini and Porfiri, 2022). As a result, transport limitations may arise even when concentration gradients appear modest, particularly under high current densities or in confined geometries. These effects manifest macroscopically as concentration polarization and voltage losses during operation.

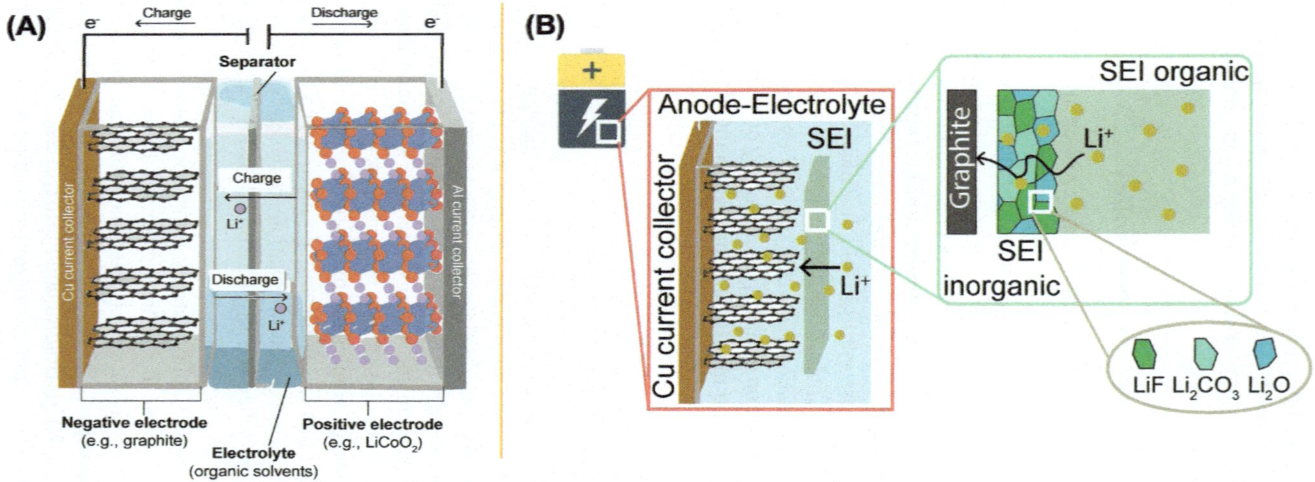

Figure 2.5 Schematic representation of ion transport processes in a representative Li-ion cell. (A) Macroscopic ion transport during charge and discharge, illustrating Li^+ migration through the electrolyte and separator and electron flow through the external circuit. (B) Microscopic ion transport at the anode-electrolyte interface, highlighting Li^+ diffusion within graphite and the role of the SEI, including inorganic and organic components.

Adapted from Cappabianca, R., De Angelis, P., Fasano, M., Chiavazzo, E., Asinari, P., 2023. An overview on transport phenomena within solid electrolyte interphase and their impact on the performance and durability of lithium-ion batteries. Energies16(13). https://doi.org/10.3390/en16135003 under the Creative Commons CC BY license.

On transport within solid electrodes introduces additional complexity, as ions must migrate through crystallographic pathways within a host lattice while overcoming site-specific energy barriers and structural constraints (Yang et al., 2025; Kulathuvayal and Su, 2023). In insertion-type materials, ionic mobility is governed by lattice topology, defect chemistry, and local compositional changes that evolve with the state of charge (Yang et al., 2025; Kulathuvayal and Su, 2023). Diffusion coefficients in solids are typically several orders of magnitude lower than those in liquid electrolytes, rendering solid-state ion transport a frequent rate-limiting step in high-power applications (Yang et al., 2025; Kulathuvayal and Su, 2023). These processes, including ion migration within active materials and across interfacial layers, are schematically represented in Fig. 2.5B.

At the electrode-electrolyte interface, ion transport is further modulated by the formation of interphase layers, such as the SEI in Li-based energy storage systems (Meda et al., 2022). The SEI, composed of both inorganic and organic components, acts as a mixed ionic-electronic barrier that selectively permits ion transport while suppressing parasitic reactions (Meda et al., 2022). Although essential for electrochemical stability, the presence of such interfacial layers introduces additional resistance to ion motion and contributes to irreversible energy losses during cycling, as highlighted by the interfacial transport pathways shown in Fig. 2.5B.

At the mesoscale, ion transport through porous electrodes involves coupled processes across electrolyte-filled pores and solid matrices (Tao et al., 2022; Gabitto and Tsouris, 2023). Tortuosity, pore size distribution, and connectivity influence effective ionic conductivity and diffusivity, creating transport pathways that are longer and more resistive than those in bulk materials (Tao et al., 2022; Gabitto and Tsouris, 2023). Inadequate ion transport through porous structures leads to localized depletion or accumulation of ionic species, increasing polarization and reducing accessible capacity (Tao et al., 2022; Gabitto and Tsouris, 2023). Thus, electrode architecture plays a decisive role in translating intrinsic material properties into practical performance.

Ion transport is also a major contributor to irreversibility in electrochemical systems. Finite transport rates require sustained chemical potential and electric field gradients, which generate entropy and dissipate energy as heat (Gabitto and Tsouris, 2023). As current density increases, transport-related losses grow disproportionately, imposing a trade-off between power output and efficiency (Gabitto and Tsouris, 2023). In extreme cases, transport limitations can trigger secondary degradation processes, such as dendritic growth, mechanical fracture, or phase separation, further compromising long-term stability.

Importantly, ion transport does not operate independently of charge-transfer kinetics. The two processes are tightly coupled: insufficient ion transport limits the supply of reactants to active sites, while sluggish charge transfer can create steep concentration gradients that hinder transport (Gabitto and Tsouris, 2023). This interdependence defines the overall electrochemical response of energy storage systems and underscores the necessity of coordinated optimization of materials, interfaces, and architectures (Gabitto and Tsouris, 2023).

In summary, ion transport in electrolytes and solids constitutes a fundamental mechanistic pillar of electrochemical energy storage. It governs the spatial and temporal redistribution of charge carriers, shapes polarization behavior, and sets practical limits on power capability and efficiency. A detailed understanding of ion transport mechanisms, their governing forces, and their coupling to electrochemical reactions is essential for interpreting device behavior and for guiding the design of advanced storage systems capable of operating efficiently under demanding conditions.

2.3.3 Interfacial phenomena and double-layer formation

Electrochemical energy storage processes are intrinsically interfacial in nature, as all charge-transfer reactions and ionic exchanges occur at boundaries between electronically conductive electrodes and ionically conductive electrolytes. These interfaces are not passive geometric surfaces but chemically and electrostatically active regions, where broken symmetry and incomplete atomic coordination give rise to excess surface charge and localized electronic states (Fig. 2.6) (Dourado, 2022). As a result, steep gradients in electric potential, charge density, and composition develop over nanometer-scale distances, rendering interfacial phenomena central to electrochemical performance, efficiency, and stability. When an electrode is brought into contact with an electrolyte, electrostatic equilibrium is established through the redistribution of electronic charge in the electrode and ionic species in the adjacent electrolyte (Vetter et al., 2009; Hausbrand and Jaegermann, 2018). This redistribution leads to the formation of an electrical double layer, consisting of a charged electrode surface and a compensating arrangement of counter-ions in the electrolyte (Vetter et al., 2009; Hausbrand and Jaegermann, 2018). In its simplest representation, the interface may be treated as a parallel-plate capacitor, as embodied in the classical Helmholtz model, where a compact counter-ion layer forms directly adjacent to the charged surface and the potential drops linearly across the interface (Fig. 2.6B)

While the Helmholtz model captures the essential concept of interfacial charge separation, it neglects thermal motion, finite ion size, and ion-ion interactions (Dourado, 2022). These limitations are addressed in the Stern model, which partitions the double layer into a compact region near the surface and a diffuse layer extending into the bulk electrolyte (Dourado, 2022). In this description, the potential decays nonlinearly with distance from the electrode, reflecting the balance between electrostatic forces and entropic contributions associated with ionic disorder (Fig. 2.6C) (Dourado, 2022). This refinement provides a more realistic account of interfacial capacitance and potential distribution under typical electrochemical conditions.

A further level of realism is introduced by Grahame's formulation, which explicitly incorporates solvent structure and distinguishes between the inner and outer Helmholtz planes (Usui, 2008). In this framework, specifically adsorbed ions and oriented solvent molecules define the inner Helmholtz plane, while solvated ions populate the outer Helmholtz plane and the diffuse layer beyond (Fig. 2.6D) (Dourado, 2022). The electrostatic potential at the boundary between the compact and diffuse regions, commonly identified with the ζ-potential, emerges as a key parameter governing electrokinetic phenomena, interfacial stability, and ion transport near the surface.

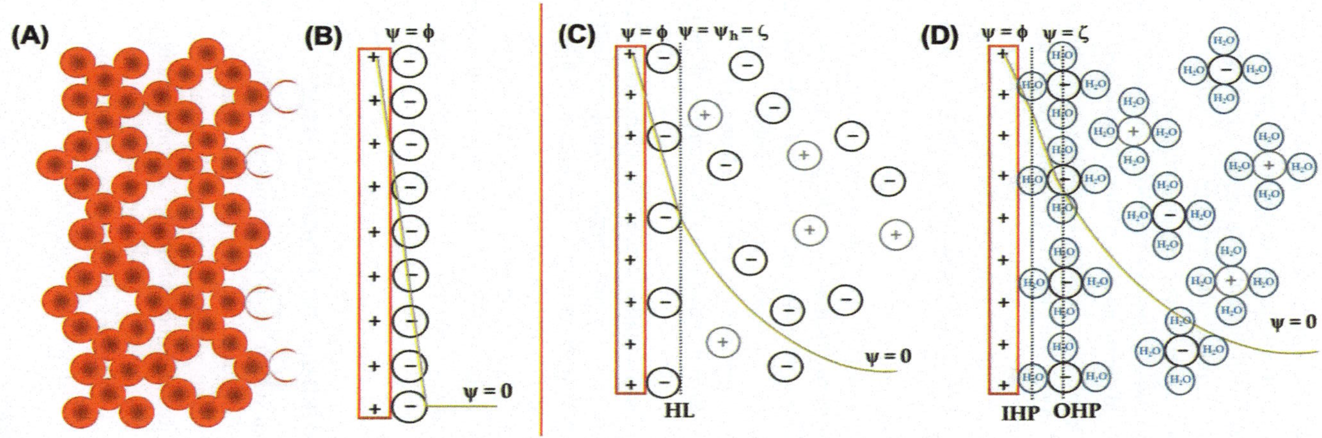

Figure 2.6 Formation and structure of the electrical double layer at electrochemical interfaces. (A) Atomic-scale representation of a solid surface, illustrating under-coordinated surface atoms that give rise to excess surface charge. (B) Helmholtz model of the electrical double layer, showing a compact counter-ion layer adjacent to a positively charged electrode (red rectangle) and a linear potential drop. (C) Stern model, in which the double layer is divided into a compact Helmholtz layer (HL) and a diffuse layer, resulting in a nonlinear decay of the electrostatic potential. (D) Grahame model incorporating solvent structure, distinguishing the inner Helmholtz plane (IHP) and outer Helmholtz plane (OHP), with oriented solvent molecules and solvated ions. *ψ denotes the local electrostatic potential, φ the electrode potential, ψ_h the Helmholtz potential, and ζ the zeta potential*; *$\psi = 0$ corresponds to the bulk electrolyte.*

Adapted from Dourado, A.H.B., 2022. Electric double layer: the good, the bad, and the beauty. Electrochem 3(4), 789–808. https://doi.org/10.3390/electrochem3040052 under the Creative Commons CC BY license.

Beyond their electrostatic role, electrical double layers strongly influence physicochemical processes at electrochemical interfaces. The intense electric fields localized within the double layer modify ion transport, alter solvation structures, and shift activation barriers for charge-transfer reactions. In energy storage systems, these effects directly impact reaction kinetics, transient current responses, and rate capability. Even in faradaic systems, double-layer charging accompanies redox reactions, contributing to nonfaradaic currents and shaping dynamic voltage behavior.

Interfaces also constitute the primary locus of irreversible processes. Large interfacial electric fields and extreme local chemical potentials can promote parasitic reactions such as electrolyte decomposition, surface reconstruction, and interphase formation. While some interfacial layers may kinetically stabilize electrode surfaces, their formation is intrinsically dissipative and entails entropy generation and irreversible consumption of active species. These processes progressively modify interfacial structure, increase resistance to ion transport and charge transfer, and ultimately degrade long-term performance.

Importantly, interfacial properties are not static but evolve continuously during operation in response to changes in potential, current density, temperature, and electrolyte composition. The structure of the double layer adapts on timescales ranging from milliseconds to hours, coupling microscopic interfacial dynamics to macroscopic performance metrics such as efficiency, power capability, and cycle life. Consequently, interfacial phenomena provide a critical mechanistic link between atomic-scale processes and system-level behavior in electrochemical energy storage devices.

In summary, electrical double-layer formation represents a fundamental organizing principle of electrochemical interfaces. It governs charge separation, modulates reaction energetics, controls ion transport in the near-surface region, and mediates both reversible energy storage and irreversible degradation. A rigorous understanding of double-layer structure and interfacial phenomena is therefore essential for interpreting electrochemical behavior and for guiding the design of advanced energy storage systems capable of approaching thermodynamic limits under practical operating conditions.

2.3.4 Polarization, overpotential, and kinetic losses

The deviation of an electrochemical system from its equilibrium potential under operating conditions is collectively described as polarization (Zalka and Péter, 2020). Polarization reflects the additional driving force required to sustain charge-transfer reactions and ionic transport at finite rates and constitutes a primary source of voltage loss in electrochemical energy storage systems (Zalka and Péter, 2020). These losses arise not from limitations imposed by thermodynamic feasibility, but from kinetic constraints associated with interfacial reactions, bulk transport, and electrical resistance within the device.

Polarization is commonly decomposed into distinct contributions, each associated with a different physical process. The total overpotential η, defined as the difference between the equilibrium electrode potential and the actual operating potential, may be expressed by Eq. (2.23).

$$\eta = \eta_{\text{act}} + \eta_{\text{ohm}} + \eta_{\text{conc}} \tag{2.23}$$

where η_{act} denotes activation overpotential, η_{ohm} represents ohmic losses, and η_{conc} corresponds to concentration (or mass-transport) overpotential. Although these terms are conceptually separable, they are strongly coupled during operation and often evolve simultaneously as current density increases. The cumulative impact of activation, ohmic, and concentration polarization on the operating cell voltage as a function of current density is schematically illustrated in Fig. 2.7 (Huang et al., 2006).

Activation polarization arises from the finite kinetics of charge-transfer reactions at the electrode-electrolyte interface. Even when a reaction is thermodynamically favorable, electrons must overcome an activation barrier associated with electronic rearrangement, bond breaking and formation, and solvent or lattice reorganization (Anantharaj et al., 2021). This barrier necessitates a shift in electrode potential to drive the reaction forward at a measurable rate. The magnitude of activation overpotential depends on the intrinsic reaction kinetics, surface electronic structure, density of active sites, and interfacial chemistry (Anantharaj et al., 2021). As current density increases, the activation barrier becomes increasingly significant, particularly in systems with sluggish redox couples or poorly optimized interfaces. This behavior is reflected in the steep initial voltage drop at low current densities shown in Fig. 2.7.

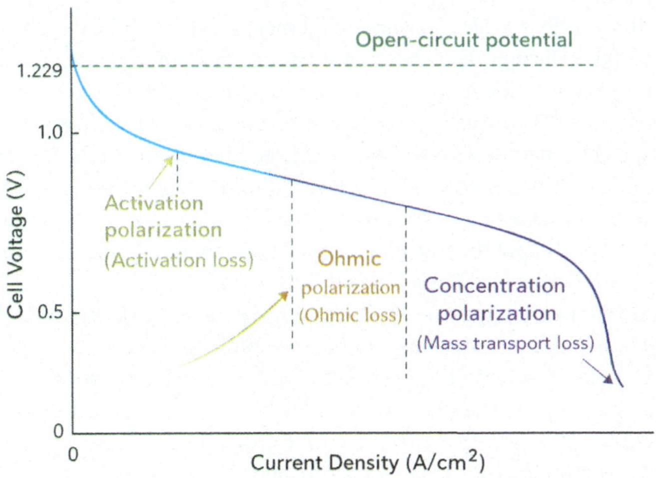

Figure 2.7 Generalized polarization behavior of electrochemical systems. Representative current–voltage polarization curve of an electrochemical energy storage system, illustrating the deviation of cell voltage from the open-circuit potential due to activation, ohmic, and concentration polarization losses. The figure highlights how interfacial charge-transfer kinetics, internal resistance, and mass transport limitations progressively reduce the operating voltage as current density increases.

Ohmic polarization originates from resistance to charge transport within the electrochemical system (Qiu et al., 2019; Guo et al., 2023). This includes electronic resistance in electrodes and current collectors, ionic resistance in the electrolyte, and contact resistance at interfaces. Ohmic losses scale linearly with current and are governed by the effective conductivities and geometrical dimensions of the constituent components. While often regarded as a purely means of energy dissipation, ohmic polarization directly influences thermal behavior and accelerates degradation by generating localized heating and amplifying other irreversible processes (Qiu et al., 2019; Guo et al., 2023). Minimizing ohmic losses is therefore essential for maintaining both efficiency and structural integrity under high-power operation.

Concentration polarization (mass transport) results from limitations in mass transport that lead to spatial gradients in ionic concentration near reactive interfaces. As ions are consumed or generated at electrode surfaces, insufficient transport rates can cause depletion or accumulation zones to form, reducing the local availability of reactants (Kim et al., 2019; Chen et al., 2024; Liu and Monroe, 2015). These concentration gradients alter the local electrochemical potential, requiring additional overpotential to sustain current. Concentration polarization becomes particularly pronounced at high current densities, in thick electrodes, or in systems with low ionic diffusivity (Barello et al., 2014; Vinoth Kumar and Khandale, 2022). In extreme cases, transport limitations can induce nonuniform reaction fronts, phase separation, or morphological instabilities that compromise reversibility and cycle life. Correspondingly, the sharp voltage decline observed at high current densities in Fig. 2.7 signifies the onset of concentration polarization under transport-limited condition (Huang et al., 2006).

The combined effect of these polarization mechanisms manifests as a reduction in operating voltage during discharge and an increase in required voltage during charging (Wojciechowski et al., 2024; Ohrelius et al., 2025; Goonetilleke et al., 2025). This voltage hysteresis represents a direct loss of usable energy and is accompanied by entropy generation and heat production. Importantly, polarization losses are rate-dependent: while minimal under near-equilibrium conditions, they grow rapidly with increasing current density, defining the fundamental trade-off between energy efficiency and power capability in electrochemical storage systems (Wojciechowski et al., 2024; Ohrelius et al., 2025; Goonetilleke et al., 2025).

From a mechanistic perspective, polarization serves as the observable consequence of finite reaction and transport rates within a complex, coupled system. Activation kinetics, ionic mobility, electronic conductivity, and interfacial structure collectively determine how closely a device can approach reversible operation. As such, polarization analysis provides a unifying framework for diagnosing performance limitations and identifying dominant loss mechanisms across different materials and architectures.

Polarization and overpotentials encapsulate the kinetic constraints that limit practical electrochemical energy storage performance. They arise from interfacial reaction barriers, resistive pathways, and transport limitations, each contributing to irreversible energy loss and reduced efficiency. Understanding the origins and interplay of these mechanisms is essential for optimizing electrochemical systems to operate closer to their thermodynamic limits while maintaining stability and durability under realistic operating conditions.

2.3.5 Coupling of thermodynamics and kinetics in electrochemical energy storage

As mentioned in previous sections, electrochemical energy storage systems operate under the simultaneous influence of both thermodynamic driving forces and kinetic constraints. Thermodynamics defines equilibrium states, feasible reaction pathways, and the upper bounds of energy conversion, whereas kinetics governs the rates at which these states are approached under practical operating conditions. The performance of an electrochemical device emerges from the continuous interaction between these two domains.

Electrochemical operation necessarily involves deviation from equilibrium to sustain finite current. This deviation activates kinetic processes associated with charge transfer, ion transport, and interfacial rearrangement, each contributing to irreversibility. The magnitude of this departure determines how effectively thermodynamic driving forces are converted into usable electrical output. Near-equilibrium operation minimizes kinetic engagement and approaches reversible behavior, while high-rate operation amplifies kinetic limitations, leading to increased polarization losses and entropy production.

The coupling between thermodynamics and kinetics is most clearly reflected in the relationship between equilibrium potential and operating voltage. Thermodynamics establishes the equilibrium voltage landscape as a function of composition and temperature, but kinetics dictates how closely the operating voltage follows this landscape during charge and discharge. Activation barriers, transport resistances, and interfacial constraints shift the system away from equilibrium without altering the underlying free-energy framework, thereby producing voltage hysteresis and efficiency loss (Panero, 2009).

Material properties critically mediate this coupling. Materials with favorable thermodynamic energy density may exhibit limited practical performance if ion mobility is low, activation barriers are large, or interfaces are unstable (Sotoudeh et al., 2024; Cao et al., 2020). Conversely, materials with rapid kinetics may operate efficiently but store limited energy due to modest thermodynamic driving forces (Sotoudeh et al., 2024; Cao et al., 2020). Effective electrochemical energy storage design therefore requires balancing thermodynamic favorability with kinetic accessibility, ensuring that energetically viable pathways remain dynamically accessible within the intended operating window (Sotoudeh et al., 2024; Cao et al., 2020).

The thermodynamics-kinetics interplay is further shaped by the evolution of materials and interfaces during cycling (Sotoudeh et al., 2024; Cao et al., 2020). Structural rearrangement, compositional heterogeneity, and interfacial modification alter both equilibrium tendencies and kinetic barriers over time (Panero, 2009). These changes reflect a feedback process in which kinetic irreversibility promotes thermodynamic relaxation toward lower-energy states, progressively reducing performance and recoverable work.

From a systems perspective, this coupling defines fundamental trade-offs between energy density, power density, efficiency, and lifetime. Conditions that favor maximal energy extraction typically require operation close to equilibrium, whereas high-power operation inherently engages kinetic losses and accelerates entropy generation. Understanding these trade-offs is essential for rational material selection, architectural design, and operational optimization.

In summary, electrochemical energy storage performance is governed by the inseparable coupling of thermodynamic limits and kinetic mechanisms. Thermodynamics establishes what is possible, while kinetics determines what is achievable within finite time and material constraints. This integrated perspective provides the conceptual foundation for interpreting electrochemical behavior and forms a natural bridge to the broader classification of storage mechanisms discussed in the subsequent section.

2.4 Storage mechanisms

Energy storage technologies can be broadly classified according to the physical or chemical form in which energy is stored and later released. While all storage systems are ultimately governed by the same thermodynamic constraints and, in electrochemical cases, by similar mechanistic processes, the nature of the storage medium fundamentally determines achievable energy density, power capability, efficiency, response time, scalability, and degradation behavior. Distinguishing storage mechanisms on the basis of how energy is embodied provides a unifying framework for comparing disparate technologies and clarifying their respective advantages and limitations.

Energy storage technologies can be systematically classified according to the primary physical or chemical mechanism by which energy is stored and later retrieved, as schematically summarized in Fig. 2.8. At the highest level, energy storage systems are commonly grouped into chemical, mechanical, electrical, and thermal categories, each encompassing distinct operating principles and performance characteristics. Chemical energy storage relies on the formation and transformation of chemical bonds, spanning molecular fuels, synthetic hydrocarbons, and electrochemical systems such as batteries. Mechanical storage exploits macroscopic motion or positional changes, including pumped hydro, compressed air, flywheels, and gravitational potential storage. Electrical storage is based on charge separation or magnetic field formation, encompassing electrostatic devices such as capacitors and supercapacitors, as well as magnetic energy storage systems. Thermal energy storage involves the accumulation of sensible, latent, or thermochemical heat, enabling energy retention through temperature change, phase transition, or reversible chemical reactions. This hierarchical classification provides a unifying framework to compare diverse storage technologies, clarify their fundamental energy storage mechanisms, and contextualize the subsequent detailed discussion of individual systems in terms of their thermodynamic limits, kinetic behavior, and application domains.

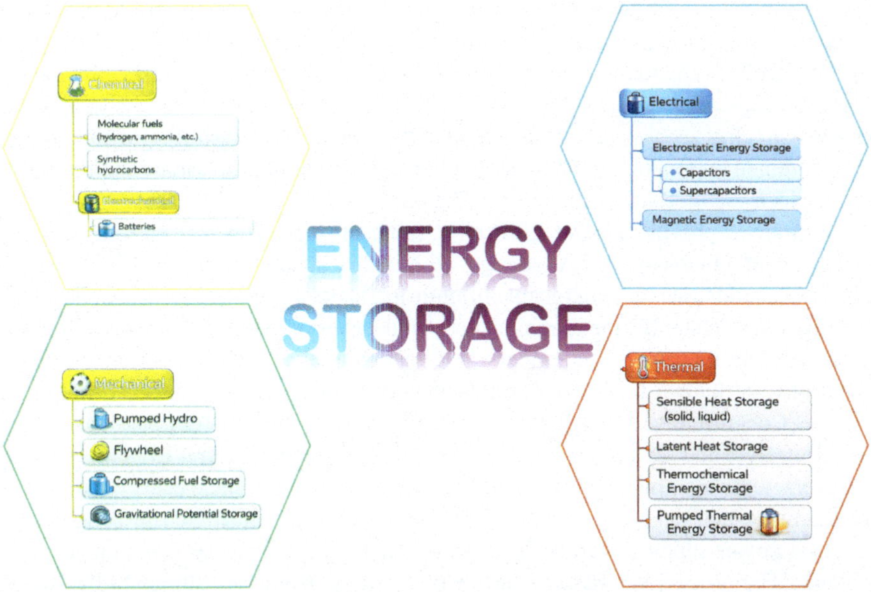

Figure 2.8 Classification of energy storage technologies. Schematic classification of energy storage technologies according to their fundamental physical and chemical mechanisms. Energy storage systems are grouped into four primary categories: chemical, electrical, mechanical, and thermal. Chemical storage includes molecular fuels, synthetic hydrocarbons, and electrochemical storage in the form of batteries. Electrical storage comprises electrostatic (capacitors and supercapacitors) and magnetic energy storage. Mechanical storage covers pumped hydro, flywheel systems, compressed fuel storage, and gravitational potential energy storage. Thermal storage includes sensible heat, latent heat, thermochemical energy storage, and pumped thermal energy storage. This classification highlights the diversity of energy storage pathways across different time scales, energy densities, and application domains.

2.4.1 Chemical energy storage

Chemical energy storage encompasses systems in which energy is stored through changes in chemical bonding or composition and released via reversible or quasi-reversible chemical reactions. In these systems, energy is embodied in the Gibbs free energy difference between reactants and products, making chemical storage inherently high in energy density relative to other storage mechanisms. This category includes electrochemical batteries, fuel-based storage, redox-active materials, and thermochemical storage systems, all of which rely on controlled chemical transformations to store and deliver energy.

In electrochemical energy storage systems, electrical energy supplied during charging is converted into stored free energy by driving the system away from thermodynamic equilibrium, resulting in increased chemical potential gradients, interfacial charge separation, and/or redox-induced changes in the electronic and ionic states of active materials (Dai et al., 2025; Li et al., 2024). Upon discharge, the system relaxes toward equilibrium, and the release of this stored free energy is converted into electrical work through the corresponding electrochemical processes (Dai et al., 2025; Li et al., 2024). The total amount of energy that can be stored is governed by the underlying free-energy change associated with the active materials and the quantity of charge-carrying species involved, whereas the practical accessibility of this energy is determined by kinetic factors, mass and charge transport limitations, and the stability of electrode-electrolyte interfaces (Dai et al., 2025; Li et al., 2024).

In order to provide a more illustrative representation, Fig. 2.9 presents a visual depiction of chemical energy storage within this specific context (Menye et al., 2025). This illustrates a rechargeable lithium-ion battery operating in two modes. During discharge (panel B), the graphite anode undergoes oxidation as a result of the separation of lithium from the interlayers, leading to the release of Li^+ ions and electrons. Electrons are known to flow externally from the anode to the cathode through the charge. Concurrently, Li^+ migrates internally toward the cathode, passing through the electrolyte and separator (e.g., $LiCoO_2$, NMC, $LiFePO_4$). Here, Li^+ enters the interlayers with incoming electrons, causing reduction, and chemical energy is thus converted into electrical work. During the charging process (illustrated in panel A), an external power supply reverses these processes by exceeding the cell's open-circuit voltage. The process entails the transfer of electrons from the cathode to the anode, concomitant with the movement of Li^+ ions in the inverse direction. Consequently, oxidation occurs at the cathode, while reduction takes place at the anode. However, the practical implementation of this process necessitates the successful navigation of ohmic, kinetic, and mass transfer losses. Consequently, in the context of chemical storage, the redox roles undergo a reversal between the charging and discharging processes, while the material labels maintain a fixed position within the designated physical locations (Fig. 2.9) (Menye et al., 2025).

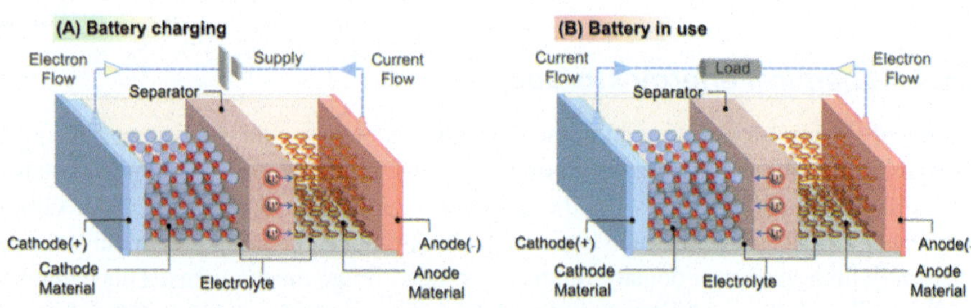

Figure 2.9 Operating principles of a rechargeable electrochemical energy storage cell. Schematic illustration of a rechargeable electrochemical energy storage cell during (A) charging and (B) discharging operation. During charging, external electrical energy drives ion transport through the electrolyte and electron flow through the external circuit, storing energy via reversible electrochemical reactions. During discharge, these processes reverse, converting stored chemical energy back into electrical energy delivered to an external load. Reprinted from Menye, J.S., Camara, M.B., Dakyo, B., 2025. Lithium battery degradation and failure mechanisms: astate-of-the-art review. Energies18(2). https://doi.org/10.3390/en18020342 under the Creative Commons CC BY license.

Beyond electrochemical cells, chemical energy storage also includes systems where energy is stored in molecular fuels such as hydrogen, ammonia, or synthetic hydrocarbons (Elalfy et al., 2024; Angenent et al., 2024; Palys and Daoutidis, 2020). In these cases, energy storage occurs through bond formation during synthesis processes, often powered by electricity or heat, and energy is later released through combustion or electrochemical conversion (Elalfy et al., 2024; Angenent et al., 2024; Palys and Daoutidis, 2020). Such systems offer exceptionally high gravimetric energy density and long-duration storage capability but typically require additional conversion steps, each introducing efficiency losses and system complexity (Elalfy et al., 2024; Angenent et al., 2024; Palys and Daoutidis, 2020).

Thermochemical energy storage represents another important subclass of chemical storage, where energy is stored via reversible endothermic reactions and released through exothermic recombination (Singh Rathore et al., 2020; Kalaiselvam and Parameshwaran, 2014). Unlike sensible or latent thermal storage, thermochemical systems store energy in altered chemical composition rather than temperature difference, enabling minimal thermal losses during storage (Singh Rathore et al., 2020; Kalaiselvam and Parameshwaran, 2014). The stored energy density is governed by reaction enthalpy, while reversibility and long-term stability depend on reaction kinetics, material compatibility, and resistance to side reactions (Singh Rathore et al., 2020; Kalaiselvam and Parameshwaran, 2014).

A defining characteristic of chemical energy storage is the decoupling of energy storage capacity from instantaneous power delivery. While energy content is set by the amount of chemically stored free energy, power capability is governed by reaction rates, transport phenomena, and system design. This decoupling enables chemical storage systems to span a wide range of applications, from high-energy, long-duration storage to high-power, short-duration operation, albeit often with trade-offs between these attributes.

Despite their advantages, chemical energy storage systems are intrinsically subject to degradation driven by irreversible chemical pathways. Side reactions, phase changes, structural reorganization, and interfacial instability gradually reduce the fraction of stored chemical energy that can be reversibly accessed. These processes reflect the system's tendency to evolve toward lower free-energy states and is a primary factor limiting cycle life and efficiency. Consequently, the design of chemical storage systems requires careful balancing of thermodynamic driving forces, kinetic control, and material stability to ensure sustained performance.

Chemical energy storage represents a cornerstone of modern energy systems by enabling energy to be stored in high-density chemical forms and released through controlled transformations. Its versatility allows decoupling of energy capacity from power delivery, supporting applications ranging from short-term, high-power operation to long-duration storage. At the same time, the reliance on chemical transformations introduces intrinsic challenges related to reversibility, stability, and degradation, as systems tend toward lower free-energy states over repeated operation. Addressing these challenges requires careful control of reaction pathways, material composition, and operating conditions to ensure that stored chemical energy remains accessible with high efficiency over extended lifetimes.

2.4.2 *Mechanical energy storage*

Mechanical energy storage systems store energy by performing work against gravitational, elastic, or pressure-related forces and recover it through controlled mechanical motion. Unlike chemical or electrochemical storage, mechanical systems do not rely on changes in chemical composition, instead, energy is embodied in macroscopic mechanical states such as elevation, rotational inertia, or compressed volume (Arabkoohsar, 2023). As a result, mechanical storage mechanisms are characterized by high durability, long cycle life, and scalability, making them particularly attractive for large-scale and grid-level applications.

One of the most widely deployed forms of mechanical storage is gravitational potential energy storage, exemplified by pumped hydroelectric systems. In such systems, electrical energy is converted into mechanical work to elevate a mass against gravity, thereby increasing its potential energy (Du et al., 2025). The stored energy is proportional to the mass, gravitational acceleration, and elevation difference and can be recovered by allowing the mass to descend through a turbine-generator assembly (Menéndez et al., 2020). The fundamental simplicity of this mechanism enables high round-trip efficiency and long operational lifetimes, although practical deployment is constrained by geographical requirements and large infrastructure demands.

Another important class of mechanical storage systems is kinetic energy storage, where energy is stored in the rotational motion of a flywheel (Koohi-Fayegh and Rosen, 2020; Kamran, 2023; Aktaş and Kirçiçek, 2021). During charging, electrical energy accelerates a rotor to high angular velocity, during discharge, the deceleration of the rotor generates electrical power (Koohi-Fayegh and Rosen, 2020; Kamran, 2023; Aktaş and Kirçiçek, 2021). The energy stored scales with the moment of inertia and the square of angular velocity, favoring lightweight, high-strength materials and low-friction environments (Koohi-Fayegh and Rosen, 2020; Kamran, 2023; Aktaş and Kirçiçek, 2021). Flywheel systems offer rapid response times and high-power density, but their total energy capacity is typically limited by material strength and safety considerations.

Compressed fluid storage represents a third category of mechanical energy storage. In CAES, energy is stored by compressing air into underground caverns or engineered vessels (Dupin and Teixeira, 2025; An et al., 2023). During discharge, the pressurized air is expanded to perform mechanical work (Dupin and Teixeira, 2025; An et al., 2023). The thermodynamic behavior of compression and expansion plays a central role in determining efficiency, as nonideal processes generate heat that must be managed or recovered (Dupin and Teixeira, 2025; An et al., 2023). Variants such as adiabatic CAES seek to improve efficiency by storing compression heat and reintegrating it during expansion, highlighting the close coupling between mechanical and thermal effects in these systems.

Mechanical energy storage systems exhibit a clear separation between energy storage capacity and power output. Energy capacity is determined by physical scale (mass, volume, or rotational inertia) whereas power capability is governed by mechanical and electrical conversion components such as turbines, generators, bearings, and control systems. This decoupling enables mechanical storage to deliver high power on demand while maintaining substantial energy reserves, albeit often at the cost of low energy density relative to chemical storage.

From a thermodynamic perspective, mechanical storage systems are subject to irreversibilities arising from friction, turbulence, material deformation, and nonideal fluid flow. These processes generate entropy and dissipate energy as heat, reducing the fraction of stored mechanical work that can be recovered. Nevertheless, because mechanical systems typically avoid complex chemical transformations, degradation mechanisms are often slower and more predictable, resulting in exceptionally long service lifetimes and stable performance over repeated cycling.

Mechanical energy storage thus occupies a distinct niche within the broader storage landscape. Its reliance on macroscopic physical processes enables robustness, scalability, and high-power capability, while its limitations in energy density and site dependence constrain applicability. These characteristics position mechanical storage as a critical complement to chemical and electrochemical systems, particularly in applications where longevity, reliability, and large-scale energy balancing are prioritized.

2.4.3 Electrical energy storage

Electrical energy storage systems store energy directly in electromagnetic fields rather than through chemical transformations or macroscopic mechanical work (Ahmed et al., 2024; Yadav and Sharma, 2025). In these systems, energy is accumulated through charge separation or magnetic field generation and is released through rapid electrical discharge (Ahmed et al., 2024; Yadav and Sharma, 2025). Because no bond breaking or large-scale mass transport is involved, electrical storage mechanisms are distinguished by exceptionally fast response times, high power density, and high round-trip efficiency, albeit typically at the expense of limited energy density (Ahmed et al., 2024; Yadav and Sharma, 2025).

The most common form of electrical energy storage is electrostatic storage, where energy is stored in the electric field established between oppositely charged conductors. Capacitive systems operate by accumulating charge on electrode surfaces separated by a dielectric or electrolyte, with stored energy proportional to capacitance and the square of the applied voltage (Azega et al., 2024). In conventional dielectric capacitors, the dielectric constant and breakdown strength of the insulating material determine storage capacity, leading to very high-power capability but modest energy density (Azega et al., 2024). Electrochemical capacitors (supercapacitors) extend this concept by exploiting interfacial charge separation at electrode-electrolyte interfaces, enabling higher capacitance through large surface area and nanoscale charge separation while preserving rapid charge-discharge behavior (Azega et al., 2024).

A second category of electrical storage involves magnetic energy storage, where energy is stored in the magnetic field generated by an electric current (Adetokun et al., 2022). Superconducting magnetic energy storage (SMES) systems exemplify this approach, using superconducting coils to circulate current with negligible resistive loss (Adetokun et al., 2022). The stored energy scales with inductance and the square of current, allowing extremely fast response and high efficiency (Adetokun et al., 2022). However, the requirement for cryogenic operation and specialized materials imposes substantial cost and complexity, limiting widespread deployment.

Electrical energy storage systems exhibit a strong coupling between voltage, field strength, and stability. The maximum energy that can be stored electrostatically is constrained by dielectric breakdown, field-induced degradation, and leakage currents, while magnetic systems are limited by critical current density, magnetic field strength, and mechanical stresses within conductors. These constraints reflect fundamental material limits rather than kinetic barriers, distinguishing electrical storage from chemical and mechanical counterparts.

A defining advantage of electrical storage mechanisms is their ability to deliver or absorb power almost instantaneously. This makes them particularly well suited for applications requiring high-frequency regulation, power smoothing, and transient load management. At the same time, the relatively low energy density of purely electrical storage restricts its role in long-duration energy supply, necessitating integration with other storage technologies for comprehensive energy management.

From a thermodynamic standpoint, electrical energy storage systems operate close to reversible limits under ideal conditions, as losses associated with charge redistribution and field establishment are minimal. Practical irreversibilities arise primarily from resistive heating, dielectric losses, and parasitic leakage, which generate entropy and reduce efficiency over time. Nevertheless, the absence of complex chemical reactions or large mechanical stresses contributes to long operational lifetimes and predictable degradation behavior.

Electrical energy storage thus occupies a specialized yet essential position within the broader storage landscape. Its unparalleled power capability and rapid response complement the high energy density of chemical storage and the scalability of mechanical systems. Together, these characteristics make electrical storage a critical enabling technology for stabilizing modern energy systems, particularly in applications requiring precise and rapid control of power flow.

2.4.4 Thermal energy storage

Thermal energy storage (TES) systems store energy in the form of heat by raising the temperature of a medium, inducing a phase transition, or driving a reversible thermochemical reaction (Koohi-Fayegh and Rosen, 2020; Rahman et al., 2022; Wolde et al., 2024). Unlike chemical or electrical storage, TES does not store energy in electrical charge or chemical bonds directly convertible to electricity without an intermediate process (Koohi-Fayegh and Rosen, 2020; Rahman et al., 2022; Wolde et al., 2024). Instead, its value lies in temporal decoupling of heat generation and heat utilization, making it particularly important for thermal management, industrial heat recovery, and the integration of variable renewable energy sources (Koohi-Fayegh and Rosen, 2020; Rahman et al., 2022; Wolde et al., 2024).

The most straightforward TES approach is sensible heat storage, in which energy is stored by increasing the temperature of a solid or fluid (Palacios et al., 2023). The amount of stored energy depends on the mass of the storage medium, its specific heat capacity, and the allowable temperature range (Palacios et al., 2023). Common materials include water, molten salts, concrete, and packed-bed solids (Palacios et al., 2023). Sensible heat storage benefits from material simplicity and low cost, but its energy density is constrained by practical temperature limits and thermal losses to the environment.

Latent heat storage exploits the energy absorbed or released during a phase transition, most commonly solid-liquid transformations (Jouhara et al., 2020; Hai et al., 2025). Phase-change materials (PCMs) store energy at nearly constant temperature, which is advantageous for applications requiring thermal regulation within narrow temperature windows (Jouhara et al., 2020; Hai et al., 2025). The stored energy density is determined by the latent heat of the transition, while performance depends on phase stability, thermal conductivity, and cycling durability (Jouhara et al., 2020; Hai et al., 2025). Although latent heat storage offers higher energy density than sensible storage within a fixed temperature range, challenges related to heat transfer rates and material degradation can limit long-term performance.

A third class of TES is thermochemical energy storage, where energy is stored through reversible endothermic reactions and released via exothermic recombination (Irwin et al., 2017; Chen and Aziz, 2025). In these systems, energy is embodied changes in chemical composition rather than temperature, enabling storage with minimal thermal loss over extended periods (Irwin et al., 2017; Chen and Aziz, 2025). The storage capacity is governed by reaction enthalpy, while operating conditions are dictated by equilibrium constraints involving temperature and pressure (Irwin et al., 2017; Chen and Aziz, 2025). Thermochemical storage offers high energy density and long-duration capability but requires precise control of reaction kinetics, material stability, and system integration to ensure reversibility and efficiency (Irwin et al., 2017; Chen and Aziz, 2025).

The effectiveness of thermal energy storage is closely linked to the quality of stored heat. High-temperature thermal reservoirs possess greater potential for conversion into useful work or high-grade heat, whereas low-temperature storage is primarily suited for space heating, cooling, or thermal buffering. As stored heat equilibrates with the environment, its ability to perform work diminishes even if total energy remains largely conserved. Consequently, TES performance cannot be fully assessed on the basis of stored energy alone; temperature level and heat-transfer pathways play a decisive role in determining practical usefulness.

Thermal energy storage systems are subject to irreversibilities arising from heat transfer across finite temperature differences, thermal gradients within storage media, and material degradation under repeated cycling. These processes generate entropy and progressively reduce the recoverable fraction of stored energy. However, because TES systems often involve relatively simple materials and operate without complex electrochemical reactions, they can achieve long service lifetimes with predictable performance when appropriately designed.

Thermal energy storage thus represents a versatile and scalable mechanism for managing energy in systems where heat is either the primary product or an unavoidable byproduct. Its integration with chemical, mechanical, and electrical storage technologies enables more flexible and efficient energy systems by matching energy form and quality to application-specific requirements.

2.5 Performance metrics

Performance metrics provide the quantitative framework for evaluating, comparing, and selecting energy storage technologies across scales and applications. While Sections 2.2–2.4 established the thermodynamic limits, mechanistic behavior, and storage modalities, performance metrics translate these principles into measurable indicators of practical capability. No single metric is sufficient to characterize a storage system; instead, performance emerges from trade-offs among energy capacity, power delivery, efficiency, lifetime, and cost. Consequently, a clear definition of key metrics is essential for meaningful comparison between fundamentally different storage mechanisms.

Among the most widely used metrics are energy density, power density, round-trip, cyclic and coulombic efficiencies, and levelized cost of storage (LCOS) (Begum Medha et al., 2026). Each metric captures a distinct aspect of system behavior and reflects different underlying constraints. Energy and power densities relate directly to how energy is stored and delivered; efficiency reflects thermodynamic and kinetic losses during cycling; and LCOS integrates technical performance with economic considerations over system lifetime (Chadly et al., 2023). Together, these metrics form a multidimensional performance space within which storage technologies must be evaluated.

2.5.1 Energy density and power density

Energy density and power density are fundamental performance metrics that describe complementary aspects of energy storage capability. The answers of how much energy can be stored and how rapidly that energy can be delivered are described by energy density and power density, respectively. Although frequently discussed together, these metrics arise from different physical constraints and must be interpreted within the context of the underlying storage mechanism and system architecture.

Energy density quantifies the total recoverable energy relative to mass or volume. In its most general form, the stored energy E_{stored} is defined as the time integral of the delivered power during discharge (Wu et al., 2012; Nagde and Dhoble, 2021; Borah et al., 2020),

$$E_{\text{stored}} = \int_{t_0}^{t_f} P(t)\,\mathrm{d}t \tag{2.24}$$

where $P(t)$ is the instantaneous power output and $t_f - t_0$ is the discharge duration. This definition is agnostic to storage mechanism and applies equally to chemical, mechanical, electrical, and thermal systems.

The gravimetric energy density is they expressed as,

$$\varepsilon_g = \frac{E_{\text{stored}}}{m} \tag{2.25}$$

where ε_g is given in Wh.kg^{-1} and m represents the total mass of the storage system or, where explicitly stated, the active storage material. Analogously, the volumetric energy density is defined as (Wu et al., 2012; Nagde and Dhoble, 2021; Borah et al., 2020),

$$\varepsilon_v = \frac{E_{\text{stored}}}{V} \tag{2.26}$$

with ε_v expressed in Wh.L^{-1} and V denoting the total system volume. In practical comparisons, system-level energy density is generally more meaningful than material-level values, as inactive components such as structural supports, current collectors, containment vessels, and thermal management systems often contribute substantially to overall mass and volume (Wu et al., 2012; Nagde and Dhoble, 2021; Borah et al., 2020).

In electrochemical energy storage systems, the recoverable energy can be conveniently expressed in terms of the voltage-capacity relationship,

$$E_{\text{stored}} \approx \int_{Q_0}^{Q_f} V(Q)dQ \tag{2.26}$$

where $V(Q)$ is the operating voltage as a function of delivered charge Q. This formulation emphasizes that energy density depends not only on capacity but also on the shape of the voltage profile, which reflects both thermodynamic driving forces and kinetic losses (Wu et al., 2012; Nagde and Dhoble, 2021; Borah et al., 2020). Consequently, two systems with identical capacities may exhibit markedly different energy densities if their voltage characteristics differ.

Power density, by contrast, characterizes the rate at which stored energy can be accessed (Elalfy et al., 2024; Zhao and Burke, 2021). The instantaneous power output of a storage system is given by Eq. (2.27).

$$P(t) = V(t)\, I(t) \tag{2.27}$$

where $V(t)$(V) and $I(t)$(A) are the time-dependent voltage and current, respectively. The gravimetric power density $\left(P_g;\ \text{W.kg}^{-1}\right)$ is defined as,

$$P_g = \frac{P_{\max}}{m} \tag{2.28}$$

and the volumetric power density (P_v; W.L^{-1}) as,

$$P_v = \frac{P_{\max}}{V} \tag{2.29}$$

where P_{max} is the maximum sustainable power output (W) under specified operating conditions. Unlike energy density, power density is governed primarily by kinetic and transport limitations, including charge-transfer rates, ionic conductivity, electronic resistance, and heat dissipation capability.

For electrochemical systems, the maximum deliverable power is often limited by internal resistance and polarization losses (Zhao and Burke, 2021; Conway and Pell, 2002). Under simplified conditions, the maximum power may be approximated as,

$$P_{max} \approx \frac{V_{oc}^2}{4R_{int}} \tag{2.29}$$

where V_{oc} is the open-circuit voltage (V) and R_{int} is the effective internal resistance (Ω). This relationship highlights the central role of transport and interfacial processes in determining power capability and illustrates why systems optimized for high power density typically sacrifice energy density.

The intrinsic trade-off between energy density and power density can be quantified through the characteristic discharge time (t_d; h),

$$t_d = \frac{E_{stored}}{P_{max}} \tag{2.30}$$

which provides a convenient measure of whether a storage system is energy-dominated or power-dominated. Large values of t_d correspond to long-duration storage applications, whereas small values indicate suitability for rapid power delivery.

The practical implications of the energy-power trade-off are illustrated in Fig. 2.10 which maps representative energy storage technologies in terms of storage capacity and characteristic discharge time (Liang et al., 2024). Systems such as PHES and thermodynamic electricity storage occupy the high-capacity, long-duration region, reflecting their suitability for bulk energy shifting and long-term balancing. In contrast, electrical storage technologies such as SMES and flywheels are confined to short discharge times, consistent with their high-power density and rapid response but limited energy capacity (Ries and Neumueller, 2001). Electrochemical storage systems span an intermediate regime, offering a balance between capacity and discharge duration that enables both energy- and power-oriented applications. This capacity-duration representation highlights that energy density and power density alone do not fully capture system functionality; instead, the characteristic discharge time provides an application-relevant metric for differentiating storage technologies across operational timescales (Liang et al., 2024; Ries and Neumueller, 2001).

Energy density and power density therefore offer complementary but incomplete descriptions of storage performance. While they capture fundamental capacity and rate limitations, they do not account for efficiency losses, lifetime degradation, or economic viability. These considerations are addressed through efficiency- and cost-based metrics, which are discussed in the following subsections.

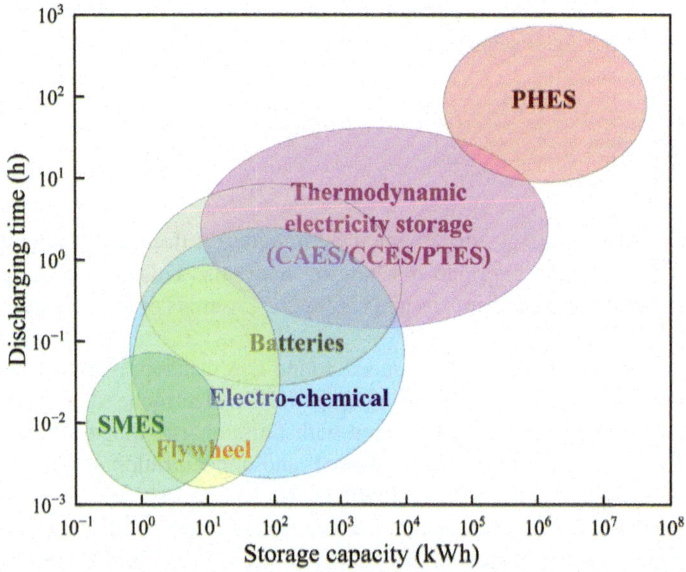

Figure 2.10 Capacity–duration map of representative energy storage technologies. Capacity–duration map of representative energy storage technologies, illustrating the relationship between storage capacity and characteristic discharge time. The figure highlights the distinct operational regimes of electrical (e.g., SMES, flywheels), electrochemical (batteries), thermodynamic electricity storage (e.g., CAES, CCES, PTES), and large-scale mechanical systems such as pumped hydro energy storage (PHES), reflecting their suitability for power-oriented or energy-oriented applications.
Reprinted with permission from Liang, Y., Peng, L., Xing, L., Su, W., Li, W., XuW., 2024. Current status of thermodynamic electricity storage: Principle, structure, storage device and demonstration. J. Energy Storage80. https://doi.org/10.1016/j.est.2023.110347 © 2023 Published by Elsevier Ltd.

2.5.2 *Efficiency metrics: Round-trip, coulombic, cycling efficiency, and energy losses*

Round-trip efficiency quantifies the fraction of input energy that can be recovered after a complete charge-discharge cycle and is a central metric for assessing the practical performance of energy storage systems. Unlike energy and power densities, which describe capacity and rate capability, efficiency reflects the cumulative impact of irreversible processes occurring during energy conversion, storage, and retrieval. As such, it provides a direct measure of how closely a storage system operates relative to ideal, reversible behavior. It is important to distinguish round-trip efficiency, which quantifies energy recovery, from coulombic efficiency, which reflects charge conservation. While closely related in electrochemical systems, the two metrics capture different loss mechanisms and are not interchangeable.

In its most general form, the round-trip efficiency η_{rt} is defined as the ratio of energy delivered during discharge to the energy supplied during charging (Bamdezh et al., 2025),

$$\eta_{\mathrm{rt}} = \frac{E_{\mathrm{discharge}}}{E_{\mathrm{charge}}} \tag{2.31}$$

where E_{charge} and $E_{\mathrm{discharge}}$ are obtained by integrating the input and output power over the respective charging and discharging intervals. This definition applies universally across storage mechanisms and captures all sources of loss, regardless of their physical origin.

Energy losses contributing to reduced round-trip efficiency arise from multiple mechanisms that act concurrently during operation. In electrochemical systems, losses include activation polarization associated with finite charge-transfer kinetics, ohmic losses due to electronic and ionic resistance, and concentration polarization arising from mass-transport limitations (Emrani et al., 2022). These losses manifest as voltage hysteresis between charge and discharge, requiring additional input energy to store a given amount of recoverable energy (Emrani et al., 2022). The electrical energy dissipated through these processes is converted into heat, increasing system entropy and reducing exergy.

Coulombic efficiency (η_{C}) quantifies the degree to which electrical charge is conserved during a charge-discharge cycle in an electrochemical energy storage system (Malekmohammadi and Mirbagheri, 2024). It reflects the fraction of injected charge that can be reversibly recovered and therefore serves as a direct indicator of parasitic reactions and irreversible charge-consuming processes (Malekmohammadi and Mirbagheri, 2024). Unlike round-trip efficiency, coulombic efficiency is independent of voltage and does not account for energy losses associated with polarization or resistive dissipation. Coulombic efficiency is defined by Eq. (2.32).

$$\eta_{\mathrm{C}} = \frac{Q_{\mathrm{discharge}}}{Q_{\mathrm{charge}}} \tag{2.32}$$

where Q_{charge} is the total electrical charge supplied during charging and $Q_{\mathrm{discharge}}$ is the charge recovered during discharge.

In ideal reversible systems, coulombic efficiency approaches unity. Deviations from unity arise from irreversible side reactions such as electrolyte decomposition, solid-electrolyte interphase growth, gas evolution, or self-discharge. Because these processes permanently consume charge carriers, coulombic efficiency is particularly sensitive to degradation mechanisms and is often used as an early diagnostic metric for long-term stability. However, a high coulombic efficiency does not guarantee high energy efficiency, as voltage losses may still be substantial.

Cyclic efficiency (η_{cyc}) is a more general and context-dependent metric that characterizes efficiency over repeated charge-discharge cycles (Liu et al., 2019; Amirante et al., 2017). Unlike coulombic efficiency, which strictly measures charge conservation, cyclic efficiency is often intended to describe the average efficiency of a storage system across multiple cycles, incorporating cumulative effects such as degradation, efficiency decay, and operational losses (Liu et al., 2019; Amirante et al., 2017). Because the term is not uniquely defined in the literature, its precise meaning must always be explicitly stated. In many electrochemical studies, cyclic efficiency is defined as the average round-trip efficiency over N cycles:

$$\eta_{\mathrm{cyc}} = \frac{1}{N} \sum_{k=1}^{N} \frac{E_{\mathrm{discharge},k}}{E_{\mathrm{charge},k}} \tag{2.33}$$

where $E_{\mathrm{charge},k}$ and $E_{\mathrm{discharge},k}$ denote the charged and discharged energies in the k-th cycle.

In electrochemical systems, round-trip efficiency can be decomposed into coulombic efficiency and voltage efficiency:

$$\eta_{\mathrm{rt}} = \eta_{\mathrm{C}} \times \eta_{\mathrm{V}} \tag{2.34}$$

where the voltage efficiency η_{V} is defined as

$$\eta_{\mathrm{V}} = \frac{\langle V_{\mathrm{discharge}} \rangle}{\langle V_{\mathrm{charge}} \rangle} \tag{2.35}$$

This relationship clarifies why systems may exhibit near-unity coulombic efficiency while still suffering from poor round-trip efficiency due to voltage polarization and resistive losses.

Efficiency can also be expressed in terms of instantaneous power and loss mechanisms. The net output power during discharge may be written as

$$P_{\mathrm{out}} = P_{\mathrm{ideal}} - P_{\mathrm{loss}} \tag{2.36}$$

where P_{ideal} represents the reversible power output and P_{loss} accounts for all dissipative contributions. Integrating this expression over time highlights how even modest instantaneous losses can significantly reduce round-trip efficiency over a full cycle.

Importantly, round-trip efficiency is not an intrinsic material constant but a system-level metric that depends on design choices, operating conditions, and environmental factors (Bamdezh et al., 2025). Temperature, current density, depth of discharge, and cycling protocol all influence the magnitude of losses and therefore the effective efficiency. Consequently, efficiency values reported for different technologies must be interpreted within the context of standardized testing conditions and application-specific requirements.

Round-trip efficiency provides a critical link between physical performance and economic viability (Bamdezh et al., 2025). Lower efficiency increases the required input energy for a given delivered output, raising operating costs and, in grid-scale applications, increasing upstream generation and transmission demands. As such, efficiency directly influences system sizing, operational strategy, and lifecycle cost, positioning it as a key metric alongside energy density and power density in the evaluation of energy storage technologies.

2.5.3 Levelized cost of storage

While energy density, power density, and round-trip efficiency describe the technical performance of energy storage systems, economic viability ultimately determines their adoption and deployment. The Levelized Cost of Storage (LCOS) provides a comprehensive metric that captures the average cost of delivering a unit of stored energy over the operational lifetime of a storage system (Khosravani et al., 2025). By integrating capital expenditure, operational costs, efficiency losses, and degradation effects, LCOS enables consistent comparison between storage technologies with fundamentally different mechanisms, scales, and lifetimes.

LCOS is defined as the ratio of the total discounted costs incurred over the lifetime of the storage system to the total discounted energy delivered during the same period (Khosravani et al., 2025). In its general form, LCOS may be expressed as

$$\mathrm{LCOS} = \frac{\sum_{t=0}^{N} \frac{C_t}{(1+r)^t}}{\sum_{t=0}^{N} \frac{E_t}{(1+r)^t}} \tag{2.37}$$

where C_t represents the total cost incurred in year t, E_t is the energy delivered to the end user in year t, r is the discount rate, and N is the system lifetime in years. This formulation accounts for the time value of money and reflects the cumulative economic burden associated with storing and releasing energy.

The cost term C_t typically includes capital investment, fixed and variable operation and maintenance costs, replacement or refurbishment expenses, and end-of-life costs where applicable. Capital expenditure is often front-loaded and dominates early years, whereas operational and maintenance costs accrue over the system's lifetime. Technologies with long service lives and low degradation rates, such as mechanical storage systems, benefit from extended cost amortization, whereas systems requiring frequent replacement of active components may exhibit higher LCOS despite favorable technical performance.

The delivered energy term E_t depends on both the nominal storage capacity and the efficiency with which energy is recovered. It may be written as

$$E_t = \eta_{\mathrm{rt}} \, E_{\mathrm{in},t} \tag{2.38}$$

where η_{rt} is the round-trip efficiency and $E_{\mathrm{in},t}$ is the energy supplied for charging in year t. Efficiency losses therefore directly increase LCOS by reducing the amount of usable energy delivered for a given input. In addition, degradation mechanisms that reduce capacity or efficiency over time lower cumulative delivered energy, further increasing LCOS even if upfront costs remain unchanged.

Cycling behavior exerts a strong influence on LCOS. Systems designed for frequent cycling with high power throughput may deliver substantial cumulative energy despite moderate energy density, while long-duration systems cycled infrequently may deliver less total energy over their lifetime. The effective utilization rate, defined by depth of discharge and cycle frequency, therefore plays a critical role in determining economic performance. LCOS inherently captures this dependence by integrating delivered energy over the full operating horizon.

Importantly, LCOS is sensitive to system boundaries and assumptions. Inclusion or exclusion of balance-of-system components, charging energy costs, auxiliary power consumption, and grid integration expenses can significantly alter calculated values. As a result, LCOS comparisons must be interpreted within a clearly defined methodological framework to ensure consistency. Despite this sensitivity, LCOS remains a powerful metric because it explicitly links physical performance to economic outcomes.

From a design perspective, minimizing LCOS requires coordinated optimization of multiple factors rather than maximizing any single performance metric. High energy density reduces capital cost per unit energy, high efficiency increases delivered energy, long cycle life extends cost amortization, and low maintenance requirements reduce operational expenses. Trade-offs among these factors are unavoidable, and LCOS provides a unified basis for evaluating their combined effect.

LCOS thus serves as a critical bridge between engineering performance and system-level decision-making. By capturing how thermodynamic limits, kinetic losses, degradation pathways, and operational strategies translate into economic cost, LCOS enables objective assessment of energy storage technologies across diverse applications. Its use is essential for guiding technology selection, policy development, and investment decisions in modern energy systems.

2.6 Efficiency and losses: Technical and economic factors

Efficiency losses in energy storage systems arise from fundamental physical processes and propagate upward to influence system design, operational strategy, and economic performance. While earlier sections quantified efficiency through performance metrics, a deeper understanding of the *origin* of losses is essential for interpreting these metrics and for identifying pathways toward improved system-level performance. Importantly, losses are not merely technical imperfections; they translate directly into higher capital requirements, increased operating costs, and reduced lifetime value, thereby shaping the levelized cost of storage and overall economic viability.

At the technical level, losses can be broadly categorized according to the physical processes responsible for energy dissipation. In electrochemical systems, ohmic losses arise from finite electronic and ionic conductivity within electrodes, electrolytes, current collectors, and interfacial contacts. These losses scale with current and manifest as resistive heating, contributing to voltage drop during discharge and increased energy input during charging. Ohmic losses not only reduce instantaneous efficiency but also impose thermal management requirements that increase system complexity and cost.

Kinetic losses, often expressed through activation overpotentials, originate from finite reaction rates at interfaces. Even when reactions are thermodynamically favorable, energy must be expended to overcome activation barriers associated with electron transfer, bond rearrangement, and solvent or lattice reorganization. These losses become increasingly significant at high current densities and are sensitive to material properties, surface chemistry, and operating temperature. From an economic perspective, kinetic losses limit achievable power density and necessitate oversizing of active materials or electrode area, increasing capital expenditure.

Mass transport losses arise when the supply or removal of reactive species cannot keep pace with reaction demand. Concentration gradients near interfaces reduce local electrochemical potential, forcing additional overpotential to sustain current. In batteries and electrochemical capacitors, such losses are exacerbated by thick electrodes, low diffusivity electrolytes, or high-rate operation. Transport-induced inefficiencies often drive nonuniform utilization of active material, accelerating degradation and reducing effective lifetime, with direct implications for replacement costs and LCOS.

Beyond electrochemical systems, thermal losses are a dominant factor in mechanical and thermal energy storage technologies. In thermal storage, heat leakage to the environment, temperature gradients within storage media, and irreversible heat transfer across finite temperature differences degrade the quality of stored energy. In mechanical systems, friction, turbulence, and nonideal fluid flow convert mechanical work into heat. Although these losses may appear modest on a per-cycle basis, their cumulative effect over long operating periods can substantially reduce recoverable energy and increase operating costs.

Parasitic consumption represents another critical but often underappreciated source of inefficiency. Auxiliary systems such as pumps, compressors, control electronics, cooling systems, and power conditioning units consume energy continuously or intermittently, reducing net delivered energy. Parasitic loads are particularly relevant in large-scale and long-duration storage, where auxiliary consumption may persist even when the system is idle. These losses directly reduce round-trip efficiency and increase LCOS by lowering delivered energy without reducing capital cost.

Closely related are standby losses, which occur when stored energy decays over time in the absence of active operation. Self-discharge in electrochemical systems, thermal leakage in heat storage, pressure loss in compressed systems, and magnetic flux decay in inductive storage all contribute to standby losses. These effects are especially consequential for applications requiring long storage durations, where energy retention rather than power capability is the dominant performance requirement. Standby losses reduce effective capacity utilization and can necessitate more frequent recharging, increasing both energy input costs and component wear.

From an economic standpoint, all loss mechanisms ultimately influence the levelized cost of storage through two primary pathways. First, losses reduce the amount of usable energy delivered over the system lifetime, increasing the denominator in LCOS calculations. Second, mitigating losses often requires additional investment in materials, infrastructure, and control systems, increasing capital expenditure (CAPEX) and operational expenditure (OPEX). For example, improving efficiency through thicker electrodes, higher-conductivity materials, advanced thermal management, or tighter system control may raise upfront cost even as it improves performance.

The relative importance of different loss mechanisms depends strongly on application context. High-power applications are dominated by kinetic and ohmic losses, while long-duration storage is more sensitive to standby and parasitic losses. Grid-scale systems prioritize lifetime efficiency and reliability, whereas mobile applications may tolerate higher losses in exchange for reduced mass or volume. As a result, efficiency optimization cannot be decoupled from economic optimization; both must be considered simultaneously within the intended operational framework.

Efficiency and losses therefore serve as the central link between physical principles and economic outcomes in energy storage systems. Understanding where and how energy is dissipated enables targeted engineering interventions that improve not only technical performance but also cost-effectiveness. This integrated perspective is essential for evaluating competing storage technologies and for guiding the development of systems capable of meeting future energy demands with both physical efficiency and economic sustainability.

2.7 Conclusion and future outlook

This chapter has established a framework for understanding energy storage systems by linking thermodynamic foundations, mechanistic processes, storage modalities, performance metrics, and efficiency losses. Rather than treating these elements in isolation, the discussion has emphasized their interdependence. Thermodynamic limits define what is theoretically possible, kinetic and transport mechanisms determine what is practically achievable, and system-level losses translate physical constraints into economic outcomes. This integrated perspective is essential for interpreting the performance of existing technologies and for guiding the rational development of future energy storage systems.

A central theme emerging from this analysis is that no single storage mechanism or metric can be universally optimized. Chemical, mechanical, electrical, and thermal storage systems each occupy distinct regions of the energy-power-efficiency-cost landscape, shaped by the form in which energy is stored and retrieved. Performance metrics such as energy density, power density, efficiency, and LCOS do not merely rank technologies; they encode the underlying physics and engineering trade-offs that govern system behavior. Meaningful comparison therefore requires application-specific evaluation rather than absolute benchmarking.

Efficiency losses have been shown to play a pivotal role in bridging fundamental physics and economic performance. Ohmic, kinetic, transport, thermal, parasitic, and standby losses arise from distinct physical origins, yet converge in their impact on delivered energy, lifetime utilization, and cost. Importantly, mitigating these losses often introduces new constraints or costs, underscoring the need for holistic optimization strategies that balance performance gains against system complexity and economic viability.

Looking forward, advances in energy storage will increasingly depend on cross-scale integration rather than incremental improvements within isolated components. At the materials level, progress will hinge on stabilizing high-energy states while maintaining kinetic accessibility and interfacial robustness. At the device level, architectures that decouple energy and power functions, manage thermal and mechanical stresses, and suppress parasitic processes will be critical. At the system level, intelligent control, hybridization of complementary storage technologies, and application-aware operation will play a growing role in maximizing usable energy and minimizing cost.

Equally important is the evolution of evaluation frameworks. As energy systems diversify, standardized metrics must be complemented by context-sensitive assessment methods that account for duty cycle, storage duration, degradation pathways, and integration costs. Metrics such as LCOS will remain indispensable, but their interpretation must be grounded in a clear understanding of the physical processes that govern efficiency and loss.

In conclusion, energy storage is inherently a multidisciplinary challenge, situated at the intersection of thermodynamics, electrochemistry, mechanics, heat transfer, and economics. Progress in this field will not be driven by singular breakthroughs but by coordinated advances that align fundamental principles with engineering design and economic reality. The framework presented in this chapter provides a foundation for such integration, supporting informed comparison of technologies and guiding the development of energy storage systems capable of meeting the performance, reliability, and sustainability demands of future energy infrastructures.

AI disclosure

During the preparation of this work the author(s) used ChatGPT 5.0 in order to During the preparation of this work the authors used ChatGPT to improve the readability and language of the manuscript. After using this tool/service, the authors reviewed and edited the content as needed and take full responsibility for the content of the publication. After using this tool/service, the author(s) reviewed and edited the content as needed and take(s) full responsibility for the content of the publication.

References

Adetokun, B.B., Oghorada, O., Abubakar, S.J.a., 2022. Superconducting magnetic energy storage systems: prospects and challenges for renewable energy applications. J. Energy Storage 55. https://doi.org/10.1016/j.est.2022.105663.

Agbaje, M.A., Akkaya, A.V., 2023. Electrochemical-energy- exergy analysis of reversible solid oxide cell-based small-scale stand-alone energy storage system. Case Stud. Therm. Eng. 52. https://doi.org/10.1016/j.csite.2023.103732.

Ahmed, M.M.S., Hasan, M.J., Chowdhury, M.S., Rahman, M.K., Islam, M.S., Hossain, M.S., Islam, M.A., Hossain, N., Mobarak, M.H., 2024. Prospects and challenges of energy storage materials: a comprehensive review. Chem. Eng. J. Adv. 20. https://doi.org/10.1016/j.ceja.2024.100657.

Aktaş, A., Kirçiçek, Y., 2021. Hybrid Energy Storage and Innovative Storage Technologies. Elsevier BV, pp. 139–152. https://doi.org/10.1016/b978-0-323-88499-0.00007-0.

Alsaç, E.P., Nelson, D.L., Yoon, S.G., Cavallaro, K.A., Wang, C., Sandoval, S.E., Eze, U.D., Jeong, W.J., McDowell, M.T., 2025. Characterizing electrode materials and interfaces in solid-state batteries. Chem. Rev. 125 (4), 2009–2119.

AlShafi, M., Bicer, Y., 2021. Thermodynamic performance comparison of various energy storage systems from source-to-electricity for renewable energy resources. Energy 219, 119626. https://doi.org/10.1016/j.energy.2020.119626.

Amir, M., Deshmukh, R.G., Khalid, H.M., Said, Z., Raza, A., Muyeen, S.M., Nizami, A.S., Elavarasan, R.M., Saidur, R., Sopian, K., 2023. Energy storage technologies: an integrated survey of developments, global economical/environmental effects, optimal scheduling model, and sustainable adaption policies. J. Energy Storage 72. https://doi.org/10.1016/j.est.2023.108694.

Amirante, R., Cassone, E., Distaso, E., Tamburrano, P., 2017. Overview on recent developments in energy storage: mechanical, electrochemical and hydrogen technologies. Energy Convers. Manag. 132, 372–387. https://doi.org/10.1016/j.enconman.2016.11.046.

An, D., Li, Y., Lin, X., Teng, S., 2023. Analysis of compression/expansion stage on compressed air energy storage cogeneration system. Front. Energy Res. 11. https://doi.org/10.3389/fenrg.2023.1278289.

Anantharaj, S., Noda, S., Driess, M., Menezes, P.W., 2021. The pitfalls of using potentiodynamic polarization curves for tafel analysis in electrocatalytic water splitting. ACS Energy Lett. 6 (4), 1607–1611. https://doi.org/10.1021/acsenergylett.1c00608.

Angenent, L.T., Casini, I., Schröder, U., Harnisch, F., Molitor, B., 2024. Electrical-energy storage into chemical-energy carriers by combining or integrating electrochemistry and biology. Energy Environ. Sci. 17 (11), 3682–3699. https://doi.org/10.1039/d3ee01091k.

Arabkoohsar, A., 2023. 1 - classification of energy storage systems. In: Future Grid-Scale Energy Storage Solutions Mechanical and Chemical Technologies and Principles, pp. 1–30.

Azega, R.K., Smith, A.D., Roy Chowdhury, N., Vyas, A., Qi, L., Haque, M., Xun, Q., Zhang, X., Thurakkal, S., Thiringer, T., Enoksson, P., Lundgren, P., 2024. Supercapacitors and rechargeable batteries, a tale of two technologies: past, present and beyond. Sustain. Mater. Technol. 41, e01111. https://doi.org/10.1016/j.susmat.2024.e01111.

Bae, J., Kim, D., Jung, J.H., Ihm, J., 2017. A computational study on hydrogen storage in potential wells using K-intercalated graphite oxide. RSC Adv. 7 (54), 33953–33960. https://doi.org/10.1039/c7ra05173e.

Bamdezh, M.A., Molaeimanesh, G.R., Mohammadi Saman Kand, R., Mostajeran, M., 2025. Effects of temperate, cold, and warm climate conditions on electric vehicle and battery thermal management system performance: energy usage, battery aging, and carbon footprints. Energy Convers. Manag. 339, 119935. https://doi.org/10.1016/j.enconman.2025.119935.

Bao, H., Ma, Z., 2022. Thermochemical Energy Storage Storing Energy: With Special Reference to Renewable Energy Sources. Elsevier, United Kingdom, pp. 651–683. https://doi.org/10.1016/B978-0-12-824510-1.00028-3.

Barbir, F., 2012. Fuel Cell Basic Chemistry and Thermodynamics PEM Fuel Cells: Theory and Practice. Elsevier, pp. 17–32. https://doi.org/10.1016/B978-0-12-387710-9.00002-3.

Barello, M., Manca, D., Patel, R., Mujtaba, I.M., 2014. Operation and modeling of RO desalination process in batch mode. Comput. Aided Chem. Eng. 33, 451–456. https://doi.org/10.1016/B978-0-444-63456-6.50076-4.

Barrera, L., Layne, B.W., Chen, Z., Watanabe, K., Kudo, A., Esposito, D.V., Ardo, S., Bala Chandran, R., 2024. Revealing the role of redox reaction selectivity and mass transfer in current-voltage predictions for ensembles of photocatalysts. Energy Environ. Sci. 17 (21), 8254–8273. https://doi.org/10.1039/d4ee02005g.

Begum Medha, M., Mahmud, I., Hasanuzzaman, Md, 2026. Chapter 14 - Electrical energy storage technology. In: Advanced Renewable Energy Conversion Technologies and Power Plant Engineering, pp. 503–549.

Bianciardi, C., Ulgiati, S., 2004. Encyclopedia of EnergyEntropy, pp. 459–470. https://doi.org/10.1016/B0-12-176480-X/00092-9.

Boampong, R., Brown, D.P., 2020. On the benefits of behind-the-meter rooftop solar and energy storage: the importance of retail rate design. Energy Econ. 86, 104682. https://doi.org/10.1016/j.eneco.2020.104682.

Boettcher, S.W., Oener, S.Z., Lonergan, M.C., Surendranath, Y., Ardo, S., Brozek, C., Kempler, P.A., 2021. Potentially confusing: potentials in electrochemistry. ACS Energy Lett. 6 (1), 261–266. https://doi.org/10.1021/acsenergylett.0c02443.

Boldini, A., Porfiri, M., 2022. A non-ideal solution theory for the mechanics and electrochemistry of charged membranes. npj Comput. Mater. 8 (1). https://doi.org/10.1038/s41524-022-00827-2.

Borah, R., Hughson, F.R., Johnston, J., Nann, T., 2020. On battery materials and methods. Mater. Today Adv. 6, 100046. https://doi.org/10.1016/j.mtadv.2019.100046.

Cao, W., Zhang, J., Li, H., 2020. Batteries with high theoretical energy densities. Energy Storage Mater. 26, 46–55. https://doi.org/10.1016/j.ensm.2019.12.024.

Cao, E., Shi, X., Inoue, H., Oshikiri, T., Liu, Y.E., Misawa, H., 2025. Activation energy in the electron transfer process and water oxidation intermediate generation under plasmon-nanocavity strong coupling. J.Phys. Chem. C 129 (3), 1590–1597.

Cappabianca, R., De Angelis, P., Fasano, M., Chiavazzo, E., Asinari, P., 2023. An overview on transport phenomena within solid electrolyte interphase and their impact on the performance and durability of lithium-ion batteries. Energies 16 (13), 5003. https://doi.org/10.3390/en16135003.

Carrasco, J., 2024. A theoretical perspective on solid-state ionic interfaces. Philos. Trans. R. Soc. A Math. Phys. Eng. Sci. 382 (2281). https://doi.org/10.1098/rsta.2023.0313.

Cesaro, Z., Nayak-Luke, R.M., Bãnares-Alcántara, R., 2020. Energy Storage Technologies: Power-To-X Techno-Economic Challenges of Green Ammonia as an Energy Vector. Elsevier, United Kingdom, pp. 15–26. https://doi.org/10.1016/B978-0-12-820560-0.00002-3.

Chadly, A., Urs, R.R., Wei, M., Maalouf, M., Mayyas, A., 2023. Techno-economic assessment of energy storage systems in green buildings while considering demand uncertainty. Energy Build. 291, 113130. https://doi.org/10.1016/j.enbuild.2023.113130.

Chen, X., Aziz, M., 2025. Solid-gas thermochemical energy storage materials for renewable energy integration in power grids. Energy Storage Mater. 80, 104340. https://doi.org/10.1016/j.ensm.2025.104340.

Chen, W., Wang, K., Li, Y., Chen, J., Wang, H., Li, L., Li, H., Ren, X., Ouyang, X., Liu, J., Pan, F., Xiao, B., Zhang, Q., Hu, J., 2024. Minimize the electrode concentration polarization for high-power lithium batteries. Adv. Funct. Mater. 34 (52). https://doi.org/10.1002/adfm.202410926.

Conway, B.E., Pell, W.G., 2002. Power limitations of supercapacitor operation associated with resistance and capacitance distribution in porous electrode devices. J. Power Sources 105 (2), 169–181. https://doi.org/10.1016/s0378-7753(01)00936-3.

Dai, Y., Zhang, C., Zhang, X., Jiang, P., Chen, J., Zong, W., Zheng, S., Gao, X., Macdonald, T. J., He, G., 2025. Interfacial energy storage in aqueous zinc-ion batteries. Energy Environ. Sci. https://doi.org/10.1039/d5ee03741g.

Ding, Y., 2021. Chapter 1. Thermodynamics for Thermal Energy Storage. Royal Society of Chemistry (RSC), pp. 1–14. https://doi.org/10.1039/9781788019842-00001.

Dourado, A.H.B., 2022. Electric double layer: the good, the bad, and the beauty. Electrochem 3 (4), 789–808. https://doi.org/10.3390/electrochem3040052.

Du, Y., Pei, W., Yan, H., Zhang, P., Wang, C., Yang, Y., Pang, M., Zhang, L., 2025. Life cycle assessment of the pumped hydro energy storage system in China. Energy 331, 137102. https://doi.org/10.1016/j.energy.2025.137102.

Dulgerbaki, C., Nohut Maslakci, N., Komur, A.I., Uygun Oksuz, A., 2016. Chemical synthesis and characterization of poly(3,4-ethylenedioxythiophene)/tungsten composite materials in the presence of ionic liquids. Compos. Interfaces 23 (4), 297–307. https://doi.org/10.1080/09276440.2016.1137707.

Dupin, V., Teixeira, D., 2025. Advanced adiabatic compressed air energy storage systems dynamic modelling: impact of the heat storage device. Heliyon 11 (1), e40730. https://doi.org/10.1016/j.heliyon.2024.e40730.

Elalfy, D.A., Gouda, E., Kotb, M.F., Bureš, V., Sedhom, B.E., 2024. Comprehensive review of energy storage systems technologies, objectives, challenges, and future trends. Energy Strategy Rev. 54. https://doi.org/10.1016/j.esr.2024.101482.

Enasel, E., Dumitrascu, G., 2025. Storage solutions for renewable energy: a review. Energy Nexus 17, 100391. https://doi.org/10.1016/j.nexus.2025.100391.

Emrani, A., Berrada, A., Ameur, A., Mohamed, B., 2022. Assessment of the round-trip efficiency of gravity energy storage system: Analytical and numerical analysis of energy loss mechanisms. J. Energy Storage 55, 105504. https://doi.org/10.1016/j.est.2022.105504.

Fang, M., Cheng, J., Li, X., Zhang, C., 2025. Degradation mechanisms of redox-active molecules toward long-cycle aqueous organic flow batteries. ACS Appl. Energy Mater. 8 (19), 14001–14013. https://doi.org/10.1021/acsaem.5c01782.

Ferguson, S.J., 2012. 8.1 Ion Electrochemical Gradients, Roles and Measurements, vol 8. Elsevier BV, pp. 1–8. https://doi.org/10.1016/b978-0-12-374920-8.00801-8.

Gabitto, J.F., Tsouris, C., 2023. A review of transport models in charged porous electrodes. Front. Chem. Eng. 4. https://doi.org/10.3389/fceng.2022.1051594.

Gedde, U.W., 2020. Essential Classical ThermodynamicsAn Introduction to Thermodynamics and the First Law SpringerBriefs in Physics. Springer, Cham, pp. 1–12. https://doi.org/10.1007/978-3-030-38285-8_1.

Goonetilleke, D., Silvan, B., Gonzalo, E., Galcerán, M., Casas-Cabanas, M., Avdeev, M., Fauth, F., Rojo, T., Sharma, N., Saurel, D., 2025. Rate-dependent structure-electrochemistry relationships and origins of capacity fading in P2-type Na2/3Fe2/3Mn1/3O2. Inorg. Chem. Front. 12 (7), 2731–2746. https://doi.org/10.1039/d4qi02804j.

Guo, J., Guo, Q., Liu, J., Wang, H., 2023. The polarization and heat generation characteristics of lithium-ion battery with electric–thermal coupled modeling. Batteries 9 (11). https://doi.org/10.3390/batteries9110529.

Gupta, S., Narajczyk, M., Sawczak, M., Bogdanowicz, R., 2025. Perspectives on electron transfer kinetics across graphene-family nanomaterials and interplay of electronic structure with defects and quantum capacitance. Sci.Rep. 15 (1). https://doi.org/10.1038/s41598-025-04357-x.

Hadidi, A., 2024. Thermodynamic design of the novel energy storage system based on liquid carbon dioxide for a 17 MW concentrated solar thermal power plant. J. Energy Storage 83, 110761. https://doi.org/10.1016/j.est.2024.110761.

Hai, T., Omar, I., Alizadeh, A., Varshney, N., Dixit, S., Sultan, A.J., Anqi, A.E., Bhatnagar, S., Rajab, H., Singh Sawaran Singh, N., 2025. Optimization of nano-finned enclosure-shaped latent heat thermal energy storage units using CFD, RSM, and enhanced hill climbing algorithm. Sci.Rep. 15 (1). https://doi.org/10.1038/s41598-025-96599-y.

Hammes-Schiffer, S., Stuchebrukhov, A.A., 2010. Theory of coupled electron and proton transfer reactions. Chem. Rev. 110 (12), 6939–6960. https://doi.org/10.1021/cr1001436.

Hassan, Q., Algburi, S., Sameen, A.Z., Salman, H.M., Jaszczur, M., 2023. A review of hybrid renewable energy systems: solar and wind-powered solutions: challenges, opportunities, and policy implications. Results Eng. 20. https://doi.org/10.1016/j.rineng.2023.101621.

Hausbrand, R., Jaegermann, W., 2018. Reaction Layer Formation and Charge Transfer at Li-Ion Cathode-Electrolyte Interfaces: Concepts and Results Obtained by a Surface Science Approach Encyclopedia of Interfacial Chemistry: Surface Science and Electrochemistry. Elsevier, Germany, pp. 232–245. https://doi.org/10.1016/B978-0-12-409547-2.14195-2.

Haynes, C., 2001. Clarifying reversible efficiency misconceptions of high temperature fuel cells in relation to reversible heat engines. J. Power Sources 92 (1–2), 199–203. https://doi.org/10.1016/S0378-7753(00)00541-3.

Honig, J.M., 2007. Thermodynamics (Third Edition) Principles Characterizing Physical and Chemical ProcessesChapter 1 - Fundamentals, 3, pp. 1–110. https://doi.org/10.1016/B978-012373877-6/50003-4.

Horlock, J.H., 2003. Advanced Gas Turbine CyclesChapter 4 - Cycle Efficiency With Turbine Cooling (Cooling Flow Rates Specified), pp. 47–69.

Horzela-Miś, A., Semrau, J., 2025. The role of renewable energy and storage technologies in sustainable development: simulation in the construction industry. Front. Energy Res. 13. https://doi.org/10.3389/fenrg.2025.1540423.

Huang, X., Zhang, Z., Jiang, J., 2006. Fuel cell technology for distributed generation: an overview. IEEE Int. Symp. Ind. Electron. 2, 1613–1618. https://doi.org/10.1109/ISIE.2006.295713.

Huang, S., Lv, P., Duan, H., 2019. Morphology evolution of liquid–gas interface on submerged solid structured surfaces. Extreme Mech. Lett. 27, 34–51. https://doi.org/10.1016/j.eml.2019.01.005.

Huang, K., 2024. Principles of Electrochemical Conversion and Storage Devices. Wiley. https://doi.org/10.1002/9783527851638.

Huo, Y., Shen, S., Zhou, H., Wang, S., Rao, Z., 2023. Using entropy generation as evaluating indicator of charging completion in a latent thermal energy storage system. Int. Commun. Heat Mass Tran. 146, 106858. https://doi.org/10.1016/j.icheatmasstransfer.2023.106858.

Irwin, L., Stekli, J., Pfefferkorn, C., Pitchumani, R., 2017. Thermochemical Energy Storage for Concentrating Solar Thermal (CST) Systems. Elsevier BV, pp. 247–267. https://doi.org/10.1016/b978-0-08-100516-3.00011-3.

Iwase, M., 2014. Treatise on Process Metallurgy Volume 1: Process FundamentalsChapter 3.1 - First, Second, and Third Laws of Thermochemistry, pp. 399–493. https://doi.org/10.1016/B978-0-08-096986-2.00030-8.

Jeon, D.H., Kim, S., Hempelmann, R., 2025. State-of-the-art review of degradation mechanisms of commercial lithium-ion batteries. J. Power Sources 646, 237242. https://doi.org/10.1016/j.jpowsour.2025.237242.

Jouhara, H., Żabnieńska-Góra, A., Khordehgah, N., Ahmad, D., Lipinski, T., 2020. Latent thermal energy storage technologies and applications: a review. Int. J. Thermofluids (5–6), 100039. https://doi.org/10.1016/j.ijft.2020.100039.

Jung, J., Onen, A., Russell, K., Broadwater, R.P., 2015. Local steady-state and quasi steady-state impact studies of high photovoltaic generation penetration in power distribution circuits. Renew. Sustain. Energy Rev. 43, 569–583. https://doi.org/10.1016/j.rser.2014.11.018.

Kalair, A., Abas, N., Saleem, M.S., Ali, R.K., Khan, N., 2021. Role of energy storage systems in energy transition from fossil fuels to renewables. Energy Storage 3 (1). https://doi.org/10.1002/est2.135.

Kalaiselvam, S., Parameshwaran, R., 2014. Chapter 6 - Thermochemical energy storage. In: Thermal Energy Storage Technologies for Sustainability Systems Design, Assessment and Applications, pp. 127–144.

Kamran, M., 2021. Renewable Energy Conversion SystemsChapter 7 - Fuel Cell, pp. 221–242. https://doi.org/10.1016/B978-0-12-823538-6.00005-1.

Kamran, M., 2023. Energy storage in smart grids. In: Fundamentals of Smart Grid Systems. Elsevier, pp. 393–429. https://doi.org/10.1016/B978-0-323-99560-3.00001-6.

Kandhasamy, M., Duvaragan, B.K., Kamaraj, S., Shanmugam, G., 2024. An overview on classification of energy storage systems. ACS Symp. Ser. 1484, 1–25. https://doi.org/10.1021/bk-2024-1484.ch001.

Kellett, C.W., Swords, W.B., Turlington, M.D., Meyer, G.J., Berlinguette, C.P., 2018. Resolving orbital pathways for intermolecular electron transfer. Nat.Commun. 9 (1). https://doi.org/10.1038/s41467-018-07263-1.

Khosravani, A., Sieving, J.K., Billings, B.W., Powell, K.M., 2025. Techno-economic analysis of long-duration energy storage integrated with high-penetration renewable energy systems. Energy Rep. 14, 4086–4110. https://doi.org/10.1016/j.egyr.2025.11.037.

Kim, D.-H., Hwang, S., Cho, J.-J., Yu, S., Kim, S., Jeon, J., Ahn, K.H., Lee, C., Song, H.-K., Lee, H., 2019. Toward fast operation of lithium batteries: ion activity as the factor to determine the concentration polarization. ACS Energy Lett. 4 (6), 1265–1270. https://doi.org/10.1021/acsenergylett.9b00724.

Kiyabu, S., Shkatulov, A., Ahmed, A., Greene, S.M., Huinink, H.P., Siegel, D.J., 2026. Materials for thermochemical energy storage and conversion: attributes for low-temperature applications. Mater. Horiz. https://doi.org/10.1039/d5mh01794g.

Koohi-Fayegh, S., Rosen, M.A., 2020. A review of energy storage types, applications and recent developments. J. Energy Storage 27, 101047. https://doi.org/10.1016/j.est.2019.101047.

Kömür, A.İ., Kızıl, Ç., Karaman, C., 2025. Nature's blueprint for energy: biomass-derived heteroatom-doped graphene materials for advanced energy applications. Carbon Lett. 35 (3), 919–961. https://doi.org/10.1007/s42823-025-00892-9.

Kulathuvayal, A.S., Su, Y., 2023. Ionic transport through the solid electrolyte interphase in lithium-ion batteries: a review from first-principles perspectives. ACS Appl. Energy Mater. 6 (11), 5628–5645. https://doi.org/10.1021/acsaem.3c00287.

Lan, P., Chen, S., Li, Q., Li, K., Wang, F., Zhao, Y., Wang, T., 2024. Comparison of different hydrogen-ammonia energy conversion pathways for renewable energy supply. Renew. Energy 227, 120602. https://doi.org/10.1016/j.renene.2024.120602.

Li, S.K., Hao, J., Liddell, M., 2013. Electrotransport across Membranes in Biological Media: Electrokinetic Theories and Applications in Drug Delivery Transport in Biological Media. Elsevier Inc., United States, pp. 417–454. https://doi.org/10.1016/B978-0-12-415824-5.00011-4.

Li, S., Peng, J.C.H., Yazami, R., 2024. Out-of-equilibrium thermodynamics analysis of lithium-ion batteries upon non-linear voltammetry fast charging. J. Power Sources 624. https://doi.org/10.1016/j.jpowsour.2024.235595.

Li, W., Yanyachi, A., Xia, J., Kuppan, S., Zhou, J., Wang, Y., Li, Y., Pianetta, P., Zhao, K., Mitlin, D., Ezekoye, O., Xiao, J., Wang, J., Liu, Y., 2025. Mechanistic considerations for battery charging protocol design. Adv. Energy Mater. https://doi.org/10.1002/aenm.202504499.

Liang, Y., Peng, L., Xing, L., Su, W., Li, W., Xu, W., 2024. Current status of thermodynamic electricity storage: principle, structure, storage device and demonstration. J. Energy Storage 80, 110347. https://doi.org/10.1016/j.est.2023.110347.

Lijesh, K.P., Khonsari, M.M., 2024. A thermodynamic approach for characterizing the degradation of Li-ion batteries. J. Energy Storage 82, 110565.

Liu, J., Monroe, C.W., 2015. On the characterization of battery electrolytes with polarization cells. Electrochim. Acta 167, 357–363. https://doi.org/10.1016/j.electacta.2015.03.104.

Liu, C., Neale, Z.G., Cao, G., 2016. Understanding electrochemical potentials of cathode materials in rechargeable batteries. Mater. Today 19 (2), 109–123. https://doi.org/10.1016/j.mattod.2015.10.009.

Liu, C.F., Liu, Y.C., Yi, T.Y., Hu, C.C., 2019. Carbon materials for high-voltage supercapacitors. Carbon 145, 529–548. https://doi.org/10.1016/j.carbon.2018.12.009.

Long, M., Charoen-amornkitt, P., Suzuki, T., Tsushima, S., 2026. Thermodynamics-informed design in electrochemical systems: entropy generation as a performance metric. Int. J. Heat Mass Tran. 256, 127952. https://doi.org/10.1016/j.ijheatmasstransfer.2025.127952.

Mahadevan, V., Raja, S., Rusho, M.A., Yishak, S., 2025. Critical review of energy storage systems: a comparative assessment of mechanisms, advantages, challenges, and integration with renewable energy. Results Eng. 26. https://doi.org/10.1016/j.rineng.2025.105589.

Maier, J., 2013. Thermodynamics of electrochemical lithium storage. Angewandte Chemie - Int. Ed. 52 (19), 4998–5026. https://doi.org/10.1002/anie.201205569.

Malekmohammadi, S., Mirbagheri, S.A., 2024. Scale-up single chamber of microbial fuel cell using agitator and sponge biocarriers. Environ.Tech. 45 (15), 2935–2943. https://doi.org/10.1080/09593330.2023.2197126.

Matsui, Y., Kawase, M., Suzuki, T., Tsushima, S., 2022. Electrochemical cell recharging by solvent separation and transfer processes. Sci.Rep. 12 (1). https://doi.org/10.1038/s41598-022-07573-x.

Meda, U.S., Lal, L., Sushantha, M., Garg, P., 2022. Solid Electrolyte Interphase (SEI), a boon or a bane for lithium batteries: a review on the recent advances. J. Energy Storage 47, 103564. https://doi.org/10.1016/j.est.2021.103564.

Menéndez, J., Fernández-Oro, J.M., Galdo, M., Loredo, J., 2020. Efficiency analysis of underground pumped storage hydropower plants. J. Energy Storage 28, 101234. https://doi.org/10.1016/j.est.2020.101234.

Meng, Z., Wang, K., Di, J., Lang, Z., He, Q., 2024. Energy analysis and exergy analysis study of a novel high-efficiency wind-hydrogen storage and power generation polygeneration system. Int. J. Hydrogen Energy 57, 338–355. https://doi.org/10.1016/j.ijhydene.2024.01.013.

Mentel, K.K., Serra, A., Abreu, P.E., Arnaut, L.G., 2018. Higher activation barriers can lift exothermic rate restrictions in electron transfer and enable faster reactions. Nat.Commun. 9 (1). https://doi.org/10.1038/s41467-018-05267-5.

Menye, J.S., Camara, M.B., Dakyo, B., 2025. Lithium battery degradation and failure mechanisms: a state-of-the-art review. Energies 18 (2). https://doi.org/10.3390/en18020342.

Mitali, J., Dhinakaran, S., Mohamad, A.A., 2022. Energy storage systems: a review. Energy Storage Sav. 1 (3), 166–216. https://doi.org/10.1016/j.enss.2022.07.002.

Mohandas, N., Bawari, S., Shibuya, J.J.T., Ghosh, S., Mondal, J., Narayanan, T.N., Cuesta, A., 2024. Understanding electrochemical interfaces through comparing experimental and computational charge density-potential curves. Chem.Sci. 15 (18), 6643–6660. https://doi.org/10.1039/d4sc00746h.

Müller, M., Pfeifer, M., Holtz, D., Müller, K., 2024. Comparison of green ammonia and green hydrogen pathways in terms of energy efficiency. Fuel 357, 129843. https://doi.org/10.1016/j.fuel.2023.129843.

Nagde, K.R., Dhoble, S.J., 2021. Li-S Ion Batteries: A Substitute for Li-Ion Storage Batteries Energy Materials: Fundamentals to Applications. Elsevier, India, pp. 335–371. https://doi.org/10.1016/B978-0-12-823710-6.00008-X.

Njema, G.G., Ouma, R.B.O., Kibet, J.K., Ponnusami, V., 2024. A review on the recent advances in battery development and energy storage technologies. J.Renew. Energy 2024, 1–35. https://doi.org/10.1155/2024/2329261.

Ohrelius, M., Lindbergh, G., Wreland Lindström, R., 2025. Polarization effects and aging characterization of lithium-ion batteries for second life applications. J. Power Sources 655, 237931. https://doi.org/10.1016/j.jpowsour.2025.237931.

Osara, J.A., Bryant, M.D., 2021. Performance and degradation characterization of electrochemical power sources using thermodynamics. Electrochim. Acta 365. https://doi.org/10.1016/j.electacta.2020.137337.

Palacios, A., Navarro-Rivero, M.E., Zou, B., Jiang, Z., Harrison, M.T., Ding, Y., 2023. A perspective on phase change material encapsulation: guidance for encapsulation design methodology from low to high-temperature thermal energy storage applications. J. Energy Storage 72, 108597. https://doi.org/10.1016/j.est.2023.108597.

Palys, M.J., Daoutidis, P., 2020. Using hydrogen and ammonia for renewable energy storage: a geographically comprehensive techno-economic study. Comput. Chem. Eng. 136, 106785. https://doi.org/10.1016/j.compchemeng.2020.106785.

Pascoe Bilibana, M., 2023. Electrochemical properties of MXenes and applications. Adv. Sens. Energy Mater. 2 (4), 100080. https://doi.org/10.1016/j.asems.2023.100080.

Panero, 2009. Electrochemical theory | kinetics. Encycl. Electrochem. Power Sources 14–22. https://doi.org/10.1016/B978-044452745-5.00034-4.

Qin, Z., Ma, J., Zhu, M., Khan, T., 2025. Advancements in energy storage technologies: implications for sustainable energy strategy and electricity supply towards sustainable development goals. Energy Strategy Rev. 59, 101710. https://doi.org/10.1016/j.esr.2025.101710.

Qiu, C., He, G., Shi, W., Zou, M., Liu, C., 2019. The polarization characteristics of lithium-ion batteries under cyclic charge and discharge. J. Solid State Electrochem. 23 (6), 1887–1902. https://doi.org/10.1007/s10008-019-04282-w.

Qu, J., Buckingham, M.A., Lewis, D.J., 2023. High-entropy materials for electrochemical energy storage devices. Energy Adv. 2 (10), 1565–1590. https://doi.org/10.1039/d3ya00319a.

Rahman, M.M., Khan, I., Alameh, K., 2022. The Role of Energy Storage Technologies for Sustainability in Developing Countries Renewable Energy and Sustainability: Prospects in the Developing Economies. Elsevier, Australia, pp. 347–376. https://doi.org/10.1016/B978-0-323-88668-0.00004-8.

Ramírez-Minguela, J., Mendoza-Miranda, J., Rodríguez-Muñoz, J., Pérez-García, V., Alfaro-Ayala, J., Uribe-Ramírez, A., 2018. Entropy generation analysis of a solid oxide fuel cell by computational fluid dynamics: influence of electrochemical model and its parameters. Therm. Sci. 22 (1 Part B), 577–589. https://doi.org/10.2298/tsci151221127r.

Randriamahazaka, H.N., 2010. Information entropy production for steady-state electrochemical systems. J. Electroanal. Chem. 647 (2), 222–229. https://doi.org/10.1016/j.jelechem.2010.05.006.

Ries, G., Neumueller, H.-W., 2001. Comparison of energy storage in flywheels and SMES. Phys. C Supercond. 357–360 (1), 1306–1310. https://doi.org/10.1016/s0921-4534(01)00484-1.

Rosen, M.A., Dincer, I., 2003. Exergy methods for assessing and comparing thermal storage systems. Int. J. Energy Res. 27 (4), 415–430. https://doi.org/10.1002/er.885.

Rouholamini, M., Wang, C., Nehrir, H., Hu, X., Hu, Z., Aki, H., Zhao, B., Miao, Z., Strunz, K., 2022. A review of modeling, management, and applications of grid-connected li-ion battery storage systems. IEEE Trans. Smart Grid 13 (6), 4505–4524. https://doi.org/10.1109/TSG.2022.3188598.

Saha, B.B., Hasan Rupam, T., 2023. Specialty grand challenge: thermal energy storage and conversion. Front. Thermal Eng. 3. https://doi.org/10.3389/fther.2023.1157794.

Salunkhe, P.B., Jaya Krishna, D., 2017. Investigations on latent heat storage materials for solar water and space heating applications. J. Energy Storage 12, 243–260. https://doi.org/10.1016/j.est.2017.05.008.

Sánchez, A.L., Williams, F.A., 2014. Recent advances in understanding of flammability characteristics of hydrogen. Prog. Energy Combust. Sci. 41 (1), 1–55. https://doi.org/10.1016/j.pecs.2013.10.002.

Singh, H., Mozafari, H., Kavianipour, M., Ghamami, M., Zockaie, A., Jackson, R., 2025. Enhancing urban sustainability through optimizing distributed energy resources for electric vehicles' fast charging. Int. J. Sustain. Transp. 19 (4), 344–362. https://doi.org/10.1080/15568318.2025.2479622.

Singh Rathore, P.K., Shukla, S.K., Gupta, N.K., 2020. Potential of microencapsulated PCM for energy savings in buildings: a critical review. Sustain. Cities Soc. 53. https://doi.org/10.1016/j.scs.2019.101884.

Sotoudeh, M., Baumgart, S., Dillenz, M., Döhn, J., Forster-Tonigold, K., Helmbrecht, K., Stottmeister, D., Groß, A., 2024. Ion mobility in crystalline battery materials. Adv. Energy Mater. 14 (4). https://doi.org/10.1002/aenm.202302550.

Steinmann, W.D., 2022. Overview of the section on mechanical energy storage. Encycl. Energy Storage 1–4, 1–4. https://doi.org/10.1016/B978-0-12-819723-3.00152-9.

Struchtrup, H., 2020. Entropy and the second law of thermodynamics—the nonequilibrium perspective. Entropy 22 (7), 793. https://doi.org/10.3390/e22070793.

Tao, H., Chen, G., Lian, C., Liu, H., Coppens, M.O., 2022. Multiscale modeling of ion transport in porous electrodes. AIChE J. 68 (4). https://doi.org/10.1002/aic.17571.

Tournassat, C., Steefel, C.I., Gimmi, T., 2020. Solving the Nernst-Planck equation in heterogeneous porous media with finite volume methods: averaging approaches at interfaces. Water Resour. Res. 56 (3). https://doi.org/10.1029/2019WR026832.

Tripathi, B.M., Shukla, S.K., Rathore, P.K.S., 2023. A comprehensive review on solar to thermal energy conversion and storage using phase change materials. J. Energy Storage 72. https://doi.org/10.1016/j.est.2023.108280.

Usui, S., 2008. Electrical double-layer interaction between oppositely charged dissimilar oxide surfaces with charge regulation and Stern–Grahame layers. J. Colloid Interface Sci. 320 (1), 353–359. https://doi.org/10.1016/j.jcis.2007.12.016.

Vetter, J., Winter, M., Wohlfahrt-Mehrens, M., 2009. Secondary Batteries - Lithium Rechargeable Systems - Lithium-Ion | Aging Mechanisms Encyclopedia of Electrochemical Power Sources. Elsevier, Germany, pp. 393–403. https://doi.org/10.1016/B978-044452745-5.00922-9.

Vinoth Kumar, R., Khandale, A.P., 2022. A review on recent progress and selection of cobalt-based cathode materials for low temperature-solid oxide fuel cells. Renew. Sustain. Energy Rev. 156, 111985. https://doi.org/10.1016/j.rser.2021.111985.

Wang, Q., Zhao, C., Hu, X., Wang, J., Ganapathy, S., Eustace, S., Bai, X., Li, B., Li, H., Aurbach, D., Wagemaker, M., 2024. Grain-boundary-rich interphases for rechargeable batteries. J.Am. Chem. Soc. 146 (46), 31778–31787.

Wang, Y., Yang, H., Hu, T., Piao, N., Li, F., Cheng, H.M., 2025. Designing electrolytes by thermodynamics. Natl.Sci. Rev. 12 (5). https://doi.org/10.1093/nsr/nwaf100.

Wojciechowski, J., Szwabińska, K., Fic, K., Lota, G., 2024. Interfacial insights into the polarization protocol: toward reducing corrosion and improving the cycle life of electrochemical capacitors. ACS Appl. Mater. Interfaces 16 (21), 27242–27253. https://doi.org/10.1021/acsami.4c00767.

Wolde, I., Cardemil, J.M., Escobar, R., 2024. Compatibility assessment of thermal energy storage integration into industrial heat supply and recovery systems. J. Clean. Prod. 440, 140932. https://doi.org/10.1016/j.jclepro.2024.140932.

Wu, Z.S., Zhou, G., Yin, L.C., Ren, W., Li, F., Cheng, H.M., 2012. Graphene/metal oxide composite electrode materials for energy storage. Nano Energy 1 (1), 107–131. https://doi.org/10.1016/j.nanoen.2011.11.001.

Xiao, J., Zeng, X., Chen, H., Yang, L., 2023. An investigation into the multi-parameter identification and model optimization strategy of the PCT curves of hydrogen storage alloys by multiple intelligent algorithms. Int. J. Hydrogen Energy 48 (5), 1943–1955. https://doi.org/10.1016/j.ijhydene.2022.09.281.

Xu, Y.Y., 2026. Thermodynamic laws and efficiency of quantum heat engines under strong coupling. Phys. Stat. Mech. Appl. 682, 131168. https://doi.org/10.1016/j.physa.2025.131168.

Yadav, S., Sharma, Y., 2025. Unravelling the potential of magnetic field in electrochemical energy storage: a review. Appl. Energy 383, 125347. https://doi.org/10.1016/j.apenergy.2025.125347.

Yang, B., Li, D., Fu, X., Wang, H., Gong, R., 2024a. Energy and exergy analysis of a novel pumped hydro compressed air energy storage system. Energy 294, 130737. https://doi.org/10.1016/j.energy.2024.130737.

Yang, Y., Liu, W., Wang, X., Tang, T., Zhu, L., Zhao, L., Zhu, K., Wang, J., 2024b. Phase modulation leads to ultrahigh energy storage performance in AgNbO3-based ceramics and multilayer capacitors. ACS Appl.Mater. Interfaces 16 (50), 69608–69616. https://doi.org/10.1021/acsami.4c17212.

Yang, P., Wu, Z., Liang, Y., Chen, H., Lin, C., Qiu, J., Meng, J., He, Y., Zhang, S., 2025. Engineering ion transport in all-solid-state sodium-ion batteries: fundamentals, strategies, and perspectives. Prog. Mater. Sci. 154, 101503. https://doi.org/10.1016/j.pmatsci.2025.101503.

Yaseen, M., Bartwal, R., Shafiq, A., Rawat, S.K., Al-Mdallal, Q.M., Khanduri, U., Sindhu, T. N., 2025. Analysis of entropy generation in tri-hybrid nanofluid flow towards a spinning disk under the influence of hall effect and variable viscosity. Int. J. Thermofluids 27. https://doi.org/10.1016/j.ijft.2025.101252.

Zalka, D., Péter, L., 2020. On the evolution and application of the concept of electrochemical polarization. J. Solid State Electrochem. 24 (11–12), 2595–2602. https://doi.org/10.1007/s10008-020-04682-3.

Zerrahn, A., Schill, W.P., Kemfert, C., 2018. On the economics of electrical storage for variable renewable energy sources. Eur. Econ. Rev. 108, 259–279. https://doi.org/10.1016/j.euroecorev.2018.07.004.

Zhang, X., Ràfols-Ribé, J., Mindemark, J., Tang, S., Lindh, M., Gracia-Espino, E., Larsen, C., Edman, L., 2024. Efficiency roll-off in light-emitting electrochemical cells. Adv.Mater. 36 (15). https://doi.org/10.1002/adma.202310156.

Zhang, Y., Wu, P., Chen, C., Liu, Y.J., Cai, X., Liang, W., Li, M., Zhuang, X., Li, Y., Chen, X., Sun, M., Lan, W., Xiang, H., Wen, Z., 2025. Electrochemical power sources enabled by multi-ion carriers. Chem. Soc. Rev. 54 (21), 9685–9806. https://doi.org/10.1039/d5cs00785b.

Zhao, J., Burke, A.F., 2021. Electrochemical capacitors: performance metrics and evaluation by testing and analysis. Adv. Energy Mater. 11 (1). https://doi.org/10.1002/aenm.202002192.

Zohuri, B., 2018. The Beginning and Concept of Cryogenics, Basic Principles. Elsevier BV, pp. 387–473. https://doi.org/10.1016/b978-0-12-814519-7.00014-8.

Zou, X., Liu, S., Qiu, G., Qian, H., Gao, S., Liu, Y., Lyu, Y., Jin, L., 2024. Improved energy storage performance at the phase boundary in BaTiO3-based film capacitors. J. Power Sources 619, 235201. https://doi.org/10.1016/j.jpowsour.2024.235201.

Battery technologies and electrochemical storage systems

3

Mona O. Albalawi[1], *Mohamed Abdel Salam*[2] and *Deepanraj Balakrishnan*[3]
[1]Department of Chemistry, Faculty of Science, University of Tabuk, Tabuk, Saudi Arabia; [2]Department of Chemistry, College of Science and Technology, Florida A&M University, Tallahassee, FL, United States; [3]Department of Mechanical Engineering, College of Engineering, Prince Mohammad Bin Fahd University, Al-Khobar, Saudi Arabia

3.1 Introduction

Energy storage technologies are vital to infrastructure, enabling the global transition toward sustainable and renewable-powered grids. Solar and wind projects have experienced unprecedented growth, but their inherent intermittency creates a basic mismatch between supply and demand. Electrochemical storage systems, such as lithium-ion batteries, flow batteries, and emerging solid-state technologies, are now the primary means of disconnecting renewable energy supply from demand. Grid-scale battery energy storage system (BESS) deployments reached 156 GWh worldwide through October 2025, marking a 38% increase over the previous year. The electrochemical energy storage industry, valued at around $104.3 billion in 2024, is projected to grow at a compound annual rate of 15%–25% through the 2030s (Energy-Storage.news, 2025).

Lithium-ion batteries have achieved technological and economic dominance, reducing costs by about 89% since their commercialization in 1991. However, this rapid adoption has revealed fundamental limitations: their specific energy density caps at around 260 Wh/kg, and geopolitical issues related to lithium and cobalt supply chains threaten future scalability. These challenges drive investment in alternative solutions: solid-state batteries (SSBs) aim for energy densities over 400 Wh/kg and improved safety, while flow batteries provide cycle lives spanning decades, suitable for long-duration grid storage (Xiao et al., 2025).

Contemporary battery engineering has shifted from empirical trial-and-error methods to physics-based, model-driven design. Advanced electrochemical modeling techniques, such as pseudo-two-dimensional (P2D) models, single-particle models (SPMs), and single-particle models with electrolyte (SPMe), now incorporate sophisticated parameterization methods that use electrochemical impedance spectroscopy (EIS) and the galvanostatic intermittent titration technique (GITT). Recognizing degradation mechanisms like solid-electrolyte interphase (SEI) growth, lithium plating, and changes in active material is crucial for predicting cycle life and enhancing recycling processes (Renewable Energy Hub, 2024).

Energy Storage Technologies. https://doi.org/10.1016/B978-0-443-45858-3.00004-8

Multiphysics integration is now leading the way, recognizing that electrochemical, thermal, and chemomechanical phenomena are highly interconnected. Real-world battery performance, affected by dynamic thermal changes, mechanical stress from lithiation expansion, and varying cycling rates, cannot be accurately modeled with single-physics methods. Integrated frameworks that combine electrochemical transport with thermal management, structural mechanics, and degradation mechanisms enable the predictive design of safer, longer-lasting batteries and systematic optimization of cell-to-pack (CTP) architectures. Seamless integration with battery management systems (BMSs) translates microscopic electrochemical processes into macroscopic control strategies for state-of-charge (SOC) estimation, thermal regulation, and grid support services.

Emerging physics-informed neural networks (PINNs), combined with electrochemical fundamentals, speed up materials discovery and reduce characterization timelines from years to months. The merging of artificial intelligence and mechanistic battery science allows for physics-based diagnostics and adaptive control strategies that improve operational lifespan and ensure safety.

This chapter summarizes modern battery technologies, electrochemical modeling methods, parametrization and validation strategies, degradation prediction, multiphysics coupling, and emerging computational intelligence techniques for next-generation electrochemical energy storage systems at the forefront of sustainable energy transition (Energy-Storage.news, 2025).

3.2 Lithium-ion systems: Architecture, materials, and modeling approaches

Lithium-ion batteries consist of four main components that enable reversible electrochemical energy storage: the cathode, anode, electrolyte, and separator. During discharge, lithium ions migrate from the anode through the electrolyte to the cathode, while electrons travel through an external circuit, generating electrical current. This intercalation-deintercalation process underpins all commercial lithium-ion chemistries and establishes basic performance limits based on material selection and electrode design (Krishnan et al., 2026) (Fig. 3.1).

3.2.1 Cathode materials and architecture

The cathode determines a battery's capacity, voltage, and thermal stability. Leading commercial cathode materials include layered transition metal oxides: lithium cobalt oxide ($LiCoO_2$, LCO), which provides high energy density around 140 mAh/g; lithium nickel manganese cobalt oxide ($Li(Ni_xMn_yCo_k)O_2$, NMC), offering balanced energy and power performance with variants like NMC-111, NMC-532, NMC-622, and nickel-rich NMC-811; and lithium iron phosphate ($LiFePO_4$, LFP), which delivers excellent thermal stability and cycle life exceeding 3000 cycles. Nickel-rich cathodes maximize specific energy by reducing cobalt content, addressing

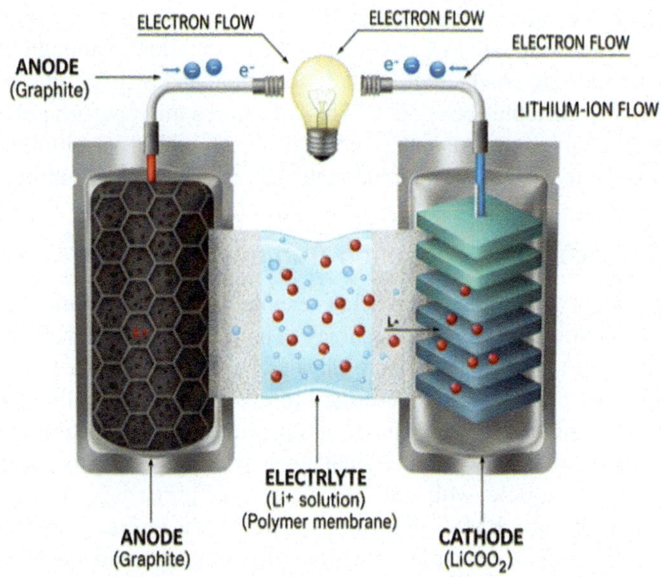

Figure 3.1 Lithium-ion systems architecture.

geopolitical and environmental concerns while reaching energy densities near 200 mAh/g. The cathode active material is deposited on aluminum foil current collectors because of aluminum's resistance to lithiation and its favorable passivation behavior in lithium hexafluorophosphate ($LiPF_6$) electrolytes (Koech et al., 2024; Krishnan et al., 2026).

3.2.2 Anode materials and emerging silicon integration

Graphite remains the main commercial anode material, with a theoretical capacity of 372 mAh/g achieved through lithium intercalation between graphene layers to form LiC_6 compounds. Copper foil is used as the current anode because lithium does not alloy with copper, helping to prevent structural degradation. However, silicon has become a revolutionary additive for anodes, offering a theoretical capacity of around 4200 mAh/g—nearly 10 times that of graphite—by alloying with lithium. Silicon/graphite composite anodes containing 9–32 wt% silicon can achieve reversible capacities between 800 and 1200 mAh/g. Still, volumetric expansion during lithiation can reach nearly 400%, requiring advanced binder systems and nanostructured architectures to maintain structural integrity and SEI stability (Krishnan et al., 2026).

3.2.3 Electrolyte and separator systems

The electrolyte facilitates ionic transport while preventing electrical conduction between electrodes. Standard formulations consist of 1 M $LiPF_6$ dissolved in organic carbonate solvents, typically a mixture of ethylene carbonate (EC) with dimethyl carbonate (DMC) or diethyl carbonate (DEC) at 3:7 weight ratios, along with additives like vinylene carbonate (VC) to enhance SEI formation. Although $LiPF_6$ offers a good balance of ionic conductivity, salt dissociation, and SEI-forming properties, it remains thermally unstable above 50°C. The separator, a microporous polyethylene or polypropylene membrane, prevents electrical shorts while permitting lithium-ion diffusion through its porous structure (Jongia Mixing Technology, 2025).

3.2.4 Electrochemical modeling frameworks

Three hierarchical modeling approaches enable battery simulation at different levels of complexity and physical accuracy. The P2D model, developed by Newman and colleagues, serves as the core physics-based framework, solving coupled partial differential equations that describe lithium diffusion in solid electrode particles and the liquid electrolyte phases along the electrode-separator stack. The SPM significantly reduces computational load by representing each electrode as a single spherical particle with uniform electrolyte properties, allowing for fast simulations suitable for BMS applications. The SPMe improves on the SPM by incorporating electrolyte transport equations that consider salt concentration and electrolyte potential while still maintaining single-particle electrode representations. SPMe achieves much better accuracy than SPM at moderate to high C-rates, where concentration polarization becomes important, with root-mean-square voltage errors below 7 mV compared to full P2D predictions. The choice of model depends critically on the application: P2D models support mechanistic studies of degradation and transport limitations, SPM offers computational efficiency for real-time SOC estimation, and SPMe provides an intermediate level of fidelity suitable for design optimization and parameter identification workflows (Laue et al., 2021) (Fig. 3.2).

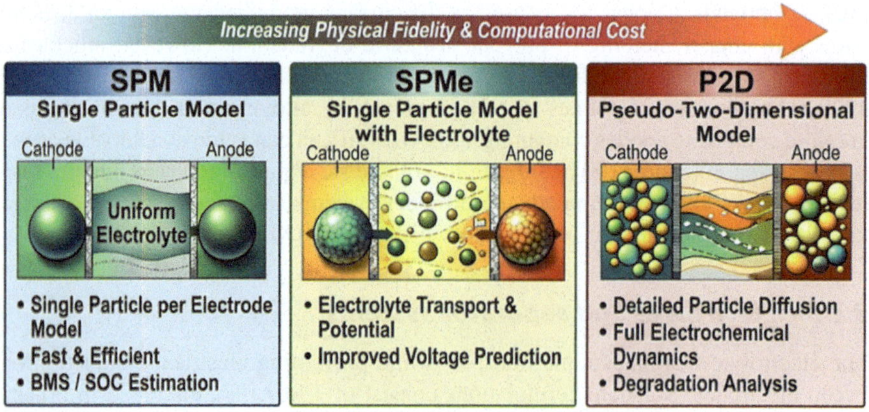

Figure 3.2 Electrochemical modeling frameworks for batteries.

3.3 Beyond lithium-ion: Solid-state, sodium-ion, and emerging chemistries

While lithium-ion batteries dominate modern energy storage uses, core issues related to energy density, safety, and resource sustainability have driven extensive research into alternative electrochemical options. SSBs, sodium-ion systems, and emerging chemistries, including lithium-sulfur and multivalent metal-ion technologies, serve as complementary approaches to overcoming these limits, each providing unique benefits for specific application areas (Bashiir, 2025).

3.3.1 Solid-state batteries

SSBs replace flammable liquid electrolytes with nonflammable solid electrolytes, greatly improving safety and energy density potential. There are three main types of solid electrolytes: oxide ceramics (e.g., garnet-type $Li_7La_3Zr_2O_{12}$, LLZO), sulfide glasses (e.g., Li_6PS_5Cl, argyrodites), and polymer electrolytes (e.g., polyethylene oxide-based systems). Sulfide electrolytes reach room-temperature ionic conductivities of 10^{-2}–10^{-3} S/cm—similar to liquid electrolytes, while oxide ceramics offer better electrochemical stability but require higher processing temperatures (Bolt. Earth, 2025) (Fig. 3.3).

3.3.1.1 Energy density and safety advantages

The solid electrolyte allows the use of lithium-metal anodes, significantly increasing the theoretical specific capacity from graphite's 372 mAh/g to lithium metal's 3860 mAh/g. SSBs achieve gravimetric energy densities of 300–500 Wh/kg and volumetric densities of 700–900 Wh/L, representing 50%–100% improvements over traditional lithium-ion batteries. Samsung SDI has demonstrated a volumetric energy

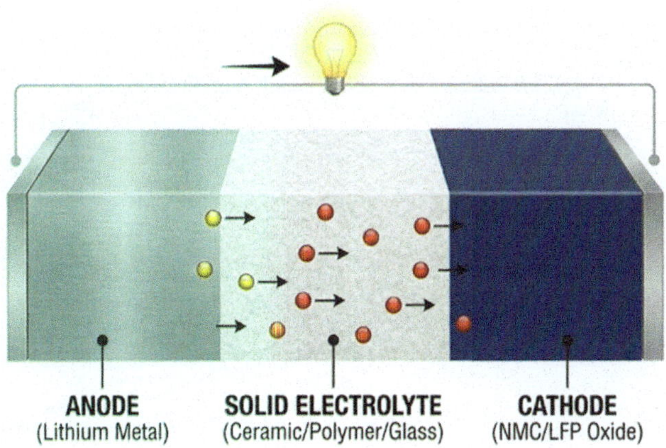

Figure 3.3 Solid-state battery structure.

density of 900 Wh/L, while QuantumScape's QSE-5 cells reach 844 Wh/L with a 12.2-minute charge from 10% to 80% state of charge. The rigid solid electrolyte physically prevents lithium dendrite growth, eliminating the primary failure mode in lithium metal anode batteries and enabling cycle lives of over 2000 cycles and safe fast charging (Bashiir, 2025).

3.3.1.2 Commercialization timeline

The SSB industry is transitioning from pilot lines to early commercialization between 2025 and 2030. Toyota plans to bring SSB-powered electric vehicles to market by 2027–28, with a 1000 km range and 10-minute fast charging. QuantumScape integrated its Cobra separator manufacturing process in June 2025 and expects to ship samples in 2026, with commercial production beginning between 2027 and 2028 (CarbonCredits.com, 2025). Gotion started pilot production in May 2025, while BMW has begun road testing with solid-state prototype vehicles. The global solid-state electrolytes market, valued at $23.7 million in 2024, is projected to grow to $61.7 million by 2034 (Simon-Kucher, 2025).

3.3.2 Sodium-ion batteries

Sodium-ion batteries (SIBs) use Earth's abundant sodium, with about 180 billion tons in seawater, instead of relying on geopolitically concentrated lithium resources. This provides clear cost and supply chain advantages. Cathode materials include layered transition metal oxides (Na_xMO_2), polyanionic compounds (e.g., $Na_3V_2(PO_4)_3$), and Prussian blue analogs (PBAs), offering theoretical capacities of up to 170 mAh/g and featuring three-dimensional open frameworks that enable sodium-ion diffusion. Hard carbon is mainly used as the anode material, providing disordered graphitic structures that can accommodate larger sodium ions (Yu et al., 2023) (Fig. 3.4).

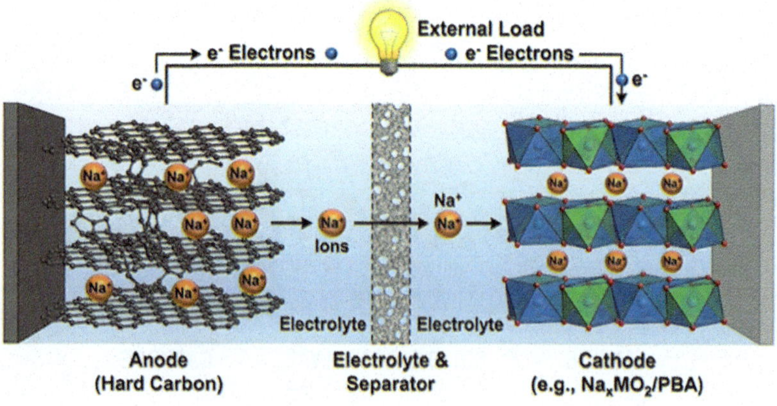

Figure 3.4 Sodium-ion battery architecture.

3.3.2.1 Performance and cost characteristics

CATL's first-generation SIB reached 160 Wh/kg energy density, with second-generation systems expected to surpass 200 Wh/kg by 2027. Current commercial products achieve approximately 165–175 Wh/kg, providing excellent low-temperature performance down to −40°C and a cycle life of over 10,000 cycles at 2C rates (Apricum, 2024). CATL has quoted $77/kWh for first-generation cells and $40/kWh for second-generation, potentially 70% cheaper than lithium-ion costs. BYD projects sodium-ion costs will match lithium iron phosphate (LFP) by 2025, with mass production costs estimated at 0.3–0.5 yuan/Wh. SIBs use aluminum instead of copper for anode current collectors, further reducing costs and eliminating copper supply chain dependencies (Simon-Kucher, 2025).

3.3.2.2 Commercialization status

CATL and BYD began mass production in 2023, with CATL launching the world's first large-scale sodium battery in April 2025. CATL's Freevoy hybrid pack combines sodium-ion and lithium-ion cells for extended-range vehicles, achieving over 400 km of range with 4C fast charging and working in extreme cold down to −40°C. BYD started building a 30 GWh sodium-ion factory in 2024, with products expected to be delivered in Q3 2025. Industry experts forecast 2026 as the milestone for large-scale commercial applications, with sodium-ion potentially replacing up to 50% of the LFP market (Simon-Kucher, 2025).

3.3.3 Emerging chemistries

Beyond solid-state and sodium-ion platforms, various emerging chemistries target specific performance needs and sustainability goals.

3.3.3.1 Lithium-Sulfur batteries

Lithium-sulfur (Li–S) batteries have theoretical specific energies exceeding 2600 Wh/kg—approximately five times that of traditional lithium-ion batteries—owing to the conversion reactions between lithium and sulfur cathodes. Sulfur's abundance (costing roughly $0.10/kg compared to over $20/kg for cobalt) and its environmental friendliness provide notable sustainability advantages. However, technical challenges such as polysulfide dissolution causing capacity fade, limited cycle life (typically 200–500 cycles), and lower volumetric energy density need to be addressed through the development of solid-state electrolytes and enhanced cathode designs (Guo et al., 2024) (Fig. 3.5).

3.3.3.2 Multivalent metal-ion systems

Magnesium-ion, calcium-ion, and zinc-ion batteries utilize divalent or trivalent charge carriers, which can potentially double or triple the capacity per intercalated ion. Zinc-air batteries reach theoretical energy densities of 1350 Wh/kg and use aqueous

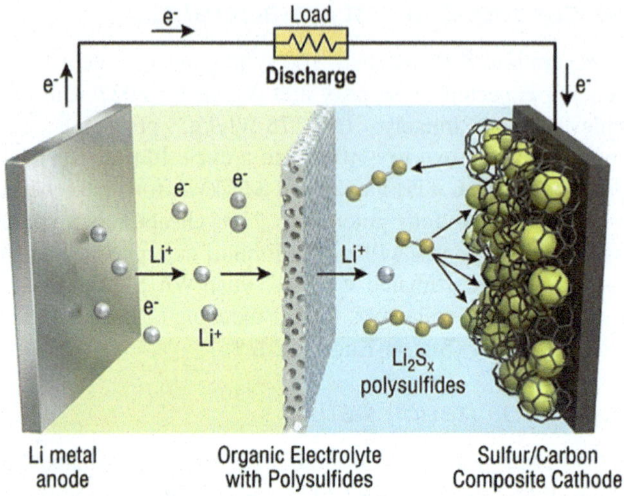

Figure 3.5 Schematic architecture of lithium–Sulfur (Li–S) battery system.

electrolytes for inherent safety. Aqueous metal batteries with zinc, aluminum, and magnesium benefit from nonflammable water-based electrolytes and simpler manufacturing processes, making them appealing for grid-scale applications where volumetric constraints are less strict (Bashiir, 2025) (Fig. 3.6).

3.3.3.3 *Metal-air systems*

Lithium-air ($Li–O_2$) batteries theoretically deliver 3500 Wh/kg through oxygen reduction reactions, but practical development relies on advances in air-electrode catalysis, electrolyte stability, and reversibility. Current research addresses critical issues such as irreversible oxidation of organic electrolytes during discharge and dendrite growth at the lithium anode, which limit cycle life to 50–100 cycles. Zinc-air systems are closer to commercialization for grid storage, benefiting from

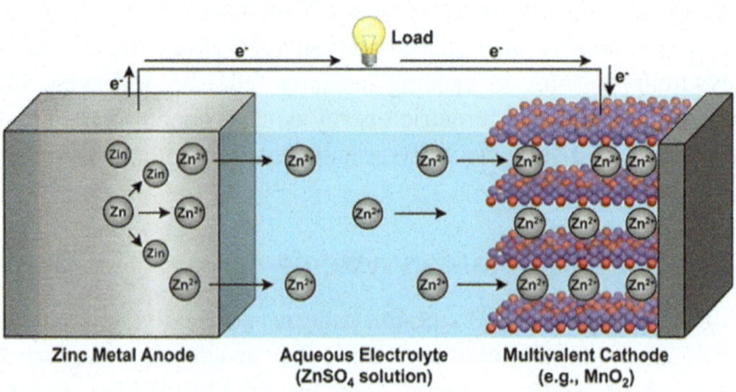

Figure 3.6 Multivalent zinc-ion battery system.

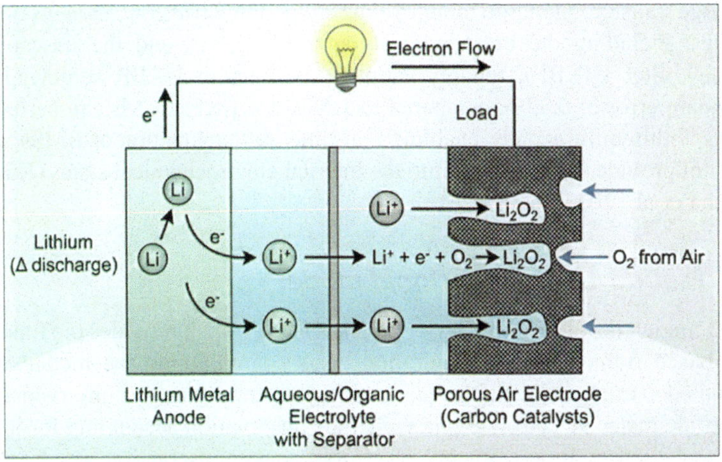

Figure 3.7 Lithium-air (Li–O_2) battery system.

established zinc electrochemistry and atmospheric oxygen as the limitless cathode reactant. Zinc-air batteries achieve volumetric energy densities of 200–400 Wh/L with aqueous electrolytes that provide inherent safety, making zinc-air technology an appealing choice for large-scale stationary energy storage (Olabi et al., 2021; Ouakki, 2024) (Fig. 3.7).

3.4 Electrochemical modeling foundations: ECM, P2D, and SPM/SPMe

Electrochemical modeling provides a mathematical framework for understanding, predicting, and optimizing battery performance across different operating conditions. Three main modeling approaches range from simple empirical methods to physics-based precision: equivalent-circuit models (ECMs) for real-time battery management, P2D models for detailed mechanistic analysis, and SPM/SPMe that balance computational speed with electrochemical accuracy (Kurt, 2022).

3.4.1 Equivalent circuit models

ECMs represent battery behavior as electrical network analogs composed of passive components that capture the overall electrochemical processes, without detailing specific transport phenomena. The Thevenin model, the most widely used ECM topology, describes terminal voltage as $Vt = VOC(SOC) - IR_0 - \sum VRC,i$, where $VOC(SOC)$ indicates the open-circuit voltage (OCV) based on the state of charge, R_0 is the series ohmic resistance, and VRC,i refers to the voltage across each RC pair that exhibits first-order dynamics. Each RC pair introduces a characteristic time constant $\tau i = RiCi$, which captures various relaxation phenomena, from milliseconds-long charge transfer kinetics to minutes-long solid-state diffusion. Variants of the Kalman filter, including the extended Kalman filter (EKF) and the unscented Kalman–Bucy filter (UKBF), reliably estimate SOC, with UKBF achieving a root-mean-square error of 0.33%, compared to 1.88% for EKF. ECMs can perform computations within milliseconds, enabling real-time implementation of BMSs, although they do not provide direct insight into the internal electrochemical states (Kurt, 2022; Kushwaha et al., 2023).

3.4.2 Pseudo-two-dimensional model

The P2D model, developed by Doyle, Fuller, and Newman, provides the fundamental physics-based framework for lithium-ion battery simulation. The model combines five coupled partial differential equations: solid-phase charge conservation within the electrode matrix, electrolyte-phase charge conservation accounting for ionic conduction and concentration gradients, electrolyte mass conservation based on concentrated solution theory, solid-phase diffusion within spherical electrode particles following Fick's law, and Butler–Volmer kinetics that couple electrochemical reactions at electrode–electrolyte interfaces. The P2D model enables mechanistic investigation of transport limitations, concentration gradients, and degradation pathways, although its computational cost typically ranges from seconds to minutes per simulation, far exceeding the capabilities of ECMs (Han et al., 2021; Laue et al., 2021).

3.4.3 Single-particle models (SPM/SPMe)

SPMs simplify computations while preserving key electrochemical physics by representing each electrode as a single spherical particle. The SPM models electrode behavior by assuming a uniform current and ignoring electrolyte dynamics, enabling speeds of over 3600× real time, suitable for embedded BMS hardware. Parameter grouping reduces the number of parameters from over 17 to 6–9, facilitating estimation. The SPMe incorporates electrolyte transport equations into the SPM, achieving RMS voltage errors below 7 mV relative to full P2D predictions at moderate-to-high C-rates, where concentration polarization is significant. Model choice depends on the application: ECM for real-time SOC estimation; SPM for low C-rates (<0.5C); SPMe for moderate-to-high C-rates (0.5–5C) state estimation; and P2D for detailed design and degradation analysis (Chambers et al., 2024; Kurt, 2022).

3.5 Parameterization and validation: OCV, EIS, GITT/PITT, and Bayesian estimation

Electrochemical modeling demands careful parameterization of material properties and kinetic factors based on experimental data (Liu et al., 2022). This section outlines the primary methods for determining model parameters: OCV measurement to establish thermodynamic basis, EIS to analyze frequency-dependent transport processes, galvanostatic and potentiostatic intermittent titration techniques (GITT/PITT) to measure diffusion kinetics, and Bayesian parameter estimation frameworks to quantify uncertainty in inferred material properties.

3.5.1 Open-circuit voltage characterization

The OCV indicates the thermodynamic equilibrium potential of a battery when no external current is flowing, and all electrochemical transients have settled. The OCV-SOC relationship is a key input for electrochemical models, providing material-specific stoichiometric details and acting as the electrochemical "fingerprint" of battery chemistry. Standard OCV characterization includes cycles of constant-current pulses at low C-rates (typically C/20 to C/10), followed by long relaxation periods—usually 1–3 h—that allow voltage to reach equilibrium until $dEdt \approx 0$ (Accrue Battery Intelligence, 2025).

Complete OCV characterization spanning 0%–100% SOC typically requires 7–14 days, with hourly relaxation periods at each SOC step. Specific battery chemistries, particularly LFP, exhibit hysteresis, wherein the charging and discharging OCV curves differ by 50–100 mV in the plateau regions, requiring characterization of both branches. Recent accelerated protocols, including the intermittent current interruption (ICI) method, achieve an 84% time reduction by optimizing relaxation durations and extrapolating equilibrium values via linear regression of voltage transients on time. Incremental capacity analysis (ICA, dQ/dV) and differential voltage analysis (DVA, dV/dQ) enable detailed resolution of OCV curve features correlated with degradation modes and phase transitions (Accrue Battery Intelligence, 2025).

3.5.2 Electrochemical impedance spectroscopy

EIS measures frequency-dependent battery impedance using sinusoidal AC voltage perturbations, allowing for the quantification of charge transfer resistance, interfacial capacitances, and ionic transport limitations (Karmakar, 2024). EIS applies small alternating current signals over a frequency range from millihertz to megahertz, and the resulting impedance data are displayed through Nyquist plots showing imaginary impedance ($-Z''$) versus real impedance (Z').

Nyquist plot features reveal distinct frequency-dependent electrochemical processes. The high-frequency intercept indicates ohmic resistance from electrolyte and contact resistances. High-frequency semicircles (1 kHz–10 MHz) correspond to lithium-ion migration through electrolyte and SEI. Mid-frequency semicircles (1 Hz–1 kHz) represent charge transfer kinetics at electrode–electrolyte interfaces, with diameter quantifying charge transfer resistance *Rct*. The low-frequency tail (<1 Hz) reflects Warburg impedance from semiinfinite solid-state diffusion within electrode particles. The charge transfer resistance follows Arrhenius temperature dependence, enabling extraction of activation energies for interfacial kinetics (Karmakar, 2024; Magar et al., 2021).

EIS data analysis uses equivalent circuit fitting, usually Randles cells consisting of series resistance, parallel charge transfer resistance, and double-layer capacitance, along with Warburg elements, employing nonlinear least-squares optimization. EIS offers nondestructive characterization of SOC, state-of-health (SOH), and degradation mechanisms without disassembling the battery, though interpretation requires careful consideration of equivalent circuit uniqueness and spatial averaging limitations (Magar et al., 2021).

3.5.3 Galvanostatic intermittent titration technique

The GITT determines lithium-ion diffusion coefficients within electrode active materials through the application of constant-current pulses interspersed with relaxation periods. GITT rests upon Fick's second law of diffusion, permitting extraction of chemical diffusion coefficients from electrochemical measurements of surface concentration gradients.

A GITT measurement consists of repeated cycles: apply constant current at low C-rate (typically C/10-C/20) for pulse duration t_1 where $t_1 \ll L^2/D$ (usually 5–30 min), interrupt current and record voltage response during relaxation time t_2 (typically 30 min–3 h) until equilibrium, then repeat across the full SOC range. The diffusion coefficient D is extracted through:

$$D = \frac{4}{\pi\tau}\left(\frac{i_0 V_m}{zFA}\right)^2 \left(\frac{\Delta E_s}{\Delta E_t}\right)^2$$

where i_0 is the applied current, V_m is the molar volume, z is the charge number, F is Faraday's constant, A is the electrode area, ΔE_s is the steady-state voltage change, and ΔE_t is the transient voltage change during the pulse. Typical diffusion coefficients for lithium-ion cathode materials range from 10^{-9} to 10^{-12} cm^2/s, with values varying significantly across SOC due to structural and thermodynamic factors. Classical GITT requires 8–14 days for complete characterization; recent extended GITT formulations applying convolution techniques enable rapid diffusion coefficient determination from general current profiles (Kim et al., 2021).

3.5.4 Potentiostatic intermittent titration technique

The potentiostatic intermittent titration technique complements GITT by fixing the electrode potential rather than the current, thereby enabling kinetic characterization with finer potential resolution. PITT applies stepwise potential increments (typically 10–50 mV) at each potential plateau until equilibrium current decays to zero, providing state-of-potential-dependent rather than SOC-dependent diffusion coefficients (Li et al., 2025b).

PITT offers advantages, including fine potential resolution for investigating phase boundaries and structural transitions, applicability to multiphase equilibria, and faster equilibration than GITT. Current decay profiles during each potential step encode diffusion kinetics, which can be extracted via Cottrell analysis for short times or via exponential decay fitting for long times. Comparative studies demonstrate that EIS provides the most accurate diffusion time constants due to the explicit separation of ohmic, capacitive, and diffusion contributions, whereas GITT and PITT measurements may overestimate time constants by 20%–80% when ohmic and charge -transfer resistances are nonnegligible (Richter and McCalla, 2025).

3.5.5 Bayesian parameter estimation

Classical least-squares parameter estimation provides point estimates without rigorous uncertainty quantification, limiting confidence in model predictions and downstream design decisions. Bayesian inference frameworks address this limitation by estimating parameters probabilistically, yielding posterior probability distributions that quantify parameter uncertainty rather than deterministic values (Miyake et al., 2021).

Bayes' theorem combines prior distributions $p(\theta)$ reflecting available knowledge with experimental likelihood distributions $p(D|\theta)$ representing measurement error to generate posterior distributions:

$$p(\theta|D) = \frac{p(D|\theta) \cdot p(\theta)}{p(D)}$$

where θ represents model parameters, D represents experimental observations, and $p(D)$ is the marginal likelihood serving as normalization. Posterior distributions are sampled using Markov chain Monte Carlo (MCMC) algorithms, typically adaptive Metropolis–Hastings or ensemble samplers, enabling exploration of high-dimensional parameter spaces. At each iteration, trial parameters are proposed, the forward model is evaluated, and proposals are accepted with probability proportional to the likelihood ratio, generating Markov chains converging to posterior distributions.

Bayesian methods have been applied to ECMs for battery-string diagnostics, revealing bimodal distributions when cell-to-cell parameter coupling creates identifiability challenges. The application of electrochemical models enables simultaneous estimation of transport parameters (diffusion coefficients, conductivities) and kinetic parameters (exchange current densities), with credible intervals quantifying parametric uncertainty that is critical for safety-critical applications. Hybrid physics-informed approaches combine principle-based models with variational Bayesian neural networks, capturing both model-form uncertainty and battery-to-battery aleatory uncertainty from manufacturing variability (Miyake et al., 2021; Wang et al., 2024a).

3.6 Degradation and lifetime modeling: SEI growth, plating, and active material loss

Battery degradation fundamentally limits the economic viability and safety of electrochemical energy storage systems. Understanding degradation mechanisms, SEI growth, lithium plating, and active material loss enables predictive lifetime modeling, which is essential for warranty estimation, second-life assessment, and optimal operating strategy design (Xu, 2022). This section synthesizes the principal degradation mechanisms and their mathematical treatment within electrochemical modeling frameworks.

3.6.1 Solid electrolyte interphase growth

The SEI forms during initial battery cycling when the anode potential drops below the electrolyte reduction potential, causing electrolyte molecules to react with electrons and lithium ions to produce a nanometer-thick passivation layer on the anode surface. While the SEI protects the electrolyte from subsequent reduction, it continues growing throughout battery life, irreversibly consuming cyclable lithium and increasing cell impedance (Li et al., 2023).

SEI growth kinetics follow characteristic time dependencies of the form t^{β} where $0 \leq \beta \leq 1$, with the scaling coefficient determined by the rate-limiting transport mechanism. Initially, the formation reaction of neutral lithium atoms limits growth; as the SEI thickens, diffusion through the layer becomes rate-limiting, producing classical $\sqrt{t}$ parabolic growth; eventually, electron migration through coordinated lithium ions dominates at very long time. SEI growth exhibits strong asymmetry between charging and discharging growing faster during charging due to low electrode potentials and high intercalation currents with capacity fade accelerated at high SOC and elevated charging rates.

Recent modeling advances reveal that SEI composition significantly influences growth kinetics. Inorganic SEI components (Li_2CO_3, LiF) follow mixed 2D/3D nucleation models dependent on overpotential, while organic SEI strictly obeys 2D instantaneous nucleation regardless of formation conditions. Large current pulses during battery formation promote 2D inorganic SEI growth and improve long-term capacity retention, offering practical strategies for manufacturing optimization (Li et al., 2023; Takenaka et al., 2021).

3.6.2 Lithium plating

Lithium plating occurs when the anode potential drops below 0 V versus Li/Li^+, causing lithium ions to deposit as metallic lithium on the electrode surface rather than intercalating into graphite. This degradation mechanism is particularly critical during fast charging and low-temperature operation, where limited lithium-ion transport causes surface accumulation exceeding intercalation kinetics (Lin et al., 2021).

Plating occurs when ionic conductivity through the SEI decreases at low temperatures or when high charging currents overwhelm the intercalation rate, creating conditions where surface deposition becomes thermodynamically favorable. Simulation studies demonstrate that at 0°C with 3–5C charging rates, lithium plating induces rapid capacity loss of nearly 10% within early cycles, stabilizing as plating-dissolution equilibrium is reached. The temperature dependence differs from other degradation mechanisms: plating severity increases at both low temperatures (<10°C) and elevated temperatures (>45°C) when poor capacity balance promotes metal deposition.

Beyond capacity loss, lithium plating poses severe safety hazards. Plated lithium can form dendritic structures penetrating the separator, causing internal short circuits and thermal runaway. Machine learning approaches now enable real-time plating detection from voltage signatures, providing early warning capabilities for BMSs. Effective mitigation strategies include optimized charging protocols limiting C-rate at high SOC, active thermal management maintaining cells above 15°C during charging, and electrolyte additives promoting stable SEI formation (Lin et al., 2021).

3.6.3 Active material loss

Loss of active material (LAM) results from mechanical degradation of electrode particles due to repeated volumetric expansion and contraction during lithiation/delithiation cycles. Mechanical stresses induce particle cracking, electrode delamination, and loss of electronic contact with current collectors and conductive additives (Li et al., 2025a).

The mechanical stress evolution follows continuum mechanics frameworks where radial and tangential stresses develop from concentration gradients within particles during (de)intercalation. When stress exceeds critical thresholds, which are approximately six times lower in graphite negative electrodes than NMC positive electrodes, crack propagation initiates irreversible capacity loss. Low temperatures dramatically accelerate LAM by reducing solid diffusion coefficients, creating steeper concentration gradients and higher mechanical stresses; simulations predict that decreasing temperature from 25°C to 5°C causes greater capacity fade from mechanical cracking than from lithium plating (Shi et al., 2025).

Silicon-containing anodes exhibit particularly severe active material loss due to ~300% volumetric expansion during full lithiation compared to ~10% for graphite. Studies on silicon-graphite composite electrodes demonstrate 80% silicon capacity loss after 400 equivalent cycles at 40°C and low SOC operation, compared to only 10% graphite capacity loss under identical conditions. The silicon/SEI interface evolves from nanometer-scale passivation into micron-scale silicon-electrolyte composite (SEC) structures, trapping lithium and underutilizing active material. Fluorine enrichment on aged silicon particles, up to 5× fresh anode concentrations, indicates preferential SEI accumulation on high-expansion regions (Bonkile et al., 2024).

3.6.4 Integrated degradation modeling

Comprehensive lifetime models must capture interactions between degradation mechanisms, as SEI growth, lithium plating, and active material loss exhibit coupled dynamics. The PyBaMM framework enables systematic comparison of degradation model combinations, revealing that enabling all four principal mechanisms (SEI growth, lithium plating, negative electrode LAM, positive electrode LAM) predicts low temperatures to cause the most irreversible capacity fade (O'Kane et al., 2022).

Capacity fade contributions evolve across battery life: lithium plating dominates early cycles under aggressive conditions, SEI growth provides approximately linear degradation in middle cycles, and active material loss accelerates nonlinearly in late life due to positive feedback between cracking and increased local current density. SOH estimation algorithms must therefore adapt degradation models to operating history, with calendar aging (storage) dominated by SEI growth and cycle aging sensitive to temperature, C-rate, and depth-of-discharge window (Alkhalil, 2025).

3.7 Multiphysics and scaling: Thermal, chemomechanical, cell-to-pack, and BMS integration

Battery systems operate as coupled multiphysics environments where electrochemical, thermal, and mechanical phenomena interact across multiple length scales. Accurate prediction of real-world performance requires integrated modeling frameworks that capture these couplings, from microscale electrode particle mechanics through cell-level thermal gradients to pack-level system integration (FFD Power, 2024). This section synthesizes multiphysics coupling approaches and scaling methodologies essential for translating fundamental electrochemical understanding into deployable energy storage systems.

3.7.1 Thermal modeling and management

Heat generation during battery operation results from reversible entropic contributions and irreversible resistive losses, with power levels spanning 10–100 W per cell, depending on C-rate and SOC. Thermal management keeps cells within optimal operating ranges (typically 15–35°C), limits cell-to-cell temperature gradients (<5°C), and prevents thermal runaway propagation.

Lumped parameter thermal models represent batteries as isothermal masses with convective and radiative heat dissipation, providing computational efficiency suitable for real-time thermal management in BMSs[41]. The governing energy balance equates heat generation to thermal capacitance accumulation and convective-radiative losses:

$$mC_p\frac{dT}{dt} = Q_{\text{gen}} - hA(T - T_{\text{amb}}) - \varepsilon\sigma A\left(T^4 - T_{\text{amb}}^4\right)$$

where m is battery mass, C_p is specific heat capacity, h is convective heat transfer coefficient, A is surface area, ε is emissive, and σ is the Stefan–Boltzmann constant. Distributed thermal models resolve internal temperature gradients through finite element discretization, capturing thermal heterogeneities critical for degradation prediction and safety analysis. Cooling strategies include passive air cooling for low-power applications, forced air convection achieving heat transfer coefficients of 25–100 W/m^2K, and liquid cooling systems reaching 500–3000 W/m^2K for high-performance applications (He et al., 2023; Nema et al., 2024).

3.7.2 Chemomechanical coupling

The strong coupling between electrochemical kinetics and mechanical stress governs electrode particle degradation and capacity fading. Lithiation induces volumetric expansion—approximately 10% for graphite and 300% for silicon—generating internal stresses that influence both electrochemical behavior and mechanical integrity (ATS Industrial Automation, 2025).

Continuum chemomechanical models couple lithium transport equations with mechanical equilibrium through concentration-dependent strain tensors:

$$\varepsilon_{ch} = \beta(c - c_0)$$

where β is the partial molar volume coefficient and c_0 is the reference concentration. The resulting stress field feeds back into electrochemical kinetics through stress-modified chemical potentials, creating bidirectional coupling. Compressive stress retards lithiation by increasing chemical potential at the reaction interface, while tensile stress accelerates lithiation explaining asymmetric swelling observed in bent electrode structures. Under mechanical constraint, the force evolution during charging follows nonlinear profiles dominated by anode swelling, with inflection points reflecting transitions between structural relaxation and lithiation-induced expansion (De Gol et al., 2024).

Silicon anodes exhibit particularly severe chemomechanical coupling, with atomistic simulations revealing that compressive stresses destabilize amorphous silicon and accelerate capacity fading beyond predictions from continuum models that neglect large volumetric deformation effects. These findings underscore the necessity of multiscale approaches linking atomic-level lithiation behavior to continuum performance predictions (De Gol et al., 2024).

3.7.3 Cell-to-pack and cell-to-body integration

CTP technology eliminates intermediate module assemblies, integrating cells directly into battery packs to improve volumetric efficiency and reduce manufacturing costs. CTP designs achieve 10%–15% increases in energy density by repurposing module casing volume for additional cells, with CATL's latest CTP technology claiming 15% energy density improvement and 20% cost reduction. The CTP battery market, valued at \$18.5–25.4 billion in 2023–24, is projected to reach \$66–77 billion by 2032 at 15% compound annual growth (Nigel, 2023).

Cell-to-body (CTB) designs extend integration further by incorporating batteries into vehicle structural elements. BYD's e-Platform 3.0 pioneered production CTB implementation, using the battery pack as a structural floor member contributing to body rigidity, aerodynamics, and crash energy absorption. The BYD Blade Battery achieves volumetric energy densities of 439–450 Wh/L through LFP chemistry and CTP/CTB integration. However, module-free architectures require enhanced thermal management strategies because cells in closer proximity without modular barriers are more susceptible to thermal runaway propagation. Thermal interface materials (TIMs) and active cooling systems have become critical for maintaining safe operating temperatures across densely packed cells (Cell to Body—Battery Design, 2023).

3.7.4 Battery management system integration

BMSs translate electrochemical models and sensor measurements into real-time state estimation and control decisions. Core BMS functions include SOC estimation indicating remaining capacity, SOH estimation tracking degradation progression, thermal management coordination, cell balancing ensuring uniform aging, and safety interlock activation (Integra Sources, 2021).

SOC estimation methodologies range from simple coulomb counting (current integration) to model-based Kalman filtering and data-driven neural network approaches. EKFs and unscented Kalman filters (UKFs) combine equivalent circuit or electrochemical models with recursive measurement correction, achieving SOC estimation accuracies of 1%–3% under dynamic conditions. Adaptive Kalman formulations augment state vectors with battery parameters including internal resistance, enabling simultaneous SOC and SOH estimation critical for lifetime prediction (Integra Sources, 2021).

Modern BMS architecture implements hierarchical control spanning cell-level monitoring, module-level aggregation, and pack-level coordination with power conversion systems (PCSs) and energy management systems (EMSs) through industrial protocols. The integration of electrochemical model insights with BMS algorithms enables physics-informed state estimation that extrapolates beyond calibration conditions while maintaining real-time computational feasibility, bridging the gap between fundamental electrochemical understanding and deployed energy storage system performance (FFD Power, 2024).

3.8 Recycling and second-life applications

The sustainability of electrochemical energy storage depends critically on establishing circular economic pathways that recover valuable materials and extend battery service life beyond initial applications. As electric vehicle adoption accelerates, the volume of end-of-life batteries is projected to surge dramatically, creating both environmental challenges and economic opportunities for recycling and repurposing industries (Srinivasan et al., 2025).

3.8.1 Recycling technologies

Three principal recycling methodologies have emerged for lithium-ion battery processing: pyrometallurgical, hydrometallurgical, and direct recycling. Pyrometallurgical recycling employs high-temperature smelting (>1400°C) to extract recyclable materials, offering simplicity in handling diverse battery chemistries but suffering from high energy consumption and limited lithium recovery. In 2023, over 50% of Europe's 160,000-ton recycling capacity relied on pyrometallurgical technology (Tembo et al., 2024).

Hydrometallurgical recycling employs aqueous chemistry, using acid leaching followed by concentration and purification to recover metals with higher selectivity and lower energy consumption. This approach achieves superior recovery rates: lithium at 70%–95%, cobalt at 90%–98%, and nickel at 85%–95%, while generating lower greenhouse gas emissions than pyrometallurgy. Direct recycling preserves cathode crystal structures through relithiation and regeneration, bypassing costly remanufacturing steps and offering the lowest environmental footprint, though scalability challenges remain (Tembo et al., 2024).

3.8.2 Market growth and regulatory drivers

The global battery recycling market, valued at $25.4 billion in 2024, is projected to reach $69.7 billion by 2032 at 12.5% compound annual growth. The circular economy battery recycling segment is expected to grow from $1.5 billion in 2024 to $4.2 billion by 2034. Regulatory frameworks are accelerating adoption: the EU Battery Regulation mandates cobalt recovery rates of 90% by 2027 and 95% by 2031, with recycled content requirements driving closed-loop manufacturing systems (Emerging Research, 2024).

By 2040, locally recycled battery materials in Europe could avoid the need for 12 new mines globally: 4 lithium, 3 nickel, 4 cobalt, and 1 manganese mine of average size. Recycled cobalt could meet 19% of European demand by 2030, rising to 53% by 2040 as legacy high-cobalt chemistries reach end-of-life (Emerging Research, 2024).

3.8.3 Second-life applications

Electric vehicle batteries typically reach automotive end-of-life at 70%–80% remaining capacity, insufficient for demanding transportation applications but adequate for less rigorous stationary storage. Second-life applications include grid backup storage, peak shaving, frequency regulation, rooftop solar storage, and microgrid systems.

Repurposed batteries offer 30%–70% cost advantages over new purpose-built storage systems, making grid-scale energy storage economically accessible for wider applications. Redwood Materials, founded by Tesla cofounder JB Straubel, has pioneered dual-pathway approaches: batteries are deployed for grid-scale energy storage when economically sensible—approximately 95% of cases—before proceeding to materials recycling, maximizing both energy value and metals recovery. When ultimately recycled, 95% of battery minerals can be reused in new batteries (Eco-ESS, 2025).

However, second-life deployment faces challenges including proprietary BMS access, lack of standardized assessment protocols, and the need for cell-level diagnostics and potential reassembly before repurposing. Connected Energy and similar companies are developing solutions that address battery-to-battery variability across diverse makes, models, and degradation states. As regulatory frameworks mature and assessment technologies improve, second-life batteries represent a critical pathway for extending resource utilization while recycling technologies continue advancing toward higher recovery efficiencies (Zhao et al., 2021).

3.9 Future directions and research frontiers

The convergence of artificial intelligence, computational methods, and materials science is fundamentally accelerating electrochemical energy storage innovation. Machine learning integration spans materials discovery, degradation prediction, and real-time state estimation. Google DeepMind's GNoME algorithm identified 2.2 million stable crystalline materials with 528 novel lithium-ion conductors for solid electrolyte development, representing 800 years of traditional discovery equivalent progress. The AI materials discovery market is projected to reach $11.7 billion by 2034 at 26.4% compound annual growth, with battery materials commanding 32.4% of this segment (Deshsorn et al., 2025).

PINNs represent transformative progress bridging data-driven machine learning with electrochemical fundamentals. NREL researchers demonstrated PINNs achieving battery health diagnostics 1000 times faster than P2D models while maintaining physical interpretability. PINNs for SOH estimation achieve mean absolute percentage errors of 0.87% across 387 batteries with diverse chemistries, demonstrating superior robustness in transfer learning scenarios compared to pure neural networks. Hybrid PINN architectures integrating Transformer networks enable feature extraction from truncated charge-discharge segments, enabling onboard battery diagnostics and adaptive fast-charge protocols that extend lifespan by 15%–25% (Wang et al., 2024b).

Generative AI accelerates cathode and electrolyte optimization through multiobjective design frameworks. Graph neural networks trained on composition-structure-property datasets propose high-entropy sodium-ion cathode materials with experimentally validated improved cycling stability, reducing discovery timelines from years to months. Recent advances in quantum computing, demonstrated by IBM's synthesis of novel photoacid generators, further compress materials-screening cycles (Zhu et al., 2022).

Critical future research priorities include: (1) scaling PINNs to capture coupled thermal-electrochemical-mechanical phenomena, (2) developing interpretable uncertainty quantification enabling confident safety-critical deployment, (3) integrating mechanistic degradation insights for causal rather than correlative AI models, (4) establishing standardized datasets enabling cross-laboratory validation, and (5) deploying physics-informed diagnostics across heterogeneous battery populations spanning multiple chemistries and manufacturers. Physics-informed computational intelligence creates unprecedented acceleration pathways for battery technology advancement toward sustainability imperatives (Zhu et al., 2022).

3.10 Conclusion and future outlook

This chapter has systematically examined battery technologies and electrochemical storage systems across the full spectrum from fundamental materials chemistry through physics-based modeling to system-level integration and lifecycle management. The synthesis of these domains reveals both the remarkable progress achieved and the critical challenges remaining for electrochemical energy storage to fulfill its transformative potential in sustainable energy systems.

Lithium-ion batteries have established technological dominance through 3 decades of cost reduction and performance optimization, yet fundamental limitations in energy density, safety, and resource sustainability motivate continued innovation. SSBs promise transformational improvements in energy density (300–500 Wh/kg) and safety through the elimination of flammable liquid electrolytes, with commercialization expected between 2027 and 2030 as manufacturing challenges are addressed. SIBs offer compelling cost and sustainability advantages for stationary storage applications, with CATL and BYD leading mass production at projected costs 40%–70% below lithium-ion equivalents.

The evolution from empirical battery development toward physics-informed engineering represents a paradigm shift enabling predictive design and accelerated innovation cycles. Electrochemical models spanning equivalent circuit representations through P2D formulations now integrate with sophisticated parameterization protocols—OCV characterization, EIS, GITT/PITT—and Bayesian uncertainty quantification, providing rigorous foundations for state estimation and lifetime prediction. Degradation modeling incorporating SEI growth, lithium plating, and active material loss enables mechanistic understanding of capacity fade, essential for warranty estimation and operational optimization.

Multiphysics integration coupling electrochemical, thermal, and chemomechanical phenomena increasingly captures the coupled dynamics that govern real-world battery behavior. CTP architectures, achieving 10%–15% energy density improvements, and integrated BMSs, translating electrochemical insights into real-time control decisions, bridge the gap between fundamental understanding and deployed systems.

Looking ahead, several paths will shape the progress of electrochemical storage: ongoing development of solid-state electrolytes toward achieving room-temperature ionic conductivities comparable to liquid systems; integrating machine learning to speed up materials discovery and enable predictive degradation diagnostics; employing advanced manufacturing methods such as dry electrode processing to lower costs and reduce environmental impact; and promoting a circular economy through improved recycling techniques and second-life applications to maximize resource use. The merging of materials innovation, computational modeling, and systems engineering offers unprecedented opportunities for electrochemical energy storage to support the sustainable energy transition. The emerging combination of PINNs with electrochemical fundamentals promises to speed up discovery processes, reducing materials characterization from years to months via machine learning-driven design optimization. The intersection of artificial intelligence and mechanistic battery science opens transformative pathways for next-generation electrochemical storage, offering physics-based diagnostics and adaptive control strategies that extend operational lifespans while ensuring safety and performance.

Declaration of AI assistance

The authors acknowledge the use of generative AI tools in the preparation of this manuscript. The tools were used solely to assist with language refinement, grammar correction, and improving clarity. All scientific content, analyses, interpretations, and conclusions presented in this work were developed independently by the authors.

References

Accrue Battery Intelligence, 2025. Understanding Open-Circuit Voltage. https://www.accure.net/blogs/battery-open-circuit-voltage.

Alkhalil, M.Y., 2025. Modeling Degradation in Li-Ion Batteries: Analyzing Single-Cell Degradation and Applying it to Customized Pack-Level. Durham School of Architectural Engineering and Construction: Dissertations, Theses, and Student Research. https://digitalcommons.unl.edu/archengdiss/80.

Apricum, 2024. Navigating the Future of Battery Tech: Sodium-Ion Batteries. https://apricum-group.com/sodium-ion-batteries/.

ATS Industrial Automation, 2025. Challenges and Solutions in Cell-To-Pack Battery Assembly. https://atsindustrialautomation.com/blog-posts/challenges-and-solutions-in-cell-to-pack-battery-assembly/.

Bashiir, A., 2025. Next-generation battery technologies: beyond lithium-ion. IDOSR J. Comp. Appl. Sci. 10 (2), 52–57. https://doi.org/10.59298/JCAS/2025/1025257.

Bolt.Earth, 2025. Sodium-ion vs. Lithium-Ion Batteries: Which Is Better for Electric Vehicles? https://bolt.earth/blog/sodium-ion-vs-lithium-ion-batteries-which-is-better-for-electric-vehicles.

Bonkile, M.P., Jiang, Y., Kirkaldy, N., Sulzer, V., Timms, R., Wang, H., Offer, G., Wu, B., 2024. Is silicon worth it? Modelling degradation in composite silicon–graphite lithium-ion battery electrodes. J. Power Sources 606, 234256. https://doi.org/10.1016/J.JPOWSOUR.2024.234256.

CarbonCredits.com, 2025. QuantumScape's Battery Breakthrough Powers Safer EVs – and Sends QS Stock up 20%. https://carboncredits.com/quantumscapes-battery-breakthrough-powers-safer-evs-and-sends-qs-stock-up-20/.

Chambers, C.A., Trimboli, M.S., Plett, G.L., 2024. Enhanced single particle model applied to lithium-metal battery cells using FDM. IFAC-PapersOnLine 58 (28), 192–197. https://doi.org/10.1016/J.IFACOL.2024.12.033.

De Gol, A., Dermenci, K.B., Farkas, L., Berecibar, M., 2024. Electro-chemo-mechanical degradation in solid-state batteries: a review of microscale and multiphysics modeling. Adv. Energy Mater. 14 (47), 2403255.

Deshsorn, K., Chavalekvirat, P., Deepaisarn, S., Chuang, H.C., Iamprasertkun, P., 2025. Historical data mining deep dive into machine learning-aided 2D materials research in electrochemical applications. ACS Mater. Au 6 (1), 28–56. https://doi.org/10.1021/acsmaterialsau.5c00030.

Eco-ESS, 2025. The Evolution of Energy Storage: From Basic Batteries to Smart Systems. https://www.eco-ess.co.uk/the-evolution-of-energy-storage-from-basic-batteries-to-smart-systems/.

Emergen Research, 2024. Circular Economy in Battery Recycling Market Size, Share and Growth Report (2024–2034). https://www.emergenresearch.com/industry-report/circular-economy-in-battery-recycling-market.

Energy-Storage.news, 2025. Global Energy Storage Deployments on Track for Record Year in 2025, BNEF Says. https://www.energy-storage.news/global-energy-storage-deployments-on-track-for-record-year-in-2025-bnef-says/.

FFD Power, 2024. Energy Storage BMS Architecture for Safety and Performance. https://ffdpower.com/bms-architecture-for-energy-storage/.

Guo, Y., Niu, Q., Pei, F., Wang, Q., Zhang, Y., Du, L., Zhang, Y., Zhang, Y., Zhang, Y., Fan, L., Zhang, Q., Yuan, L., Huang, Y., 2024. Interface engineering toward stable lithium–sulfur batteries. Energy Environ. Sci. 17 (4), 1330–1367. https://doi.org/10.1039/D3EE04183B.

Han, R., Macdonald, C., & Wetton, B. (2021). A Fast Solver for the Pseudo-two-dimensional Model of Lithium-Ion Batteries. arXiv. https://doi.org/10.48550/arXiv.2111.09251.

He, J., Sazzad Hosen, M., Youssef, R., Kalogiannis, T., Van Mierlo, J., Berecibar, M., 2023. A lumped electro-thermal model for a battery module with a novel hybrid cooling system. Appl. Therm. Eng. 221, 119874.

Integra Sources, 2021. Battery Management System: SoC and SoH Estimation Solutions. https://www.integrasources.com/blog/battery-management-system-bms-state-charge-and-state-health/.

Jongia Mixing Technology, 2025. Mixing Electrolyte for Ion-Lithium Batteries. https://www.jongia.com/industries/mixing-electrolyte-for-ion-lithium-batteries/.

Karmakar, S., 2024. Impedance spectroscopy for electroceramics and electrochemical system. In: Advanced Energy Conversion Materials, pp. 10–56. https://doi.org/10.37256/aecm.6120255567.

Kim, J., Park, S., Hwang, S., Yoon, W.S., 2021. Principles and applications of galvanostatic intermittent titration technique for lithium-ion batteries. J. Electrochem. Sci. Technol. 13 (1), 19–31. https://doi.org/10.33961/JECST.2021.00836.

Koech, A.K., Mwandila, G., Mulolani, F., Mwaanga, P., 2024. Lithium-ion battery fundamentals and exploration of cathode materials: a review. South Afr. J. Chem. Eng. 50, 321–339. https://doi.org/10.1016/J.SAJCE.2024.09.008.

Krishnan, R., Packirisamy, V., Palani, D., Ariyamuthu, R., 2026. Historical perspective in technological advances in lithium-ion battery development. JCIS Open 21, 100163. https://doi.org/10.1016/J.JCISO.2025.100163.

Kurt, E., 2022. Physics-Based Modeling of Lithium-Ion Batteries for Electrified Vehicle Simulations. Sabancı University. Master's thesis. https://research.sabanciuniv.edu/47045/.

Kushwaha, S.K., Reddy, P.K., Kant, S., Garg, M., 2023. Electro-thermal model using advanced ECM approach for automotive application. In: Proceedings of the 5th International Transportation Electrification Conference (ITEC-India 2023). https://doi.org/10.1109/ITEC-INDIA59098.2023.10471509.

Laue, V., Röder, F., Krewer, U., 2021. Practical identifiability of electrochemical P2D models for lithium-ion batteries. J. Appl. Electrochem. 51 (9), 1253–1265. https://doi.org/10.1007/S10800-021-01579-5.

Li, B., Chao, Y., Li, M., Xiao, Y., Li, R., Yang, K., Cui, X., Xu, G., Li, L., Yang, C., Yu, Y., Wilkinson, D.P., Zhang, J., 2023. A review of solid electrolyte interphase (SEI) and dendrite formation in lithium batteries. Electrochem. Energy Rev. 6 (1), 7. https://doi.org/10.1007/S41918-022-00147-5.

Li, D., Shen, C., Zheng, Y., Xu, J., 2025a. Electrochemo-mechanical degradation and failure of active particles in high energy density batteries: a review. Small 21 (8), 2407740.

Li, K., Wang, S., Shi, X., Huang, Y., 2025b. An analysis of the chemical stress field under potentiostatic intermittent titration techniques for interfacial reaction-controlled systems. Acta Mech. Solida Sin. 38 (3), 508–516. https://doi.org/10.1007/S10338-024-00578-1.

Lin, X., Khosravinia, K., Hu, X., Li, J., Lu, W., 2021. Lithium plating mechanism, detection, and mitigation in lithium-ion batteries. Prog. Energy Combust. Sci. 87, 100953. https://doi.org/10.1016/J.PECS.2021.100953.

Liu, K., Gao, Y., Zhu, C., Li, K., Fei, M., Peng, C., Zhang, X., Han, Q.L., 2022. Electrochemical modeling and parameterization towards control-oriented management of lithium-ion batteries. Control Eng. Pract. 124, 105176. https://doi.org/10.1016/J.CONENGPRAC.2022.105176.

Magar, H.S., Hassan, R.Y.A., Mulchandani, A., 2021. Electrochemical impedance spectroscopy (EIS): principles, construction, and biosensing applications. Sensors 21 (19), 6578. https://doi.org/10.3390/S21196578.

Miyake, T., Suzuki, T., Funabashi, S., Saito, N., Kamezaki, M., Shoda, T., Saigo, T., Sugano, S., 2021. Bayesian estimation of model parameters of equivalent circuit model for detecting degradation parts of lithium-ion battery. IEEE Access 9, 159699–159713. https://doi.org/10.1109/ACCESS.2021.3131190.

Nema, P.K., Vijaya, Muthukumar, P., Thangavel, R., 2024. Recent advancements and future prospects in lithium-ion battery thermal management techniques. Energy Storage 6 (8), e70076.

Nigel, 2023. Cell to Body. BatteryDesign.net. https://www.batterydesign.net/cell-to-body/.

O'Kane, S.E.J., Ai, W., Madabattula, G., Alonso-Alvarez, D., Timms, R., Sulzer, V., Edge, J.S., Wu, B., Offer, G.J., Marinescu, M., 2022. Lithium-ion battery degradation: how to model it. Phys. Chem. Chem. Phys. 24 (13), 7909–7922. https://doi.org/10.1039/D2CP00417H.

Olabi, A.G., Sayed, E.T., Wilberforce, T., Jamal, A., Alami, A.H., Elsaid, K., Rahman, S.M.A., Shah, S.K., Abdelkareem, M.A., 2021. Metal-air batteries—a review. Energies 14 (21), 7373. https://doi.org/10.3390/EN14217373.

Ouakki, M., 2024. Design and optimisation of metal-air batteries: a critical review of mechanisms and performance. In: Book Title. IGI Global, pp. 237–276. https://doi.org/10.4018/979-8-3373-3156-0.CH006.

Renewable Energy Hub, 2024. The Role of Battery Storage in the Quest for Renewable Energy. https://www.renewableenergyhub.co.uk/blog/the-role-of-battery-storage-in-the-quest-for-renewable-energy.

Richter, J., McCalla, E., 2025. EIS and pitt – extending the portfolio of high-throughput characterization of battery materials. ECS Meet. Abstr. 5, 535. https://doi.org/10.1149/MA2025-015535MTGABS. MA2025-01.

Shi, T., An, Z., Du, X., Ding, Y., Jia, R., Li, W., Zhang, D., 2025. Experimental study on the degradation characteristics and mechanism of lithium-ion batteries under mild mechanical deformation. J. Energy Storage 108, 115097. https://doi.org/10.1016/J.EST.2024.115097.

Simon-Kucher, 2025. Global Pricing Study 2025. www.simonkucher.com/en/insights/global-pricing-study-2025.

Srinivasan, S., Shanthakumar, S., Ashok, B., 2025. Sustainable lithium-ion battery recycling: a review on technologies, regulatory approaches and future trends. Energy Rep. 13, 789–812. https://doi.org/10.1016/J.EGYR.2024.12.043.

Takenaka, N., Bouibes, A., Yamada, Y., Nagaoka, M., Yamada, A., 2021. Frontiers in theoretical analysis of solid electrolyte interphase formation mechanism. Adv. Mater. 33 (37), 2100574.

Tembo, P.M., Dyer, C., Subramanian, V., 2024. Lithium-ion battery recycling—a review of the material supply and policy infrastructure. NPG Asia Mater. 16 (1), 43.

Wang, B.C., He, Y.B., Liu, J., Luo, B., 2024a. Fast parameter identification of lithium-ion batteries via classification model-assisted Bayesian optimization. Energy 288, 129667. https://doi.org/10.1016/J.ENERGY.2023.129667.

Wang, F., Zhai, Z., Zhao, Z., Di, Y., Chen, X., 2024b. Physics-informed neural network for lithium-ion battery degradation stable modeling and prognosis. Nat. Commun. 15 (1), 4332. https://doi.org/10.1038/s41467-024-48779-z.

Xiao, D., Xu, X., Xu, Z., 2025. Circular economy and energy storage technologies: a comprehensive approach to reduce emission and promoting sustainable growth. Energy Rep. 13, 6596–6608. https://doi.org/10.1016/J.EGYR.2025.05.063.

Xu, B., 2022. The role of modeling battery degradation in bulk power system optimizations. MRS Energy Sustain. 9 (2), 198–211. https://doi.org/10.1557/S43581-022-00047-7.

Yu, T., Li, G., Duan, Y., Wu, Y., Zhang, T., Zhao, X., Luo, M., Liu, Y., 2023. The research and industrialization progress and prospects of sodium ion battery. J. Alloys Compd. 958, 170486. https://doi.org/10.1016/J.JALLCOM.2023.170486.

Zhao, Y., Pohl, O., Bhatt, A.I., Collis, G.E., Mahon, P.J., Rüther, T., Hollenkamp, A.F., 2021. A review on battery market trends, second-life reuse, and recycling. Sustain. Chem. 2 (1), 167–205. https://doi.org/10.3390/SUSCHEM2010011.

Zhu, Z.W., Qiu, J.Y., Wang, L., Cao, G.P., He, X.M., Wang, J., Zhang, H., 2022. Application of artificial intelligence to lithium-ion battery research and development. J. Electrochem. 28 (12), 1. https://doi.org/10.13208/j.electrochem.2219003.

Further reading

Mahmood, A., Bai, Z., Wang, T., Lei, Y., Wang, S., Sun, B., Khan, H., Khan, K., Sun, K., Wang, G., 2025. Enabling high-performance multivalent metal-ion batteries: current advances and future prospects. Chem. Soc. Rev. 54 (5), 2369–2435. https://doi.org/10.1039/D4CS00929K.

PS Market Research, 2024. Cell-to-pack Battery Market Size, Growth, Analysis, and Forecast. https://www.psmarketresearch.com/market-analysis/cell-to-pack-battery-market.

Hydrogen storage and metal hydride systems

4

Rajeswaran Ramaraj[1], *Ajcharapa Chuanchai*[2], *Natthawud Dussadee*[3], *Rameshprabu Ramaraj*[3] *and Yuwalee Unpaprom*[4]
[1]Institute of Green Products, Feng Chia University, Taichung, Taiwan; [2]Department of Aeronautics and Astronautics, National Cheng Kung University, Tainan, Taiwan; [3]School of Renewable Energy, Maejo University, Chiang Mai, Thailand; [4]Program in Biotechnology, Maejo University, Chiang Mai, Thailand

4.1 Introduction

4.1.1 The carrier of energy, hydrogen

The characteristic of hydrogen in the current energy systems is that it has various primary sources, such as renewable electricity, carbon-captured natural gas, as well as biomass, which, when converted into any form of electricity, heat, or chemical feedstocks, does not emit carbon at the point of use. Hydrogen, unlike fossil fuels, does not constitute an energy source, but is an energy carrier, the overall usefulness of which is dictated by the efficiency of the production, transport, storage, and end conversion process. As a result, hydrogen storage is not a fringe logistical issue; instead, it is a core component of the hydrogen value chain and one of the key factors that dictates the cost, safety, and performance of the system (Aguey-Zinsou and Ares-Fernández, 2010).

The physical characteristics of hydrogen make it challenging to store it: its volumetric energy density at ambient temperature and pressure is very low, but this element has a high specific energy mass (120 MJ kg^{-1}). It can only be practically used, however, by densifying it physically or chemically. The storage pathway should be the correct one based on the target use, the discharge rate needed, the permissible pressure–temperature and refueling or recharge limits, and permissible system weight and volume (Jena, 2011; Schlapbach and Züttel, 2001; Züttel, 2004).

There are numerous broad categories of hydrogen storage technologies: physical, adsorption-based and material-based chemical, or metallurgically bound hydrogen storage. The most commercially developed physical options are compressed gaseous hydrogen and cryogenic liquid hydrogen. The metal hydrides, the subject of this chapter, confine hydrogen with the help of metallurgical bonding and often offer high volumetric densities with extremely low storage pressures along with a favorable safety profile (Jena, 2011; Principi et al., 2009; Schlapbach and Züttel, 2001). The main storage paths are schematically compared in Fig. 4.1, and the main features of the main storage paths are summarized in Table 4.1.

Energy Storage Technologies. https://doi.org/10.1016/B978-0-443-45858-3.00007-3

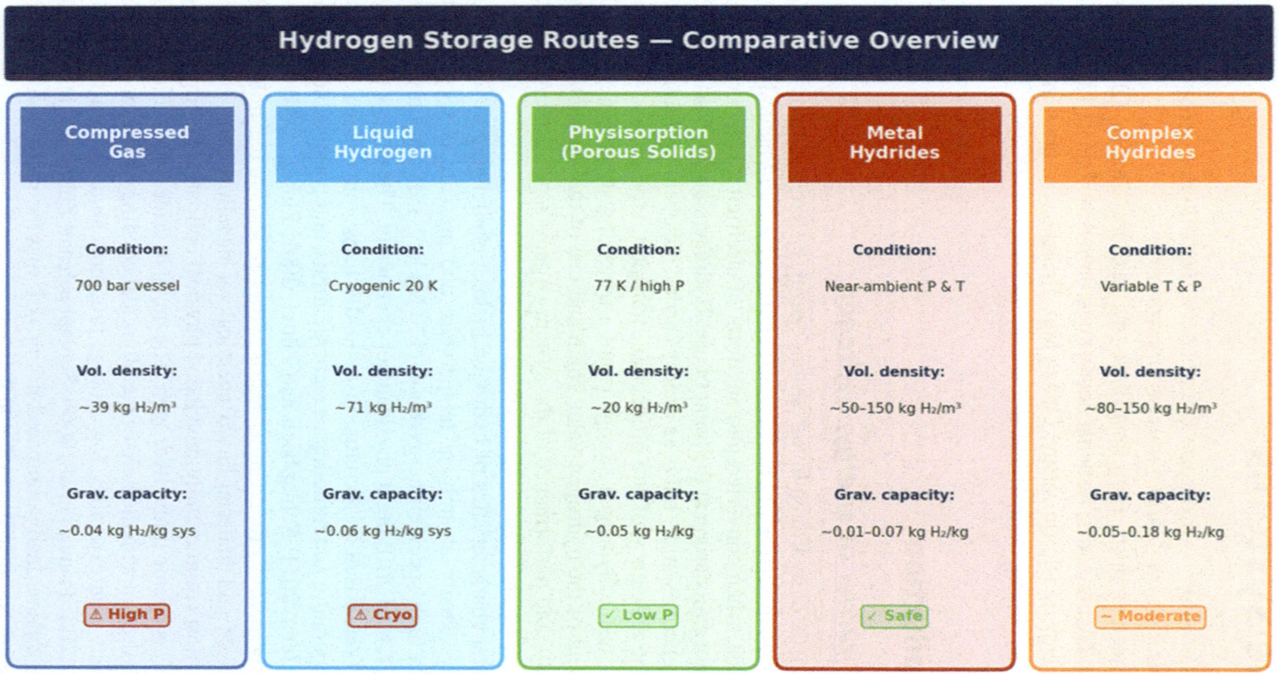

Figure 4.1 Comprehensive graphical chart of the key hydrogen storage routes, with the typical state of storage, volumetric/gravimetric density, and safety profile of compressed gas, liquid hydrogen, physisorption, metal hydrides, and complex hydrides.

Table 4.1 Comparison of key routes in hydrogen storage.

Storage route	Typical storage state	Main advantages	Main limitations	Typical application
Compressed gas	High-pressure H_2 cylinders/tanks	Mature technology; rapid filling; direct gas delivery	High-pressure vessel requirements; lower volumetric density than condensed/ solid-state routes	Transport, refueling, and distributed storage
Liquefied hydrogen	Cryogenic liquid at ~20 K (−253 °C)	High volumetric density; large throughput possible; established for rocket/ aerospace use	Liquefaction is energy-intensive (~30% of H_2 energy); boil-off losses; requires cryogenic insulation	Large-scale distribution, aerospace, bulk export terminals
Physisorption/ adsorption	H_2 adsorbed on high-surface-area solids (e.g., MOFs, activated carbon)	Potential reversibility; low chemical reactivity; moderate pressures possible	Maximum useful capacity typically requires cryogenic temperatures, low gravimetric density at ambient conditions; sorbent mass penalty	Niche and emerging storage concepts; research applications
Metal hydrides	Reversible solid-state hydride phase	High volumetric density; moderate pressure; favorable safety profile	Thermal management, sorbent mass, and kinetics can be limiting	Stationary storage, compressors, batteries, specialty systems
Chemical/ complex hydrides and carriers	Hydrogen stored in reactive compounds (e.g., NH_3, LOHC, $NaBH_4$)	Very high theoretical capacity in select systems; liquid carriers compatible with existing infrastructure	Regeneration is complex and energy-intensive; kinetics and reversibility can be challenging; handling hazards	Portable power, off-board regeneration, long-distance shipping, specialty systems

4.1.2 Scope and structure of this chapter

This chapter builds up the scientific and engineering principles of hydrogen storage in metal hydrides to a reader who has a background in the science of materials, in chemical engineering, or in the energy-technology field. Section 4.2 presents the hydride formation chemistry, pressure–composition–temperature (PCT) behavior, and key performance indicators. A review of the major classes of hydride materials, conventional intermetallic, magnesium-based, and complex hydrides is made in Section 4.3. In Section 4.4, synthesis, activation, and characterization are considered. Section 4.5 discusses the thermodynamic and kinetic constraints that are involved in real-world performance and the important approaches to overcoming. Section 4.6 discusses reactor and system design, Section 4.7 is a survey of major areas of application, and Section 4.8 is a review of current constraints and future research priorities (Griessen and Riesterer, 1988; Lototskyy et al., 2017; Principi et al., 2009; Sakintuna et al., 2007).

4.2 Fundamentals of metal hydrides

4.2.1 Chemistry of hydride formation

The key reversible storage reaction of a metal hydride system is

$$M + (x/2)\,\mathrm{H_2} \rightleftharpoons \mathrm{MH}_x \tag{4.1}$$

In which M is the host metal or alloy and MH4-6 is the related hydride phase (Griessen and Riesterer, 1988). Hydrogen atoms are interstitially located in a metallic host in the interstices of the lattice in interstitial hydrides, or are attached to ionic or covalent substructures in the complex hydrides and the storage process may be multi-step reactions and intermediate steps (Griessen and Riesterer, 1988; Orimo et al., 2007). Since absorption of hydrogen is exothermic and desorption is endothermic, the sense of reaction (1) is directly related to temperature and pressure—the basic utility and the key engineering problem of hydride systems (Edalati et al., 2018; Griessen and Riesterer, 1988; Sandrock, 1999; Lototskyy et al., 2014).

4.2.2 The mechanisms of absorption and desorption

The process of hydrogen sorption in metals and alloys occurs in a series of elementary and mesoscale reactions, coupled together. To be absorbed, they include: (i) H_2 is transported to the material surface (molecular); (ii) H_2 is dissociatively chemisorbed into atomic hydrogen (ii); (iii) H atoms penetrate the surface (subsurface); (iv) diffusion of hydrogen atoms through the host lattice (vi); and (v) nucleation and growth of the hydride phase (Lototskyy et al., 2014, 2017; Principi et al., 2009). The reverse order of desorption takes place. Fig. 4.2 shows the scheme of this mechanism of a single particle.

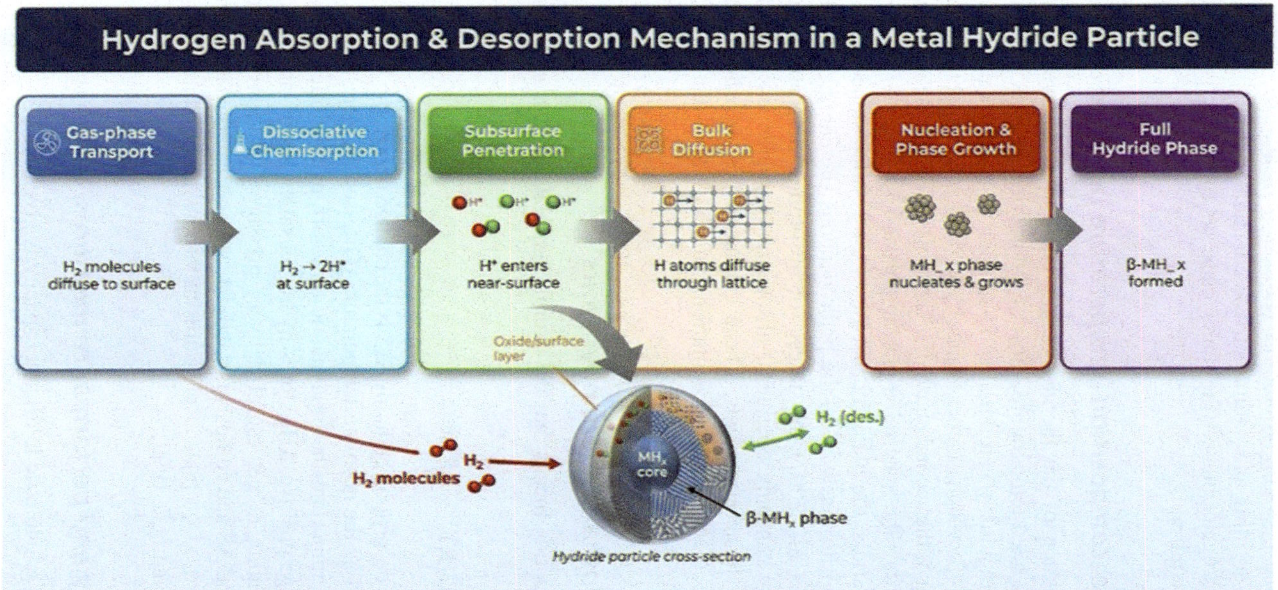

Figure 4.2 Hydrogen absorption and desorption processes in a metal hydride particle in a schematic manner. Examples of the steps are gas-phase transport (①), dissociative chemisorption (α-phase) (2), subsurface penetration α+β two-phase coexistence (③), bulk diffusion β phase, and nucleation and growth of the hydride phase (⑤–⑥). The reverse chain of reaction of desorption takes place.

The total rate can be regulated by surface dissociation, diffusion in a product layer, phase boundary motion, or heat transfer in the packed sorbent bed, depending on the material system and operating conditions (Lototskyy et al., 2017; Principi et al., 2009). Natural oxide films and natural surface contamination regularly damage the initial steps of hydrogen uptake—the major cause as to why activation treatments (Section 4.2) are required.

4.2.3 Pressure–composition–temperature behavior

The typical expression of the thermodynamic equilibrium of hydride is the PCT isotherm. Within a fixed temperature, most of the hydride systems are characterized by three different regions: (i) an initial solid-solution (α-phase) region; (ii) a plateau region (between $\alpha+\beta$ two-phase coexistence); and (iii) a region that tends to approach complete hydrating of the second phase (0). The most easily available thermodynamic quantity that describes the material is plateau pressure.

In the real materials, the plateau caused by PCT is often sloped due to the disorder in the composition as well as the heterogeneous site energies and the elastic strain (Fetcenko et al., 2007). Hysteresis of absorption–desorption—In a given pressure range, we can only operate at absorption pressures less than desorption pressures, that is, hysteresis exists. These characteristics are shown in Fig. 4.3 together with the Van Hoff representation, which relates plateau pressure to temperature.

4.2.4 Van't Hoff and thermodynamics

The Van't Hoff relation means that the equilibrium plateau pressure is temperature dependent:

$$\ln P\_eq = \Delta H/(RT) - \Delta S/R \tag{4.2}$$

in which ΔH is the desorption enthalpy, ΔS is the change in entropy, R is the universal gas constant, and T is the absolute temperature (Griessen and Riesterer, 1988; Sandrock, 1999). The important thermodynamic trade-off, as quantified in Eq. (4.2): a stable hydride (large positive ΔH) stores hydrogen safely but in large amounts, and needs a lot of heating to get useful delivery pressure. The conventional way of determining ΔH and ΔS and comparing candidate storage materials is by constructing Van't Hoff plots based on plateau pressures at various temperatures (Griessen and Riesterer, 1988; Sandrock, 1999).

4.2.5 Capacity and performance measures

The material level gravimetric hydrogen capacity is given as:

$$w\_g = (m\_H_2 \ / \ m\_MH_x) \times 100 \tag{4.3}$$

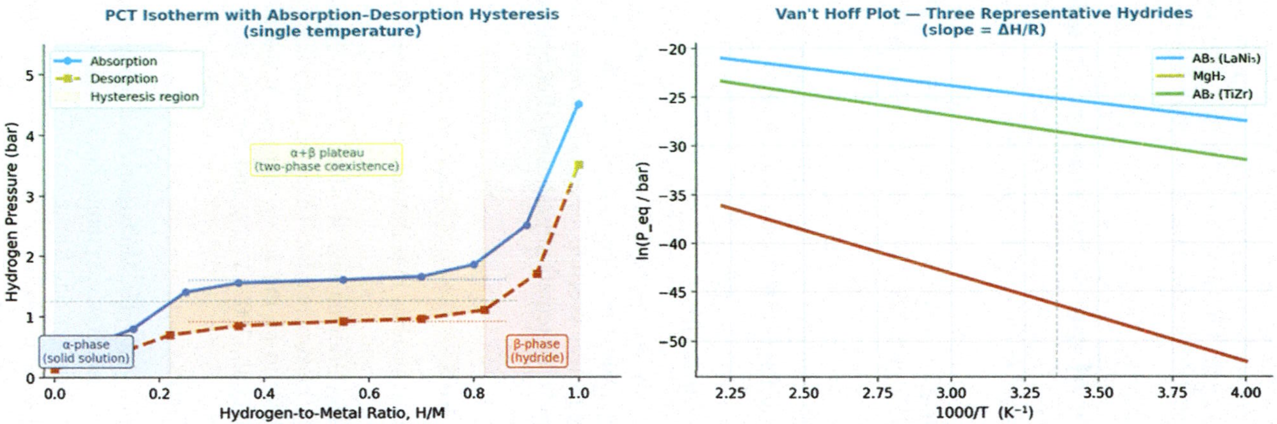

Figure 4.3 Left: Idealized PCT isotherm with the 2 plateaus of coexistence between 2 and 3, 2-hydride and absorption-desorption hysteresis. Right: Van't Hoff plot of three representative hydrides (AB_5, MgH_2, AB_2) and how their thermal (operating) stability is traded off against temperature.

and m H_2 is the mass of hydrogen stored and MH is the total mass of the hydrided material (Jena, 2011; Principi et al., 2009). Practically, pressure vessels, heat-exchange equipment, and controls are all lower capacity relative to Eq. (4.3). Other important measures are the rates of absorption/desorption, activation behavior, the magnitude of hysteresis, cycle life, impurity tolerance, thermal conductivity, and reversibility (Principi et al., 2009; Sakintuna et al., 2007; Sandrock, 1999).

4.3 Classification of metal hydride materials

4.3.1 Elementary and interstitial hydrides

The PdH 8 and TiH_2 are elemental hydrides used as model systems to study the interactions and diffusion of hydrogen and metals and the phase equilibria (Griessen and Riesterer, 1988; Sandrock, 1999). They have significant scientific importance but are not typically good in large-scale stores because of cost (Pd), or because they are too thermodynamically stable (very stable hydrides). They still offer the conceptual framework for the understanding of more complicated intermetallic and composite hydrides (Griessen and Riesterer, 1988; Sandrock, 1999).

4.3.2 Intermetallic hydrides

The most technologically advanced type of reversible hydrides is intermetallic compounds. Their main strength lies in the wide range of tunability of crystal structure and composition, allowing the plateau pressure, hysteresis, kinetics, activation properties, and impurity resistance to be systematically adjusted (Fetcenko et al., 2007; Young and Nei, 2013).

4.3.2.1 AB_5-type alloys

AB alloys (and $LaNi_5$-based systems in particular) are typified by kinetics which are fast, low activation barriers, and predictable reversibility at near-ambient temperatures (Edalati et al., 2018; Griessen and Riesterer, 1988; Sandrock, 1999). They have a modest but strong electrochemical behavior and their gravimetric capacity (~1.4 wt) has enabled them to be the active material in the negative electrode of Ni-MH battery (Fetcenko et al., 2007; Young and Nei, 2013).

4.3.2.2 AB_2-type alloys

AB_2 Laves-phase alloys (Ti, Zr, V, Cr, Mn based) have more compositional freedom and a little better capability versus AB_5 systems, but still can operate at room temperature (Edalati et al., 2018; Griessen and Riesterer, 1988; Sandrock, 1999). Elemental substitution enables adjustment of plateau pressure over a broad range, which makes the alloys promising in hydrogen compression and design of application-specific hydrogen storage (Sandrock, 1999; Lototskyy et al., 2014).

4.3.2.3 Other intermetallic families

Further design space expansion is A_2B and A_2B_7 compounds, which add to the design space (Griessen and Riesterer, 1988; Sandrock, 1999). The ability to trade-off between capacity, kinetics, plateau pressure, and cycling stability can be quite deliberate in contrast to simple binary hydrides, and the ongoing centrality of intermetallics to applied hydride engineering (Griessen and Riesterer, 1988; Sandrock, 1999).

4.3.3 Magnesium-based hydrides

MgH_2 is a highly promising anode because of its high theoretical capacity (~7.6 wt %), and the low cost and abundance of magnesium (Griessen and Riesterer, 1988; Sandrock, 1999). The main drawbacks are thermodynamic stability and kinetic slowness: hydrogen desorption generally involves temperatures of over 300 °C and untreated MgH_2 desorbs slowly (Principi et al., 2009; Sakintuna et al., 2007). These issues have been tackled by a lot of research on particle-size reduction, catalytic additives, and alloying, but are yet to be implemented in practice due to further improvements in kinetics and thermal capabilities (Fetcenko et al., 2007; Jensen and Gross, 2001).

4.3.4 Complex hydrides

Complex hydrides do not store hydrogen in interstitial sites, but store it in complex anions $[AlH_4]$—(alanates) or $[BH_4]$—(borohydrides), allowing the hydrogen density to be extremely high (Lototskyy et al., 2014).

4.3.4.1 Alanates

A breakthrough material was sodium alanate ($NaAlH_4$), as it was demonstrated that Ti-containing additives allow reversibility at much more moderate conditions than thought possible before (Bogdanović and Schwickardi, 1997; Jensen and Gross, 2001). This founded the principle that catalytic adjustment can fundamentally modify the seemingly reversible and kinetics of a complex hydride-later to be utilized in most material families (Jensen and Gross, 2001).

4.3.4.2 Borohydrides

$LiBH_4$ and other borohydrides are some of the highest theoretical hydrogen contents ($LiBH_4$ has an approximate of 18.5 wt%) of any solid material. High decomposition temperatures (>400 °C), large kinetic barriers, and incomplete reversibility are limiting factors to practical application (Edalati et al., 2018; Lototskyy et al., 2014). The additive chemistry, reactive destabilization, and nanoconfinement have been actively sought to expand the practical operating range.

4.3.5 Compare the material classes

The main families of hydride are summarized in Table 4.2. None of the classes can fulfill all the application needs at the same time. The most reliable room-temperature kinetics are those of intermetallics; those of magnesium-based materials may have very high theoretical capacity; complex hydrides may have the highest theoretical densities but may have multistep reaction pathways and may be hard to reverse (Aguey-Zinsou and Ares-Fernández, 2010; Vajo et al., 2005). Table 4.2 gives a comparison of metal hydride families under three capacities of less than 2 to approximately 18.5 wt. which indicates a definite tradeoff between hydrogen capacity and operating conditions. Storage capacity of materials, which operate at or close to ambient temperature, is usually low and higher capacity systems have higher temperature requirements, catalysts, or regeneration complexity. Consequently, each hydride is unlikely to be a high capacity, fast kinetic, mild conditions, and completely reversible hydride at the same time. The most technologically developed systems are the AB_5 alloys (e.g., $LaNi_5H_6$), which provide close to ambient operation, kinetics, and stable cycling, but with low capacity (Aguey-Zinsou and Ares-Fernández, 2010).

The advantages of AB_2 Laves-phase alloys are similar to higher flexibility in composition, but the storage capacity is small. More intermediate systems like Mg_2NiH_4 and MgH_2 have a much larger hydrogen storage; they have higher operating temperatures, but slower kinetics, which, for example, necessitate the strategies of nanostructuring and catalytic enhancement. Theoretical capacity is also cooled by using complex hydrides. $NaAlH_4$ shows reversible hydrogen storage, but it is catalytically doped, which makes its decomposition a multistep process, which decreases its usefulness. $LiBH_4$ is a highly hydrogenated material but with high thermal stability, low kinetics, and slow reversibility, which limits its efficient application in long-term research (Vajo et al., 2005). There are broadly three categories of metal hydrides: Low-capacity, near-ambient systems (AB_5, AB_2); intermediate-capacity, thermally activated systems (Mg-based hydrides); and highly capacity, complex hydrides with advanced process engineering necessary. AB_5 and AB_2 alloys are currently used with the prospects of Mg-based systems in the mid-term, with borohydrides continuing to be a priority area of future development.

Table 4.2 Representative families of metal hydride and indicative material-level properties.

Material class	Representative example	Capacity (wt%)
Elemental/interstitial hydrides	PdH_x, TiH_2	Low–moderate
AB_5 intermetallics	$LaNi_5H_6$	~1.4
AB_2 intermetallics	Ti–Zr-based Laves phases	~1.5–2.0
A_2B/related intermetallics	Mg_2NiH_4	~3.6
Magnesium-based hydrides	MgH_2	~7.6
Alanates	$NaAlH_4$	~5.6
Borohydrides	$LiBH_4$	~18.5

4.4 Material synthesis, activation, and characterization

4.4.1 Synthesis routes

The composition is not the only characteristic of hydrogen-storage behavior; microstructure, particle size, defect density, distribution of catalysts, and surface state also control the behavior; so, the direction of synthesis is thus one of the key design variables (Aguey-Zinsou and Ares-Fernández, 2010; Edalati et al., 2018). Traditional intermetallic alloys are obtained by means of arc or induction melting, homogenization, and comminution. Hydrides based on magnesium are made or modified using ball milling and reactive mechanical alloying that decrease the size of the particles, form defects, spread catalysts, and can produce valuable metastable phases (Edalati et al., 2018). Further refinement in grain structure can be done by rapid solidification (melt spinning) (Fetcenko et al., 2007; Vajo et al., 2005).

4.4.2 Activation behavior

Newly prepared hydrides normally have poor initial sorption caused by oxide films inherent on the surface or impurities. Activation-conditioning to reproducible, full capacity sorption—Emasculation with high temperature, repeated hydrogenation–dehydrogenation recycling, or exposure to high hydrogen pressure. Alloys of overtype $LaNi_5$ are fairly easily activated, and MgH_2-based systems are much more challenging, a main stimulus behind catalytic and nanostructural alteration (Aguey-Zinsou and Ares-Fernández, 2010; Edalati et al., 2018).

4.4.3 Characterization methods

To achieve reliable characterization, structural (XRD), microstructural (SEM/TEM) and thermal (DSC/TGA), and sorption (Sieverts apparatus) analyses have to be used in combination (Aguey-Zinsou and Ares-Fernández, 2010). Since the outcome is sensitive to the history of cycling and the pretreatment of the samples, to make real sense in comparing literature, it is vital to report specifically the conditions of activation, size of particles, loading of the catalyst, and thermal procedures (Edalati et al., 2018; Vajo et al., 2005). The most popular methods are summarized in Table 4.3.

The main processing and analysis methods used in metal hydride research are summarized in Table 4.3 along with their grouping based on the three major stages of the process: synthesis, activation, and characterization. Synthesis techniques are designed to synthesize materials with defined composition and structure that form the hydrogen bond. Bulk intermetallic alloys are prepared using techniques including arc or induction melting, ball milling, and reactive mechanical alloying, which refines particle size, adds defects, and enhances hydrogen diffusion routes. An additional method of performance enhancement is melt spinning, whereby fine-grained or metastable structures are formed, which may enhance reaction kinetics. There are ways of storing hydrogen that are crucial before testing.

Table 4.3 Typical methods of synthesizing, activating, and characterizing metal hydride.

Category	Method	Main purpose
Synthesis	Arc/induction melting	Intermetallic alloy preparation
Synthesis	Ball milling	Particle refinement, defect generation, and catalyst dispersion
Synthesis	Reactive mechanical alloying	Direct hydride/composite formation
Synthesis	Melt spinning/rapid solidification	Microstructure refinement
Activation	Vacuum heating/gas cycling	Surface conditioning
Characterization	XRD	Phase identification
Characterization	SEM/TEM	Microstructural observation
Characterization	DSC/TGA	Thermal analysis
Characterization	Sieverts apparatus	Sorption testing

Surface oxides and contaminants are removed by vacuum heating and repeated cycles of the hydrogen gas to result in a material that is capable of additions and removals of hydrogen in a reproducible fashion. Such a step is important since well-synthesized materials can exhibit poor performance unless they are well-conditioned on the surfaces. The techniques of characterization measure structural, thermal, and hydrogen storage. X-ray diffraction (XRD) detects crystal phases and traces the structural changes during cycling. SEM and TEM give images in detail of morphology, particle size, as well as catalyst dispersion. Thermal analysis (DSC/TGA) identifies temperatures of reactions and the stability of the material.

4.5 Thermodynamics, kinetics, and performance enhancement

4.5.1 Thermodynamic constraints

The main thermodynamic problem is the need to balance the conditions to obtain release (equilibrium against the conditions necessary to release hydrogen) as is formulated by Eq. (4.2). A stable hydride (large ΔH) would be useful in holding hydrogen, but large amounts would need to be delivered thermally. Such a trade-off is observed between AB_5 systems (near-ambient reversibility) and systems of MgH_2 (high capacity, 7.6 wt, per, yet over 300 °C to desorb) (Aguey-Zinsou and Ares-Fernández, 2010; Fetcenko et al., 2007). To achieve thermodynamic control through alloying deals a shift in the Van't Hoff slope and intercept; in complicated hydrides, a reduction in reactive destabilization can be done to decrease the effective reaction enthalpy (Edalati et al., 2018; Lototskyy et al., 2014).

4.5.2 Kinetic limitations

In cases where the thermodynamics are good, the effective rate of sorption can be unrealistically slow. One of the general expressions of phenomenology is

$$r = k_0 \exp(-E\_a / RT) \times (P / P\_eq - 1) \times f(X) \quad (4.4)$$

and r is the rate of sorption, k 0 is a preexponential term, Ea is an apparent activation energy, Pf = the equilibrium pressure [Eq. 4.2], and f(X) is the dependence on picked fraction X [5, 6, 14]. At reactor scale, heat transfer often becomes the rate-limiting mechanism in practice: exothermic desorption heat increases the temperature of the bed, boosts the equilibrium, and leads to a decrease in the driving force P/P equilibrium; conversely, a lack of heat to sustain heat payback causes the reverse (Fetcenko et al., 2007; Lototskyy et al., 2017).

4.5.3 Catalysis and doping

Catalytic modification is among the most effective tools for accelerating sorption. The seminal example is Ti-doped $NaAlH_4$, demonstrating that additive chemistry can fundamentally alter the reversibility of a complex hydride [9,10]. In Mg-based systems, transition-metal oxides, halides, and metallic nanoparticles enhance surface dissociation, reduce E_a, and promote nucleation (Aguey-Zinsou and Ares-Fernández, 2010). In intermetallics, doping optimizes plateau pressure, suppresses hysteresis, and extends cycle life.

4.5.4 Nanostructuring and nanoconfinement

Nanostructuring decreases the diffusion path, increases the interfacial region, and introduces defects to enable faster diffusion of hydrogen exchange (Aguey-Zinsou and Ares-Fernández, 2010). The most straightforward path and most often successful method to enhance the Mg-based kinetics by an order of magnitude is ball milling. Nanoconfinement within porous scaffolds (carbons, oxides) can also inhibit grain coarsening and, in certain cases, be subject to apparent thermodynamic behavior through changing the surface/interface energies (Edalati et al., 2018).

4.5.5 Cyclability and degradation

Major methods of enhancing the performance of the metals hydrogen are shown in Table 4.4. Real-world hydrides need to be able to maintain capacity and kinetics throughout numerous cycles. Capacity fade occurs due to the redistribution of the particles, coarsening of particles, oxidation, as well as segregation (Aguey-Zinsou and Ares-Fernández, 2010; Edalati et al., 2018). Mechanical effects too play important roles: volumetric expansions of 10–25 percentage due to hydriding lead to decrepitation, loss of thermal contact, and alterations in the bed packing density—all of which need to be considered when designing the system (Lototskyy et al., 2014).

Table 4.4 Major methods of enhancing the performance of the metals hydrogen.

Strategy	Primary mechanism	Typical benefits	Common trade-offs
Alloying/ substitution	Thermodynamic tuning; lattice modification	Plateau pressure adjustment; hysteresis reduction; cycling improvement	Composition complexity; possible cost increase
Catalytic additives	Faster dissociation and nucleation; lower activation barrier	Improved kinetics; reduced desorption temperature	Additive dispersion and long-term stability issues
Ball milling/ nanostructuring	Shorter diffusion lengths; higher defect density	Faster sorption; better catalyst mixing	Contamination risk; possible agglomeration on cycling
Nanoconfinement in scaffolds	Particle stabilization; interface control	Improved reversibility; reduced coarsening	Lower effective active-material fraction
Conductive matrices/ expanded graphite	Improved heat transfer	Faster bed charging/ discharging at reactor scale	Added mass and reduced net capacity
Surface modification/ coatings	Reduced oxidation and contamination	Better activation and durability	Added processing steps

4.6 Metal hydride system design and engineering

4.6.1 Reactor concepts

A metal hydride storage unit is a thermochemical reactor consisting of a single integrated unit, comprising the active component, hydrogen manifold, pressure vessel, active material, filters, valves, thermal control hardware, and control electronics (Lototskyy et al., 2014; Vajo et al., 2005). The engineering goal is to ensure that the conditions that allow rapid hydrogen exchange and as much active material used per cycle are maintained. Since the intrinsic thermal conductivity of hydride powders is low (0.1–1 $W\ m^{-1}\ K^{-1}$), the geometries of the reactor are normally created with the main constraint on heat transfer. Fig. 4.4 shows an example of a reactor cross-section.

4.6.2 Heat and mass transfer management

The most noteworthy issue of hydride reactor engineering is thermal management. The reacting bed has an energy balance.

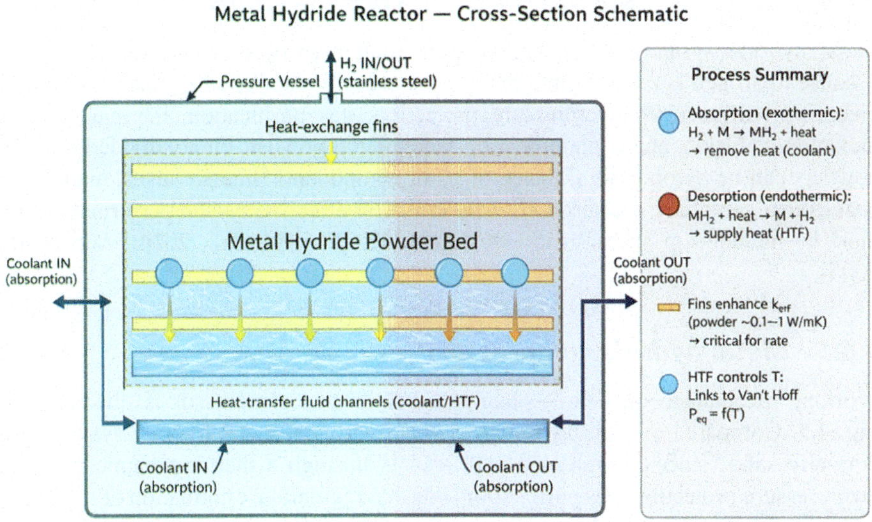

Figure 4.4 Schematic of a cross-section of metal hydride storage reactor, including internal finance schematic of heat-exchange, coolant/heat-transfer fluid (HOTF) channeling. The exothermic absorption heat is dissipated with the help of a coolant, and the endothermic desorption is motivated with the help of HTF heating. To provide effective thermal conductivity in the hydride bed, expanded natural graphite (ENG) or metal foams can be added.

$$(\rho C_p)_{eff}\partial T / \partial t = \nabla \cdot (k_{eff}\nabla T) + (-\Delta H)\partial X/\partial t \qquad (4.5)$$

and $(\rho C_p)_{eff}$, the effective volumetric heat capacity, k_{eff}, the effective thermal conductivity, 8H, the reaction enthalpy and X, the local reacted fraction (Lototskyy et al., 2014; Vajo et al., 2005). The temperature that was not controlled during and in the absorption stage elevates P_{eq} and retards uptake, whereas inadequate heat in the desorption stage inhibits the release of hydrogen. To enhance k_{eff} and thermal homogeneity, internal fins, ENG additives, metal foams, and phase-change buffers are employed.

4.6.3 Modeling and scale-up

The reactor models combine Eqs. (4.2), (4.4), and (4.5) in a numerical framework and are needed in scale-up, as the performance of beds, at the gram-scale, cannot predict the performance at the kilogram-scale. Another solid engineering literature insight is that the material with the best material properties does not necessarily result in the best system performance: a higher-capacity hydride might not perform due to poor thermal or matching the equilibrium pressure with the application (Lototskyy et al., 2014; Vajo et al., 2005).

4.6.4 Safety and balance-of-plant considerations

Metal hydride systems are generally safer than high-pressure gaseous hydrogen because hydrogen is released in a thermodynamically controlled manner. Sound engineering still requires overpressure protection, gas purification, thermal control, moisture exclusion, and containment of decrepitated powder. Finely divided hydride powders can be pyrophoric. Balance-of-plant components (filters, valves, manifolds, heat exchangers) have a disproportionate effect on effective system performance and must be included in all realistic assessments (Lototskyy et al., 2014; Vajo et al., 2005).

4.6.5 Metal hydride compressors

Working mechanism of a three-stage metal hydrogen compressor is illustrated in Fig. 4.5. Compared to high-pressure gaseous hydrogen, metal hydride systems are normally safer, since the released hydrogen is through a thermodynamic process. Overpressure protection, gas purification, thermal regulation, elimination of moisture, and removal of decrepitated powder are still needed in sound engineering. Particles of hydride that are finely divided may be pyrophoric. Balance-of-plant (filters, valves, manifolds, heat exchangers) has an inappropriate impact on effective system operation, and should always be factored into all realistic evaluations (Lototskyy et al., 2014; Vajo et al., 2005).

4.7 Applications

4.7.1 Nickel–metal hydride batteries

The NiMH battery, which utilizes an AB5617-type alloy as a negative electrode in its hydrogen storage, is the most commercially successful application of an alloy for hydrogen storage. The Ni-MH technology has shown that hydride materials don't need to have the highest possible gravimetric capacity in order to have commercial significance, but must meet reversibility, rate capability, corrosion solidity, and cycle-life objectives of the operating conditions (Edalati et al., 2018; Young and Nei, 2013).

4.7.2 Stationary hydrogen storage

One of the most apparent areas where hydride is immediately promising to be applied is in stationary storage since the mass limitation is greatly diminished. Electrolyzer-fuel cell when used as a hydrogen buffer in electrolyzer-fuels, Electrolyze Metal hydrides can serve as a backup power supply as well as renewable energy integration and distributed hydrogen networks (Lototskyy et al., 2014; Vajo et al., 2005). The safety and inherent characteristics can easily justify the penalty of the sorbent mass of moderate pressure and compact volumetric density in such applications.

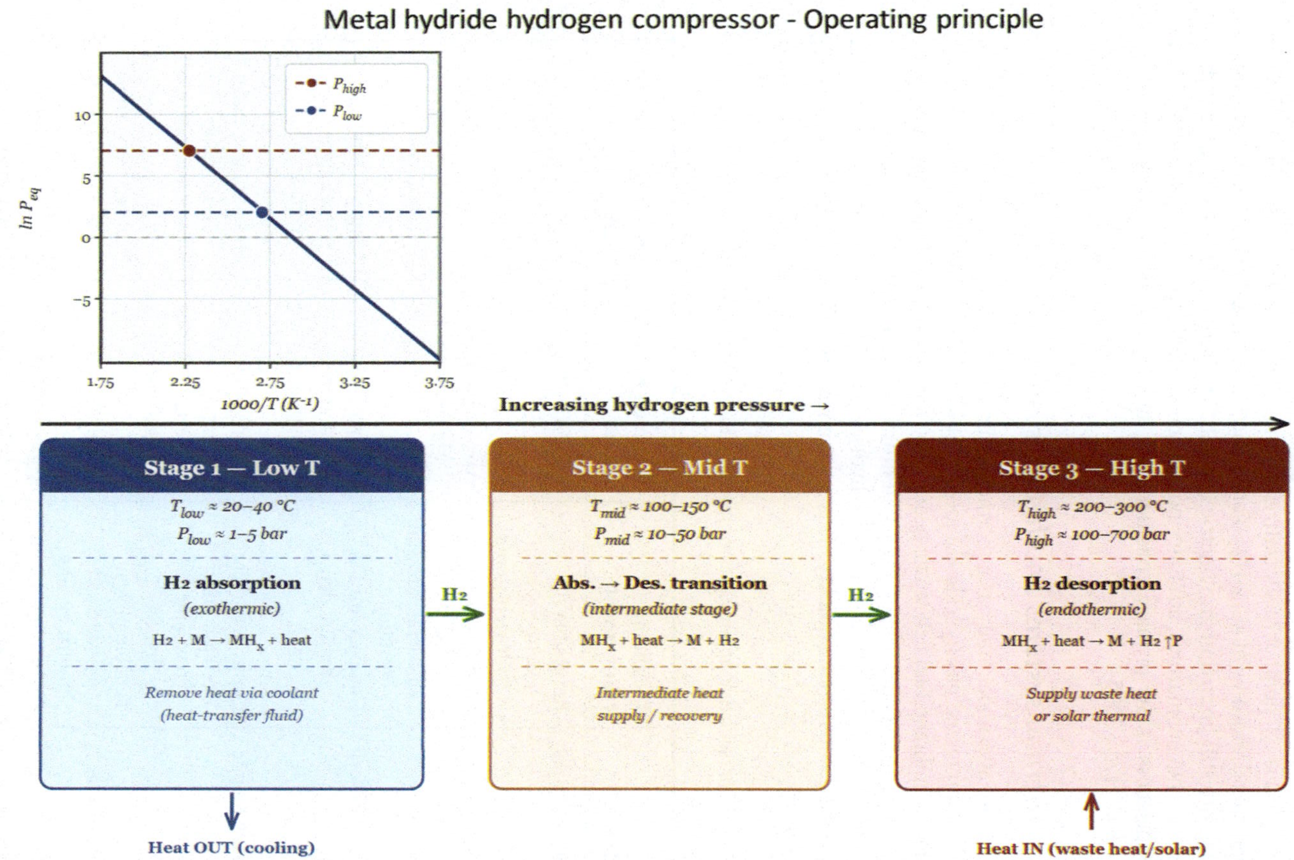

Figure 4.5 Working mechanism of a three-stage metal hydrogen compressor. The low temperature, low pressure absorption of hydrogen takes place (stage 1) and then it is transferred in stages, and lastly the hydrogen is desorbed at high temperature and high pressure (stage 3). The inset Van't Hoff plot illustrates the response of P eq to the change in temperature, as will be illustrated by Eq. (4.2).

4.7.3 Portable and mobile

Mobile platforms with low-pressure hydrogen can be applied to portable fuel-cell systems and on a small number of specialized mobile platforms of low-pressure hydrogen requirements (Lototskyy et al., 2014; Young and Nei, 2013). The current intermetallic materials are still limited in gravimetric capacity to achieve targets used in light-duty automotive, but submarines, auxiliary power units, select mobility uses, where safety or compactness is the primary consideration, have opportunities (Lototskyy et al., 2014; Vajo et al., 2005).

4.7.4 Thermal systems and hybrid systems

Hydrides are also under study as a thermochemical heat storage medium as well as in a hybrid compressed gas/hydride (Lototskyy et al., 2014; Vajo et al., 2005). Proton-exchange membrane fuel cell waste heat can be used to help with desorption with an integrated hydride bed to enhance system efficiency. The hybrid architectures are in use based on the volumetric density of hydrides at low or moderate pressures and compressed gas to respond quickly. This is summarized in Table 4.5 as the application requirements.

Table 4.5 Representative applications of metal hydride systems and their dominant design requirements.

Application	Dominant requirement	Common hydride classes	Principal engineering focus
Ni–MH batteries	Reversible electrochemical hydrogen storage	AB_5, AB_2 alloys	Cycle life, corrosion resistance, rate capability
Stationary hydrogen buffering	Safety, compactness, moderate pressure operation	AB_5, AB_2, selected Mg composites	Thermal integration, long cycle life, reliability
Portable fuel-cell supply	Compact low-pressure delivery	Intermetallics, catalyzed complex hydrides	Startup control, refueling/recharge, thermal management
Specialty mobile systems	Controlled hydrogen release under constrained volume	Intermetallics, composites, selected Mg-based systems	Mass–volume trade-off, dynamic response, safety
Hydride compressors	Tunable plateau pressure and robust cycling	AB_2, other intermetallics	Heat recovery, stage matching, gas purity
Thermochemical heat storage	Favorable reaction enthalpy and reversibility	Mg-based and selected complex hydrides	Thermal stability, integration with heat source/sink

4.8 Current limitations and future outlook

4.8.1 Gravimetric penalty and thermal burden

The gravimetric penalty to store hydrogen in a heavy host lattice is the most inadvertently persistent weakness of other hydrides (Sakintuna et al., 2007; Vajo et al., 2005). The materials have a high capacity, like MgH_2 and complex hydrides, which enhance storage density by means of Eq. (4.3) at the cost of a greater thermal load during the desorption as a direct result of Eq. (4.2). There is a redistribution of the challenge instead of an elimination.

4.8.2 Sensitivity to cost, resource constraints, and purity

The factors that determine economic viability include the cost of raw materials, the complexities of synthesizing a particular material, the use of catalysts, cycling stability, and sensitivity to purity of hydrogen (Aguey-Zinsou and Ares-Fernández, 2010; Sakintuna et al., 2007; Vajo et al., 2005). Intermetallics containing rare earths have challenges in the supply chain. Complex hydrides are moisture and oxygen-sensitive. Kinetics and reversibility can be impaired by trace impurities (CO, O_2, H_2O, H_2S) that surface poison (Edalati et al., 2018; Vajo et al., 2005), which becomes more pronounced when the field is started up compared to idealized laboratory conditions.

4.8.3 Scale, shape, and manufacture

Reproducibility on a large-scale production has remained a challenge. Nanopowders that excel on a laboratory level may be challenging to prepare at a commercial scale, to guard against oxidation in the preparation and transfer, and to add to mechanically stable, thermally conducting reactor beds (Edalati et al., 2018; Lototskyy et al., 2014; Sakintuna et al., 2007). Thermal-contact engineering, powder-shaping, and compaction are now considered to be vital in the technology-development pathway (Vajo et al., 2005; Young and Nei, 2013).

4.8.4 Emerging research directions

Future application-specific optimization and future research are shifting toward application-specific optimization by: high-throughput computational screening (DFT and machine learning potentials); surface and interface engineering to lower kinetic barriers; nanoconfined and composite hydrides with a higher reversibility; thermally conductive bed design to meet the constraint of Eq. (4.5); hybrid systems combining hydrides with compressed gas buffers or waste-heat-assisted operation (Edalati et al., 2018; Sakintuna et al., 2007). The research roadmap is macrosummarized in Fig. 4.6.

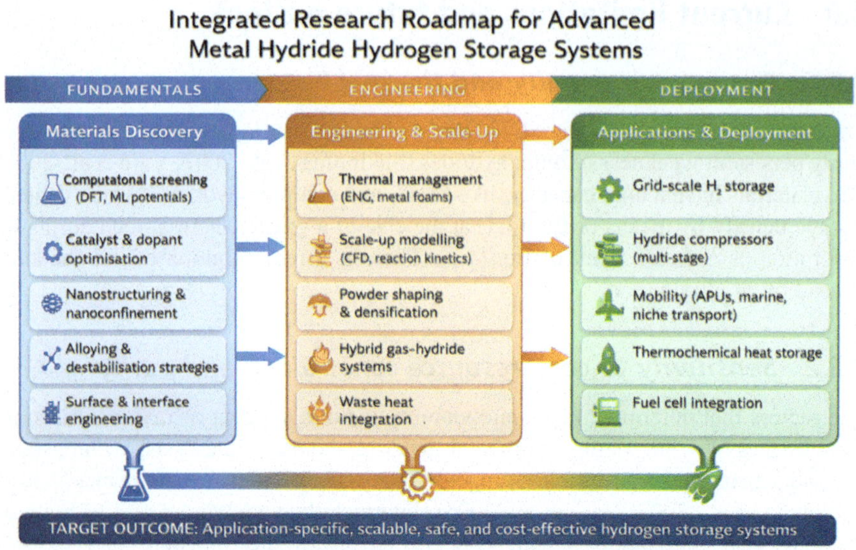

Figure 4.6 Research roadmap for metal hydride systems, linking development tracks converging on application-specific, manufacturable, safe, and cost-effective hydrogen storage solutions.

4.9 Conclusion

This chapter reviews hydrogen storage as a crucial element of the hydrogen energy system, noting that hydrogen's low volumetric density necessitates densification. Storage types include compressed gas, liquid hydrogen, adsorption, and material-based, focusing on reversible metal and complex hydrides, which offer high density and safety but reduced thermal and system simplicity. Hydrogen absorption and release are reversible, involving gas transfer, surface interactions, and diffusion; hydride systems are designed for this. Their thermodynamic relations, setting equilibrium pressure and temperature, control behavior, balancing storage stability and energy for hydrogen release. Performance factors include capacity, kinetics, cycle stability, and thermal properties, with system-level components affecting practical performance. Material classes are interstitial hydrides and intermetallic alloys (AB_5, AB_2, etc.), magnesium-based, and complex hydrides. Intermetallic are favored for reversibility and kinetics, while magnesium-based and complex hydrides offer high capacity but require higher temperatures and have limited kinetics. Balancing operating conditions and capacity is a challenge. Material processing and synthesis, like melting, ball milling, and rapid solidification, significantly impact performance. Activation procedures may enhance hydrogen uptake. Characterization techniques include X-ray diffraction, electron microscopy, thermal analysis, and sorption measurements. Performance enhancement strategies involve alloying, catalytic doping, and nanostructuring. Degradation issues like oxidation, particle growth, and structural change can reduce long-term stability. Reactor design needs to address low thermal conductivity, requiring heat management solutions like conductive additives and designed structures. Applications include nickel-metal hydride batteries, stationary energy storage, hydrogen compression, and hybrid energy systems, but weight, cost, and durability restrictions persist. Advances in material design, nanostructured composites, improved thermal integration, and hybrid storage systems are developing safe, efficient, and scalable hydrogen storage.

References

Aguey-Zinsou, K.F., Ares-Fernández, J.R., 2010. Hydrogen in magnesium: new perspectives toward functional stores. Energy Environ. Sci. 3 (5), 526–543.

Bogdanović, B., Schwickardi, M., 1997. Ti-doped alkali metal aluminium hydrides as potential novel reversible hydrogen storage materials. J. Alloys Compd. 253–254, 1–9.

Edalati, K., Akiba, E., Horita, Z., 2018. High-pressure torsion for new hydrogen storage materials. Sci. Technol. Adv. Mater. 19 (1), 185–193.

Fetcenko, M.A., Ovshinsky, S.R., Reichman, B., et al., 2007. Recent advances in NiMH battery technology. J. Power Sources 165 (2), 544–551.

Griessen, R., Riesterer, T., 1988. Heat of formation models. In: Schlapbach, L. (Ed.), Hydrogen in Intermetallic Compounds I. Springer, Berlin.

Jena, P., 2011. Materials for hydrogen storage: past, present, and future. J. Phys. Chem. Lett. 2 (3), 206–211.

Jensen, C.M., Gross, K.J., 2001. Development of catalytically enhanced sodium aluminum hydride as a hydrogen-storage material. Appl. Phys. A 72 (2), 213–219.

Lototskyy, M.V., Tolj, I., Pickering, L., Sita, C., Barbir, F., Yartys, V., 2017. The use of metal hydrides in fuel cell applications. Prog. Nat. Sci. Mater. Int. 27 (1), 3–20.

Lototskyy, M.V., Yartys, V.A., Pollet, B.G., Bowman Jr., R.C., 2014. Metal hydride hydrogen compressors: a review. Int. J. Hydrogen Energy 39 (11), 5818–5851.

Orimo, S., Nakamori, Y., Eliseo, J.R., Züttel, A., Jensen, C.M., 2007. Complex hydrides for hydrogen storage. Chem. Rev. 107 (10), 4111–4132.

Principi, G., Agresti, F., Maddalena, A., Lo Russo, S., 2009. The problem of solid state hydrogen storage. Energy 34 (12), 2087–2091.

Sakintuna, B., Lamari-Darkrim, F., Hirscher, M., 2007. Metal hydride materials for solid hydrogen storage: a review. Int. J. Hydrogen Energy 32 (9), 1121–1140.

Sandrock, G., 1999. A panoramic overview of hydrogen storage alloys from a gas reaction point of view. J. Alloys Compd. 293, 877–888.

Schlapbach, L., Züttel, A., 2001. Hydrogen-storage materials for mobile applications. Nature 414 (6861), 353–358.

Vajo, J.J., Skeith, S.L., Mertens, F., 2005. Reversible storage of hydrogen in destabilized $LiBH_4$. J. Phys. Chem. B 109 (9), 3719–3722.

Young, K., Nei, J., 2013. The current status of hydrogen storage alloy development for electrochemical applications. Materials 6 (10), 4574–4608.

Züttel, A., 2004. Hydrogen storage methods. Naturwissenschaften 91 (4), 157–172.

Thermal energy storage and thermochemical systems

5

Jayashri Narayanan Nair [1], *Dhana Raju Vallapudi* [2] *and Sachin Lotan Borse* [3]
[1]Department of Mechanical Engineering, VNR Vignana Jyothi Institute of Engineering and Technology, Hyderabad, Telangana, India; [2]Department of Mechanical Engineering, Lakireddy Bali Reddy College of Engineering, Mylavaram, Andhra Pradesh, India; [3]Department of Mechanical Engineering, JSPM Narhe Technical Campus, Pune, Maharashtra, India

5.1 Introduction

Sustainable development and energy efficiency needs of the world have driven the need for advanced storage technologies. Thermal energy storage (TES) is a significant component in enhancing the flexibility and reliability of energy systems. TES enables energy management in both renewable and conventional energy systems, contributing to reduced fuel consumption enhanced system, efficiency, and mitigation of intermittently in solar and wind energy sources. Depending on the storage mechanism, TES technologies are generally categorized into three main types based on the storage mechanism, sensible, latent, and thermochemical heat storage.

5.1.1 Sensible heat storage

In this method, energy is stored by raising or lowering the temperature of a storage medium without changing its phase (Kumar et al., 2024). The amount of heat stored depends on the material's specific heat capacity, mass, and temperature difference according to the relation:

$$Q_{stored} = m\ Cp\ \Delta T$$

where Q is the stored thermal energy (j), Cp is the specific heat capacity(J-kg/K), and ΔT is the temperature rise in K.

Sensible heat storage (SHS) materials are typically categorized as solid or liquid media. Solid media, such as natural rocks, concrete, and ceramics, are commonly used for applications requiring high-temperature operation and mechanical stability, such as in solar thermal power plants or industrial waste heat recovery. For moderate and high temperature ranges water and molten salts are preferred, as they are characterized by their high heat transfer rates and high volumetric heat capacities. SHS systems have lower energy storage density (10–100 kWh/m^3) compared to latent and thermochemical systems but is also cost-effective, reliable with good cyclic stability,

Energy Storage Technologies. https://doi.org/10.1016/B978-0-443-45858-3.00001-2

and long operational lifetimes. However, the thermal losses are more over longer durations and large volume of material is required for storage. Thermal gradients and heat dissipation limit the storage efficiency of SHS. Adding metallic foams, graphite, or nanoparticles enhances the storage performance of SHS (Islam et al., 2024). In addition to the additives, encapsulation techniques and integration with heat transfer fluids (HTFs) improves the thermal response. Comprehensively, SHS is the base for thermal energy management, contributing a durable and cost-effective solution for short- to medium-term storage in various applications like district heating and industrial process optimization.

5.1.2 Latent heat storage

Latent heat storage (LHS) is an advanced form of energy storage method, which depends upon the phase change of the storage material for storing and releasing the energy at constant temperature (Sarbu and Dorca, 2019). It utilizes the latent of fusion, vaporization, or sublimation (Yang et al., 2024). The governing energy balance for LHS can be expressed as:

$$Q = m\ \Delta h_{\text{phase}} + m\ C_p\ \Delta T$$

where Q is the total stored thermal energy (j), Δh_{phase} is latent heat of phase diagram (j/kg), C_p is the specific heat capacity (j-kg/K), and ΔT is the temperature rise in K. The main component of an LHS system is the phase change material (PCM), which stores energy through reversible melting and solidification processes (Mahdi et al., 2019). The decisive factor for the operating range of PCM is its melting temperature, hence selection of material for a specific application becomes crucial. PCMs can be classified into two main groups:

1. Organic PCMs possess high chemical stability, less supercooling, and good compatibility with container materials. However, the low thermal conductivity of organic PCMs limits the heat transfer rates. For example, paraffins, fatty acids, and esters.
2. Inorganic PCMs display higher volumetric latent heat densities and thermal conductivities but the material degrades after multiple cycles. For example, primarily salt hydrates, molten salts, and metallic alloys.

For medium temperature range applications (20–300°C), like space heating, solar water heating, and waste heat recovery, LHS systems can be used, as their energy storage density is higher than the SHS systems (Mahdi et al., 2019). As the operations are nearly at constant temperature, it provides better thermal control and efficiency in energy exchange processes. Nevertheless, a number of obstacles prevent LHS technologies from being widely used. Many PCMs have low thermal conductivity, which causes charging and discharging rates to be slow (Sarbu and Dorca, 2019). The addition of metallic foams, graphite matrices, nanoparticle dispersion, and encapsulation in high-conductivity shells are some of the methods that researchers have investigated to improve thermal conductivity. Furthermore, long-term material stability under frequent cycling and leakage prevention during the liquid phase continue to be crucial design factors. Recent developments in microencapsulation, composite PCMs, and hybrid TES systems that combine sensible and latent mechanisms have demonstrated promise in enhancing system compactness, structural stability, and thermal performance. As an intermediary step toward high-efficiency and high-density TES technologies, LHS is therefore still a major area of research interest.

5.1.3 Thermochemical energy storage

Reversible chemical reactions involving sorption or decomposition processes store and release heat. Although sensible and LHS systems are well-established, heat losses and low energy density limit their efficiency, especially for long-term applications. Thermochemical energy storage (TCES), on the other hand, has a number of clear benefits, such as increased energy storage density, reduced thermal losses during storage, and the potential for long-term and transportable energy storage. TCES is a promising technology for upcoming sustainable energy systems because of these characteristics. Additionally, integrating thermochemical systems into building heating and cooling systems, industrial waste heat recovery processes, and renewable energy plants can greatly increase overall energy efficiency. Finding appropriate thermochemical materials, enhancing reaction kinetics, and creating reactor designs that optimize heat and mass transfer efficiency are the main goals of ongoing research.

5.2 Sensible heat storage: Mechanisms and materials

5.2.1 Mechanism

SHS mechanisms involve three main stages:

a) Charging (heat absorption)
b) Storage (thermal retention)
c) Discharging (heat release)

During these stages, heat is exchanged with a storage medium (solid or liquid) through conduction, convection, and sometimes radiation, without any phase change in the material (Fig. 5.1).

a) **Charging (Heat Absorption Mechanism):** In the charging phase, thermal energy from a heat source, such as solar collectors, industrial exhaust gases, or electric heaters, is transferred to the storage medium. If a HTF like air, oil, or water flows through the system, it transfers heat to the storage medium through convective heat transfer, for example, in a rock bed, hot air passes between the rocks, heating them up. Within the solid or liquid medium, heat moves from the surface to the interior through molecular vibration and diffusion, for example, in a concrete or metal block, heat gradually penetrates the core by conduction. At higher temperatures, radiant heat exchange can also contribute, especially between solid surfaces in a packed bed (Li, 2016). The temperature of the medium rises, storing energy in the form of increased internal energy of the material.

$$Q_{\text{stored}} = mCp\,\Delta T$$

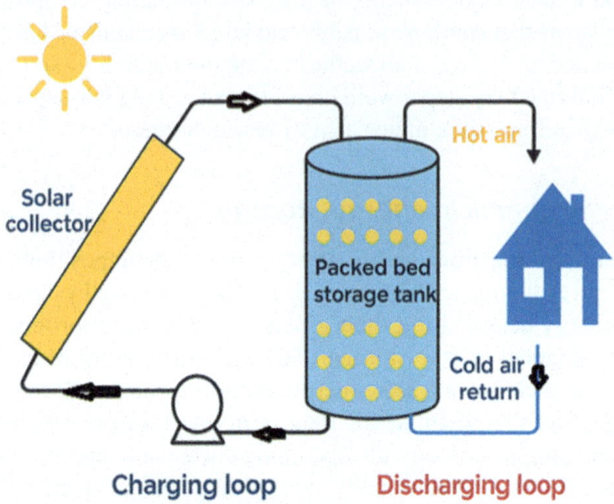

Figure 5.1 Sensible heat storage mechanism.

b) **Thermal Retention Mechanism:** Once the medium is heated, the system is insulated to minimize heat losses. The stored heat remains as increased internal energy within the medium's molecules (Sathyamurthy, 2024). The quality of insulation (e.g., polyurethane foam, vacuum panels) determines how long the heat can be retained. Thermal stratification may occur, especially in liquids like water, forming layers of different temperatures, which improves retrieval efficiency by drawing heat from the top layers first.
c) **Heat Release Mechanism:** During the discharging phase, the process is reversed (Lugolole et al., 2019). A cooler fluid (HTF) or environment draws heat from the storage medium. Heat is released by conduction from the interior of the medium to its surface, by convection from the surface to the HTF or surrounding air, for example, in a hot water tank, cooler water entering the tank gets heated by the stored hot water at the top.

5.2.2 Materials

High specific heat capacity, long-term thermal stability, and compatibility with container materials are necessary for effective SHS. The two primary types of materials that are utilized are solid and liquid heat storage materials (Dinker et al., 2017). Due to their high thermal conductivity, metals like aluminum, copper, and lead are frequently used for SHS; however, they have disadvantages such as low heat capacity, high density, and high cost. Water is a preferred medium because of its high thermal conductivity and affordability. Building materials have also been investigated for SHS to create energy-efficient and temperature-controlled buildings. Solar water tanks and solar ponds, which are constructed with materials like steel, concrete, aluminum, and

Table 5.1 Desirable properties of sensible heat storage materials (Boquera et al., 2021a).

Properties	Desirable properties
Thermophysical properties	High energy, thermal conductivity, heat capacity, density, and long-term thermal cycling stability
Chemical properties	Prolonged chemical stability with no chemical breakdown, less corrosive to HTFs, nontoxic, nonexplosive, and compatible with building materials
Economic properties	Less expensive
Mechanical properties	High compressive strength, high fracture tension, low coefficient of thermal expansion, and good mechanical stability
Environmental properties	Minimal CO_2 footprint and energy requirements for manufacturing

fiber glass to stop heat loss, can accomplish SHS (Nkhonjera et al., 2017). Salts like sodium and magnesium chloride are used in solar ponds to produce thermal stratification, with the bottom, where solar radiation is absorbed, having the highest temperature. Other liquids like molten salts, petroleum-based oils, and molten metals are also used for SHS, with petroleum-based oils being suitable due to low vapor pressure (Table 5.1).

5.2.2.1 Reinforced concrete

Concrete has attracted a lot of interest as a potential sensible heat medium. Concrete is promising storage medium for sensible TES systems for concentrating solar power (CSP) plants. Concrete is a composite made up of cement binders, aggregates, admixtures, and water. It is cheap, durable, and requires less maintenance. It is actively involved in TES research due to its special combination of high heat capacity, moderate thermal properties, affordability, and widespread availability (Rahjoo et al., 2022a). Concrete is particularly well suited for use in CSP plants and solar process heat (SPH) systems due to its natural thermal inertia and capacity to store energy through sensible heat transfer (Rahjoo et al., 2022b; Salomoni et al., 2014; Wang et al., 2023).

Concrete has high specific heat capacity, moderate thermal conductivity, and good thermal cycling stability which makes it suitable for its use as SHS material. Specific heat capacity determines the amount of thermal energy that can be stored per unit mass for a given temperature change however concrete's specific heat capacity varies depending on its composition. Siliceous aggregate concrete typically exhibits higher specific heat capacity compared to carbonate aggregate concrete (Hoivik et al., 2019). The thermal conductivity property of concrete influences the rate of heat transfer during charging and discharging cycles; however, extremely high values can lead to increased heat losses. Optimization of thermal conductivity incompliance with the application is necessary. Siliceous aggregate concrete usually has a higher thermal conductivity of about 2.5 W/mK when compared to lime stone aggregate concrete, which has a thermal conductivity of about 1.8 W/mK at room temperature (Hoivik et al., 2019).

Compact thermal storage systems are made possible by the high density of concrete, which varies from 2320 kg/m^3 for carbonate aggregate concrete to 2465 kg/m^3 for siliceous aggregate concrete with steel fibers. However, this high density also raises the cost of materials and structural requirements. For long-term durability in SHS applications, concrete must be able to endure repeated heat cycles without noticeably losing its mechanical and thermal qualities. It is crucial to look into spalling, microstructural damage, and changes in thermal characteristics during cycling.

Ordinary Portland cement (OPC)-based materials are commonly used by Sensible TES; however, their use is limited to operation temperatures below 400°C to 500 °C because of thermal degradation processes. Improved thermal properties have resulted from the addition of supplementary cementitious materials such as fly ash, blast furnace slag, and silica fume (Boquera et al., 2021b). Concrete is now a state-of-the-art material for high-temperature TES, able to function at temperatures higher than 500°C thanks to these developments. Significantly, recent research on geopolymer-based concretes has extended the possibilities of concrete-based TES systems by raising temperature thresholds to 600°C (Rahjoo et al., 2023). Additionally, its ingredients are widely available and reasonably priced worldwide. The cementitious binder typically used in both regular concrete and concrete-based TES devices is OPC. In fact, OPC concrete's performance and suitability for TES have been thoroughly examined at research facilities and verified in solar thermal applications. However, on occasion, CSP plants and SPH applications may require temperatures as high as approximately 1000 °C (Lovegrove and Stein, 2012), or at least significantly higher than what OPC-based concretes can tolerate. The most significant hydration product of OPCs, the C–S–H gel (Cement Chemist Notation C = CaO, S = SiO_2, H = H_2O), largely controls the durability and thermal properties of OPC-based composites. Unfortunately, using OPC-based TES modules at temperatures between approximately 400 and 600 °C is extremely dangerous because C–S–H gel degrades at such temperatures (Andic et al., 2008; John et al., 2010).

5.2.2.2 Cast steel

Cast steel can be employed as a sensible storage application due to its appealing thermophysical and thermomechanical characteristics. There is no phase change when thermal energy is stored (Vigneshwaran et al., 2019a). Because of its high density and moderate specific heat (450–600 J/kg K), cast steel has a high volumetric heat capacity that ranges from 3.6 to 4.5 MJm^{-3}K^{-1} (Vigneshwaran et al., 2019b). Cast steel's high thermal conductivity (25–45 W/mK) speeds up internal diffusion, leading to quicker charging and discharging. Cast steel demonstrates remarkable resistance to thermal fatigue and repeated thermal cycling, ensuring thermal stability under prolonged operating conditions and varying heat input (Jurczyk et al., 2024). Cast iron can be combined with solar thermal collectors and industrial waste heat recovery systems because it can withstand temperatures as high as 700°C. If surface treatment is

Table 5.2 Thermophysical properties of cast steel.

Property	Typical range
Density	7700–7900 kg/m^3
Specific heat	460–600 J/kg K
Thermal conductivity	25–45 W/mK
Thermal diffusivity	6–12 × 10^{-6} m^2/s^{-1}
Max service temperature	700°C

not applied, cast iron may experience convective and radiative heat losses (Paul et al., 2022). In situations where high temperatures are important, cast iron is a SHS material that is both technically sound and thermally dependable (Khare et al., 2013; Zhao et al.). Table 5.2 demonstrates the thermophysical properties of cast steel.

5.2.2.3 Cast iron

Cast iron is a sensible material for storage at medium to high temperatures. Cast iron has a volumetric heat storage capacity of approximately 3.0–3.8 MJ m^{-3} K^{-1}, which is comparable to cast steel and significantly higher than conventional materials like concrete (Tripathi and Rathore, 2025). It has a density of 6800–7300 kg m^{-3} and a specific heat capacity of approximately 450–550 J kg^{-1} K^{-1}. Particularly in air-based or gas–solid heat exchange systems, cast iron's thermal conductivity (35–55 W m^{-1} K^{-1}) is relatively high among sensible storage materials, enabling efficient internal heat diffusion and faster charging–discharging rates. Its suitability for applications requiring rapid thermal response is further supported by its thermal diffusivity, which is usually on the order of (7–12) × 10^{-1} m^2 s^{-1} (Naveenprabhu et al., 2024). Cast iron has strong resistance to thermal degradation, good dimensional stability, and excellent compressive strength from a mechanical standpoint. However, because of its brittle nature, careful design is required to prevent thermal shock and stress concentration during abrupt temperature changes (Miliozzi et al., 2025). Because cast iron can reliably operate at temperatures up to about 600–650 °C in oxidizing environments, it can be integrated with solar thermal systems, waste heat recovery units, and high-temperature air preheating applications (Olivkar et al., 2022). Cast iron is more castable and less costly than cast steel, despite having slightly less ductility and tensile strength. Sufficient insulation and surface protection are essential to lowering convective and radiative heat losses at high temperatures (Jotshi et al., 1996). All things considered, cast iron is a SHS material that is both technically sound and economically appealing.

5.2.2.4 Rocks

Rocks are an economical and useful way to store heat. Both active and passive thermal storage systems can be found in rocks (Abddaim et al., 2025). The forced convection system makes up active TES. A heat exchanger is used in passive TES to circulate the

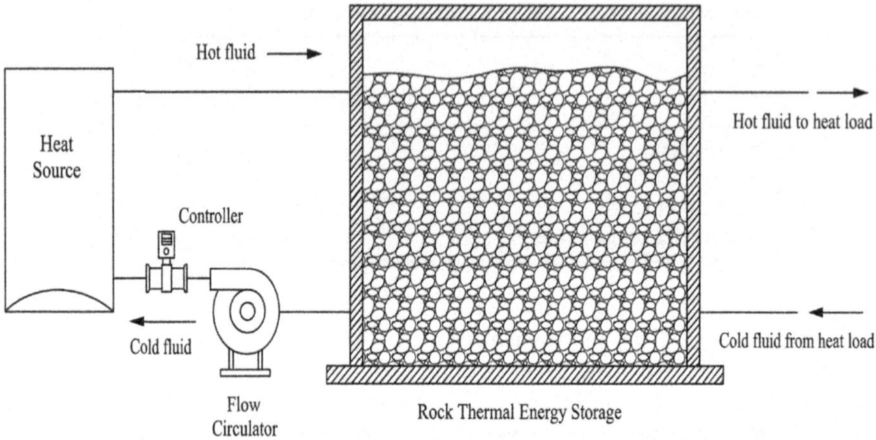

Figure 5.2 Schematic diagram of typical rock thermal energy storage (Amiri et al., 2024).

storage medium. Thermal energy can be stored in a fluidized bed by using small rocks (Abrha et al., 2024) (Fig. 5.2). Heat is transferred to and from the storage medium by the HTF that passes through passive TES (Salehudress, 2024). Rocks with specific heat values between 0.75 and 0.95 kJ/kg K, densities between 2500 and 3000 kg/m^3, and thermal conductivities between 2 and 4 W/mK include granite, basalt, and quartzite (Malinowski et al., 2025). Rocks typically have a volumetric heat storage capacity of 2–2.7 MJ/m^3K^{-1}. Rocks can compete with concrete and cast iron at temperatures below 500°C. There are embedded rocks in the packed bed thermal storage system. As hot air moves through the bed of porous rock particles, conduction takes place inside the solid matrix. Rocks show good mechanical strength, chemical inertness, and resistance to thermal shock under cyclic operation. At higher temperatures, rocks experience oxidation, microcracking, and mineral phase transformation.

5.2.2.5 Graphite

The crystalline form of carbon known as graphite is distinguished by its exceptional thermal stability, high thermal conductivity, and resistance to thermal shock. Graphite's use in medium-to-high temperature thermal storage applications has been made possible by these characteristics (Zhao and Wu, 2011). Graphite is useful in thermal energy systems that require a lot of power because of its exceptionally high heat transfer rate capability. Graphite has remarkable stability during thermal cycling. Phase transformation and thermal degradation over many cycles are insignificant (Nguyen et al., 2024). Graphite is a great material to withstand thermal shock because of its low coefficient of thermal expansion. A protective coating is necessary for long-term durability because graphite is prone to oxidation at temperatures higher than 450–500°C. Graphite and ceramic, concrete, or clay composites significantly improve charging and discharging while remaining commercially feasible (Deng et al., 2024). Graphite is best used in hybrid SHS systems, where it serves as both a secondary

Table 5.3 Thermophysical properties of graphite (Li et al., 2024).

Property	Typical range
Density	1700–2200 $kg \cdot m^{-3}$
Specific heat	0.7–0.85 $kJ/kg \cdot K$
Thermal conductivity	100–400 $W \cdot m^{-1} \cdot K^{-1}$
Max service temperature	3000 °C

storage medium and a heat transfer enhancer, due to cost and oxidation limitations. Graphite-ceramic composites, graphite-enhanced concrete or clay matrices, and embedded graphite fins or plates inside ceramic storage blocks are examples of common configurations (Li et al., 2024). While retaining economic viability, such hybrid designs greatly enhance charging–discharging kinetics and overall, TES efficiency. Table 5.3 demonstrates the thermophysical properties of graphite.

5.2.2.6 Egyptian clay

The Nile's alluvial deposits are used to make Egyptian clay. Bricks, ceramics, and thermal enclosures are all made from Egyptian clay. Egyptian clay is utilized in solar air heaters, solar dryers, and passive building applications as an eco-friendly alternative for TES where 58% montmorillonite and 48% kaolinite make up Egyptian clay (Dincer and Rosen, 2011). When thermal cycling occurs, the clay's thermos-mechanical behavior is determined by its mineral composition (Sharma et al., 2009). Heat and cooling cycles below 500°C (Cabeza, 2009; Cultrone et al., 2004) show very little microstructural degradation and very little mass loss. Due to pore collapse and dehydration during repeated heat cycles, sun-dried clay is mechanically degraded (Cengel and Ghajar, 2015). For long-term use, this calls for carefully regulated heat treatment of the clay. Egyptian clay has many benefits, including being inexpensive, widely available, and environmentally friendly. Its low thermal conductivity, mechanical brittleness under heat shock, and heterogeneous composition, however, restrict its use (Incropera and DeWitt, 2007; Neville, 2011). Clay-based composites that incorporate sand, graphite, or metallic additives to improve their thermophysical characteristics, such as mechanical strength and thermal conductivity, are the focus of recent trends (Zhang et al., 2013) (Table 5.4).

Table 5.4 Thermophysical properties of Egyptian clay (Voronin et al., 2020).

Property	Typical range
Density	1800–2200 $kg \cdot m^{-3}$
Specific heat	0.84–1.05 $kJ/kg \cdot K$
Thermal conductivity	0.6–1.2 $W \cdot m^{-1} \cdot K^{-1}$
Max service temperature	600–900 °C (fired clay)

Table 5.5 Thermophysical properties of aluminum.

Property	Typical range
Density	2700 kg/m^3
Specific heat	900 J/kg K
Thermal conductivity	205 W/m K
Max service temperature	300–400°C
Thermal diffusivity	$8.44 \times 10^{-5}\ m^2 \cdot s^{-1}$
Volumetric heat capacity	$2.43 \times 10^6\ J \cdot m^{-3} \cdot K^{-1}$

5.2.2.7 Aluminum

For high-power SHS, aluminum is a desirable option. Compared to metallic and rock media, aluminum has a high specific heat and high thermal diffusivity. Thermal equilibrium is reached rapidly, and less heat transfer space is needed. The alloy's metallurgical limitations, such as its melting point of roughly 660°C, are its practical limitations. When a high heat flux is needed, aluminum can be used. Table 5.5 demonstrates thermophysical properties of aluminum.

5.3 Latent heat storage: Phase change materials and applications

LHS makes use of the enthalpy released during the material's phase transition. When compared to SHS, it provides a volumetric energy density that is 5 to 14 times higher. This characteristic makes it suitable for use in CSP, solar drying, and waste heat recovery systems. PCMs are broadly classified as organic and inorganic PCMs (Fig. 5.3).

5.3.1 Organic PCM

The latent heat of fusion of organic PCMs enables them to absorb and release thermal energy at a temperature that is almost constant (Sharma et al., 2009). These materials are widely used in low-to-medium-temperature latent heat thermal energy storage (LHTES) applications because of their chemical stability, noncorrosiveness, and dependable performance under suitable formulation and containment (Kahwaji and White, 2021). Their qualities make them appropriate for a variety of energy storage options, advancing thermal management technologies. Paraffinic and nonparaffinic compounds are the two main categories of organic PCMs (Duquesne et al., 2021).

5.3.1.1 Paraffinic PCMs (n-alkanes)

In addition to different commercial paraffin waxes, paraffinic PCMs, specifically n-alkanes, include a number of compounds, such as n-octadecane and n-eicosane

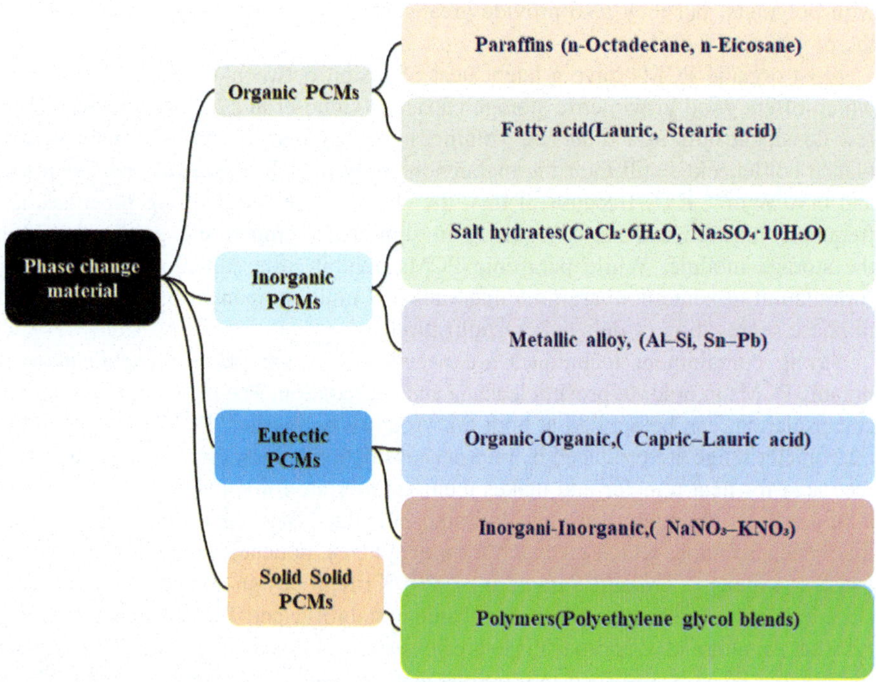

Figure 5.3 Classification of phase change material.

(Peng et al., 2018). The melting range of these materials usually ranges from 10 to 70 °C, and the precise temperature is mostly determined by the blend's composition and carbon number. The advantages of paraffinic PCMs include high levels of chemical inertness, minimal supercooling, and robust cycling stability (Himran et al., 1994). They are suitable for a variety of thermal management applications because of these characteristics. However, there are certain disadvantages to these materials. The efficiency of paraffinic PCMs in heat transfer applications may be hampered by their low thermal conductivity, which is measured at about 0.2 $W\ m^{-1}\ K^{-1}$. Additionally, flammability concerns and a moderate volumetric storage density must be taken into account when evaluating their practical applications (Alkan and Alakara, 2025).

5.3.1.2 Nonparaffinic PCMs

Nonparaffinic materials include polyethylene glycols, fatty acids, fatty alcohols, and esters. Fatty acids typically melt between 20°C and 70 °C, and their eutectic mixtures provide relatively high latent heat with sharper phase transitions. Polyethylene glycols have melting temperatures that depend on molecular weight and are commonly used in applications that need a comfortable temperature (Chung et al., 2014). These materials may have disadvantages like odor, moisture sensitivity, or compatibility issues with polymers, but they also provide greater control over melting temperature and thermophysical behavior.

Most organic PCMs have a latent heat of fusion between 150 and 250 kJ kg^{-1}, which offers good gravimetric storage capacity (Goud et al., 2023). However, their low-density results in a moderate volumetric energy density, and the main performance bottleneck is still their thermal conductivity, which is typically between 0.15 and 0.30 W m^{-1} K^{-1}. Because of this, the charging and discharging processes are frequently uneven and slow, resulting in significant temperature gradients within the storage module. While paraffinic PCMs exhibit little supercooling, improper formulation of some nonparaffinic materials and multicomponent blends may result in phase segregation or delayed crystallization.

Strong containment techniques are necessary for the practical application of organic PCMs in order to prevent leakage and degradation. The most popular method, encapsulation, can be applied at both micro- and macroscales. PCM particles in the micrometer range are produced by microencapsulation, which gives them a large surface area for heat transfer and makes them appropriate for slurries, wallboards, and textiles. Solar thermal tanks and building-integrated storage units frequently use microencapsulation, which takes the form of spheres, pouches, or tubes and is easier and less expensive (Nazari Sam et al., 2020). Critical design considerations in both situations are shell permeability, mechanical stability, and fatigue under repeated cycling. An alternative approach is to create shape-stabilized composite PCMs, where the organic PCM is absorbed into porous matrices such as expanded graphite (EG), activated carbon, silica, or biochar. Through capillary forces, these matrices maintain the liquid phase while increasing thermal conductivity. The intrinsic low thermal conductivity of organic PCMs makes heat-transfer enhancement techniques essential at the system level (Oliveski et al., 2021). These include adding radial or longitudinal fins, applying heat pipes to disperse heat more evenly, using metal or graphite foams, and adding highly conductive fillers like graphene or graphite (Singh et al., 2022). Cascaded storage systems employing several PCMs with different melting points are also being used more often in order to maintain a high temperature driving force and improve overall energy efficiency.

Accurate characterization of organic PCMs requires differential scanning calorimetry to measure melting temperature and latent heat, bulk-scale T-history methods for realistic enthalpy evaluation, and extended thermal cycling tests to assess long-term stability. While leakage and compatibility tests are performed to make sure that container materials and sealing elements are not negatively impacted during extended contact with molten PCM, chemical degradation is usually tracked using FTIR or chromatographic techniques. Applications like cold-chain packaging for food and pharmaceutical transportation, solar thermal and home hot-water systems in the 40°C–70°C range, building thermal regulation in the 20°C–35°C range, and short-term thermal buffering in electronic devices are dominated by organic PCMs. Notwithstanding their benefits, issues with fire safety, leakage, phase transition volume change, and restricted heat transfer rates still exist. In order to provide safe, effective, and long-lasting LHS systems, current research is concentrated on graphite-based shape-stabilized composites, hybrid encapsulation shells with decreased permeability, cascaded storage architectures, and bio-derived PCMs (Manar et al., 2025).

5.3.2 Inorganic PCMs

For LHTES in medium- and high-temperature applications, inorganic PCMs are widely used because of their high volumetric energy density, better thermal conductivity than organic PCMs, and comparatively low cost. Compared to organic PCMs, inorganic materials generally have higher thermal diffusivity and latent heat per unit volume, which enhances charging and discharging performance. However, the practical use of inorganic PCMs is limited by problems like corrosion, phase segregation, supercooling, and long-term chemical instability. Salt hydrates, molten salts, and metallic alloys are the three main categories into which inorganic PCMs can be broadly divided (Mohamed et al., 2017).

a) Salt hydrates

Salt hydrates are the most widely used inorganic PCMs for low- and medium-temperature applications (0–120 °C). These materials store energy through the reversible dehydration and hydration of crystalline salts like $Mg(NO_3)_2{\cdot}6H_2O$, $Na_2SO_4{\cdot}10H_2O$ (Glauber's salt), and $CaCl_2{\cdot}6H_2O$ (Purohit and Sistla, 2021). They have high latent heat values, sharp phase transition temperatures, and good thermal conductivity.

b) Molten salts

High-temperature TES systems use molten salts, usually between 150 and 600 °C. Nitrate, nitrite, and carbonate salts, such as $LiCO_3$–KCO_3 mixtures and $NaNO_3$–KNO_3 eutectics, are common examples (Zuo et al., 2024). These materials are frequently utilized in CSP facilities to generate dispatchable electricity and provide thermal buffering.

c) Metallic and alloy PCMs

Materials like aluminum–silicon and magnesium–zinc alloys are examples of metallic and alloy PCMs, which function at extremely high temperatures (>300 °C) (Höhlein et al., 2018). They have very high volumetric energy density and thermal conductivity, but they are constrained by issues with cost, corrosion, and encapsulation.

5.3.2.1 Stability and compatibility issues of inorganic PCMs

Inorganic PCMs' long-term performance is determined by their chemical, structural, and interfacial stability under repeated thermal cycling in addition to their latent heat capacity (Rathod and Banerjee, 2013). The most important problems affecting the dependability of salt-based PCMs are corrosion of containment materials, chemical degradation, and hydration instability. Inorganic PCMs are naturally hostile to common building metals, especially salts based on chloride, nitrate, nitrite, and sulfate.

Electrochemical reactions between molten or semihydrated salts and metal surfaces are the main cause of corrosion. For example, nitrate salts can cause uniform oxidation and intergranular attack at high temperatures, while chloride ions encourage pitting corrosion in stainless steels by penetrating the passive oxide layer. When exposed to molten $CaCl_2{\cdot}6H_2O$ or $NaNO_3$–KNO_3 eutectics, aluminum and copper alloys show rapid mass loss and surface embrittlement because soluble metal chlorides or nitrates are formed. The protective oxide layers of steels are dissolved by molten nitrate and carbonate salts at temperatures above 200 °C, resulting in accelerated corrosion kinetics controlled by Arrhenius-type temperature dependence. The rate of corrosion may increase by an order of magnitude for every 25–30°C increase in operating temperature (Ushak et al., 2016). Because of this, containment systems often require high-nickel alloys, stainless steel grades with high Cr and Mo content, ceramic linings, or polymeric barrier coatings to maintain structural integrity.

Salt hydrates store energy through a reversible process of hydration and dehydration. However, many salt hydrates undergo incongruent melting, where the salt melts independently of its crystallizing water. Phase segregation, in which excess water accumulates at the top and heavier anhydrous salt precipitates at the bottom, causes a gradual loss of effective latent heat over cycles. For example, when $Na_2SO_4{\cdot}10H_2O$ partially dehydrates to $Na_2SO_4 + H_2O$, the stoichiometric composition needed for reversible phase change is destroyed. Repeated thermal cycling also modifies the hydration number, producing metastable hydrate phases with different enthalpy values and melting temperatures. After a few hundred cycles, this phenomenon reduces latent heat by 10%–40% and shifts the phase change temperature by several degrees, severely impairing TES performance. Thermal decomposition reactions like

$$2NO_3- \rightarrow 2NO_2- + O_2$$

occur above 450 °C in molten nitrate and nitrite systems used in CSP plants, producing nitrites and oxygen gas (Behzadi and Farid, 2014). The released oxygen accelerates corrosion while nitrite formation increases chemical reactivity and modifies the melting temperature. Additionally, trace impurities such as Mg^{2+}, Cl^-, or moisture catalyze side reactions, accelerate degradation rates, and result in gas formation, voids, and pressure build-up in sealed TES modules. The volumetric expansion of inorganic PCMs during melting, which is typically 5–15%, causes cyclic thermomechanical stresses in encapsulation shells and container walls. Over hundreds of melt-freeze cycles, this results in coating delamination, microcracking, and ultimately leakage. These mechanical failures are exacerbated by the stiff crystalline structure of salts, which lacks the organic PCMs' ability to relax under stress. The combined effects of corrosion, phase segregation, hydration instability, and decomposition cause latent heat capacity, melting temperature accuracy, and heat recovery efficiency to progressively decline. In practical TES systems, this manifests as increased supercooling, delayed heat release, progressive storage capacity reduction, and incomplete solidification. Thus, the long-term technoeconomic viability of inorganic PCMs depends on material stabilization techniques like corrosion inhibitors, nucleating agents, thickening additives, hermetic encapsulation, and the selection of chemically compatible container materials (Fu et al., 2024).

5.3.2.2 Performance enhancement techniques

The intrinsic limitations of inorganic PCMs, such as supercooling, phase segregation, low effective heat transfer rate, and chemical reactivity with container materials, necessitate targeted performance-enhancement techniques at the material and system levels. The lack of stable crystallization nuclei during solidification causes supercooling in inorganic PCMs. Heterogeneous nucleation governs solidification in hydrated and molten salts; however, the high purity of the salts and the smooth interior surfaces of the containers inhibit nucleation, keeping the liquid PCM metastable well below its equilibrium freezing temperature. Borax ($Na_2B_4O \cdot 10H_2O$), $SrCl_2$, $Ba(OH)_2$, AgI, TiO_2 nanoparticles, or metal powders are examples of nucleating agents that offer lattice-matching surfaces that reduce the Gibbs free energy barrier for nucleation (Kundu et al., 2022). By introducing crystal planes that are crystallographically compatible with the PCM, these additives lower the nucleus' critical radius and allow solidification to start at or close to the melting temperature. This ensures consistent heat release and stable cycling performance by lowering the degree of supercooling from usual values of 10–15 °C to less than 2–3 °C.

Density differences between the hydrated salt and the water that is released during melting cause phase segregation in salt hydrates, which causes anhydrous salt to settle. Carboxymethyl cellulose, polyacrylamide, xanthan gum, or fumed silica are examples of thickening agents that are added to PCM to increase its viscosity and create a weak three-dimensional polymeric or colloidal network. By imparting pseudoplastic or gel-like rheological behavior, these additives successfully immobilize the water molecules and salt crystals, preventing gravitational separation (Liu et al., 2012). Over several melt–freeze cycles, the elevated yield stress keeps compositional homogeneity and stops sedimentation. Excessive thickener concentration, however, should be avoided since it decreases the capacity to store latent heat by taking up volume without increasing phase change enthalpy.

Even though inorganic PCMs have a higher thermal conductivity than organic PCMs, when used in bulk, they still don't transfer heat effectively enough. A low-conductivity liquid layer forms next to the heat transfer surface during melting, resulting in a thermal resistance that controls the total charging time. Metal fins, metal foams, graphite matrices, and heat pipes are added to PCM containers to get around this restriction. The total heat flux into the PCM is increased by metal fins because they increase the heat transfer area and shorten the conduction path. But the more the melted PCM region grows, the less effective they are. The PCM's effective thermal resistance is significantly reduced by open-cell metal foams (aluminum, copper, and nickel), which offer a continuous, highly conductive three-dimensional skeleton.

In comparison to pure PCM, the foam network reduces charging and discharging times by a factor of three to six by encouraging both conductive and natural convective heat transfer in the molten region. EG and graphite matrices provide a lightweight substitute with extremely high in-plane thermal conductivity (up to 200 W m^{-1} K^{-1}). While retaining a high latent heat density, PCM impregnated into EG displays nearly isotropic heat transfer behavior. Heat pipes function as thermal superhighways, removing thermal bottlenecks and guaranteeing consistent temperature distribution by quickly moving heat from the heat source deep into the PCM volume through phase-change evaporation–condensation cycles of the working fluid. Encapsulation prevents leakage, corrosion, and contamination by separating inorganic PCMs from the TES system's structural elements. In macroencapsulation, the PCM is enclosed in metal, polymer, or composite shell tubes, spherical capsules, or pouches (Cárdenas and León, 2013). Although this method is reliable and appropriate for systems operating at medium to high temperatures, the large capsule size may result in considerable thermal resistance. However, during microencapsulation, PCM droplets are encased in polymeric, ceramic, or hybrid shells that have thicknesses in the micrometer range. This greatly increases the surface area of heat transfer and lessens mechanical stress caused by volume change. For extremely corrosive molten salts, ceramic or silica-based shells are advised because of their strong thermal stability and chemical inertness. In general, nucleation control, rheological stabilization, conductive enhancement, and encapsulation work in concert to convert inorganic PCMs from lab materials into thermally efficient, chemically stable, and industrially deployable TES media.

5.3.2.3 Applications of inorganic PCMs

Inorganic PCMs are used in different temperature ranges because of their ability to accurately match the temperature requirements of different energy systems by using the phase transition temperatures, latent heat densities, and thermal conductivity. Their high volumetric storage density, relatively high thermal conductivity, and availability of stable salt systems at specific operating temperatures make them suitable for use in buildings, district heating, solar thermal, and CSP installations.

Salt hydrates are attractive for building thermal regulation because of their melting temperatures, which fall between 20°C and 35°C. Two examples are $Na_2SO_4{\cdot}10H_2O$ ($T_m \approx 32$ °C) and $CaCl_2{\cdot}6H_2O$ (Tm ≈ 29 °C). These materials can lessen HVAC load and even out indoor temperature swings by absorbing large amounts of latent heat during daytime temperature peaks and releasing it during night time cooling (Jaguemont et al., 2018). Their high volumetric latent heat, which usually exceeds 250 MJ m^{-3}, enables significant thermal buffering even when integrated into thin wallboards or ceiling panels. The higher thermal conductivity of salt hydrates ensures a faster thermal response than organic PCMs in dynamic indoor environments with rapid cycles of heat gain and loss.

Inorganic PCMs are used in solar water heating and district heating systems to extend energy availability during off-sun hours and stabilize outlet temperature. Shell-and-tube or encapsulated storage modules that are connected to solar collectors frequently incorporate hydrated nitrates and sulphates with melting temperatures between 40 and 80 °C (Wu et al., 2020). By effectively capturing intermittent solar heat and delivering it at a nearly constant temperature, their higher heat of fusion and quicker heat transfer characteristics increases collector efficiency and lower the need for auxiliary heating. By temporally separating generation and demand, PCM-based buffer tanks in district heating networks reduce load fluctuations and enable peak-shaving.

As molten nitrate can function at temperatures above 300 °C with considerable thermal stability, it is preferably used for CSP plants. Common binary nitrate eutectics, like $NaNO_3$–KNO_3, melt at around 220 °C and stay stable until about 565 °C (Banerjee et al., 2022). LHS is used in conjunction with SHS in these systems to boost energy density and sustain turbine inlet temperature during cloud transients or at night. Molten salts' comparatively high density and thermal conductivity allow for quick heat exchange between the steam generator, storage tanks, and receiver, enabling dispatchable electricity production.

Metallic and alloy PCMs are preferred in applications where operating temperatures surpass 500 °C, such as industrial waste heat recovery and aerospace thermal protection systems, metallic and alloy PCMs are becoming more popular (Mika et al., 2025). Compact thermal buffers that can absorb short-duration, high-intensity heat loads are made possible by the exceptionally high volumetric latent heat and thermal conductivity of aluminum–silicon or magnesium-based alloys. Metallic PCMs manage aerodynamic heating during atmospheric re-entry in aerospace systems by acting as temporary heat sinks; in the metallurgical and glass industries, they recover high-grade waste heat that would otherwise dissipate, increasing overall process efficiency. Therefore, the ability of inorganic PCMs to provide high-density, temperature-specific LHS with superior heat transfer performance is the scientific basis for their widespread use. This makes them essential in high-temperature renewable and industrial energy systems as well as low-temperature building applications.

5.4 Thermochemical storage: Reversible reactions and cycles

The highest-density type of TES is TCES, which uses reversible reactions to transform heat into chemical potential. Since the stored energy is maintained in the chemical bonds of separated reactants rather than at a high temperature, TCES permits loss-free long-term storage, in contrast to sensible and LHS (André and Abanades, 2020). As a result, TCES are particularly appealing for next-generation CSP systems, high-temperature industrial heat recovery, and seasonal solar energy storage. Generally, TCES reaction is written as:

$$AB + Q_{in} \leftrightarrow A + B$$

Heat causes compound AB to break down endothermically into A and B during charging. These goods are kept apart in storage. Heat equal to the reaction enthalpy is released when A and B recombine during discharging. Theoretically, the stored energy can be stored for an infinite amount of time because it is solely dependent on the chemical conversion rather than temperature gradients.

5.4.1 Classification of reversible thermochemical cycles

5.4.1.1 Hydration–dehydration systems

Hydration–dehydration thermal energy is stored and released through the reversible reaction between a solid material and water vapor. The principle is expressed as

$$MO(s) + H_2O(g) \rightleftharpoons M(OH)_2(s),$$

The dehydration is an endothermic reaction where heat is absorbed to dissociate the hydroxide into its corresponding oxide and water vapor and stored in chemical form. The hydration reaction is an exothermic releasing heat when the oxide reacts with water vapor to reform the hydroxide. The operating range of system (100°C–600°C) suits for solar thermal applications, industrial waste heat recovery, and space heating. ($CaO/Ca(OH)_2$) is extensively investigated due to its high reaction enthalpy, quick reaction kinetics, and availability and reversibility. Hydration–dehydration system is characterized by display faster reaction kinetics and lower operating temperature. However, there are some critical issues like agglomeration, pore blockage, and volume expansion during hydration.

For medium-temperature applications, hydration–dehydration systems like $Ca(OH)_2/CaO$ and $Mg(OH)_2/MgO$ function between 350 and 550 °C. Although they offer volumetric storage densities of 1500–2000 MJ m^{-3}, they have pore blockage over repeated cycles and slow hydration kinetics (Afflerbach and Trettin, 2019).

5.4.1.2 Carbonation–decarbonation cycles

The reversible gas solid reactions between metal oxide and carbon dioxide enable the storing and releasing of the high temperature thermochemical energy in carbonation–decarbonation cycles. This system offers high energy storage densities and prolonged storage periods with minimum thermal loss (Han et al., 2022). This feature allows the integration of the system with CSP systems and industry applications at higher temperatures. The principle is expressed as

$$MO(s) + CO_2(g) \rightleftharpoons MCO_3(s).$$

The charging phase is known as decarbonation and discharging phase is known as carbonization. During decarbonation the high temperature thermal energy is absorbed and the metal carbonate decomposes into oxide and carbon dioxide. The endothermic reaction stores energy as chemical potential. During carbonation exothermic reaction takes place between metal oxide and carbon dioxide and forms metal carbonate which releases the stored heat. For CSP integration, carbonation–decarbonation cycles in particular, $CaCO_3/CaO$, are thoroughly investigated due to their cost effectiveness and availability. While carbonation releases heat at 600–700 °C, charging takes place at 800–900 °C through calcination (Liu et al., 2020). These systems have some of the highest storage densities (>2500 MJ m^{-3}) that have been reported, but they are constrained by CaO sintering and a gradual loss of carbonation reactivity.

5.4.1.3 Metal oxide redox cycles

Metal oxide redox cycles, like MnO_3/MnO_4 and CoO_3O_4/CoO, store energy through reversible oxygen release and function above 800 °C. Although they need sophisticated oxygen separation and high-temperature containment, they are ideal for solar power reactors (Wokon, 2020). Ammonia dissociation–synthesis cycles offer a very high energy density and the extra benefit of storing hydrogen as a by-product by using the reversible reaction between NH_3, H_2, and N_2. However, stringent gas management and catalytic reactors are required.

Despite their favorable thermodynamics, TCES systems are severely constrained by kinetics. Reaction rates progressively decrease due to sintering, diffusion resistance through product layers, and porosity loss. The transfer of heat through crowded beds, which frequently leads to temperature gradients and incomplete conversion, is another major disadvantage, especially in large-scale reactors. In TCES, material durability is a major problem. Hydration systems experience pore clogging due to volume expansion, whereas carbonate and redox systems suffer from crystal growth and agglomeration at high temperatures. These phenomena reduce reactive surface area over a number of cycles, which reduces storage capacity and conversion efficiency.

5.4.2 System integration challenges

The main obstacle to the large-scale deployment of TCES systems is the difficulty of combining high-temperature heat sources, auxiliary subsystems, and solid–gas reactors into a cohesive and thermodynamically effective architecture. TCES depends on reversible heterogeneous reactions like,

$$\mathrm{MOx} \rightleftharpoons \mathrm{MOx} - 1 + 1/2\mathrm{O_2} \quad \text{or} \quad \mathrm{Ca(OH)_2} \rightleftharpoons \mathrm{CaO} + \mathrm{H_2O}$$

which require exact control of temperature, pressure, and reactant partial pressures during repeated charging–discharging cycles, in contrast to sensible or LHS.

5.4.2.1 High-temperature reactor design

The majority of TCES reactions require operating temperatures between 400 and 900°C, although some metal-oxide redox systems require temperatures above 1000°C. Designing reactors that can sustain these conditions under cyclic thermal stresses is extremely difficult (Rodriguez-Pastor et al., 2024). High thermal gradients between the reactor's core and shell cause thermomechanical fatigue, which can lead to reactive gas leakage, structural integrity loss, and microcracking. Additionally, solid reactants often exhibit volume change, sintering, and particle agglomeration during cycling, which leads to nonuniform reaction fronts and bed compaction that reduce mass and heat transfer coefficients.

5.4.2.2 Gas handling and pressure management

The functioning of TCES depends on efficient gas–solid interaction. Dehydration–hydration or carbonation–decarbonation reactions require exact control of reactant gas flow rates, humidity levels, and system pressure (Rodriguez-Pastor et al., 2024). In large reactors, uneven gas distribution causes concentration polarization, localized reaction zones, and incomplete conversion. High-temperature reactive gases like steam, CO_2, or oxygen are also challenging to seal and contain because leaks not only reduce round-trip efficiency but also increase the risk of oxidation and safety hazards for adjacent components.

5.4.2.3 Corrosion and material compatibility

Corrosion is a significant barrier to TCES integration. Reactive gases accelerate the oxidation, carburization, and corrosion of metallic components due to sulfate or chloride at high temperatures. For example, redox oxide systems release oxygen that promotes the formation of scale on steel alloys, while moisture and CO_2 can create extremely corrosive environments in $CaO/Ca(OH)_2$ or $CaO/CaCO_3$ systems (Qiu and Galbraith, 2013). This necessitates the use of expensive high-nickel alloys, ceramic coatings, or inert liners, which significantly increases capital costs and complicates fabrication.

5.4.2.4 Integration with CSP receiver and heat exchangers

TCES reactors must be thermally coupled with solar receivers in CSP plants in order to minimize energy losses. Thermal inefficiencies result from the mismatch between the relatively slow kinetics of solid-state reactions and the extremely transient nature of solar flux. Additionally, fouling, thermal expansion mismatch, and the small heat transfer area inside packed beds make conventional heat exchangers ineffective at these temperatures (Codd et al., 2020). One of the main unsolved engineering challenges is to achieve uniform heat flux into the reactor while preventing thermal shock during cloud transients.

5.4.2.5 Cycling stability and system control

Long-term TCES operation requires that reaction enthalpy, kinetics, and material morphology remain stable over thousands of cycles. However, repeated phase transformations lead to pore collapse, sintering, and loss of active surface area, reducing reaction rates and storage capacity.

In order to compensate for degradation and maintain safe operation under varying solar input and load demand, system-level control strategies must dynamically modify operating temperature, gas flow, and pressure.

5.5 Integration with concentrated solar power systems

The integration of TCES with CSP facilities represents a paradigm shift from conventional sensible and LHS technologies toward high-exergy-density and nearly lossless long-term storage (Codd et al., 2020). Because TCES systems store solar energy as reversible chemical bonds, they are naturally suited for dispatchable solar power generation under high-temperature operating conditions typical of central receiver and parabolic dish CSP configurations.

In CSP facilities, TCES reactors may be heated directly or indirectly. In directly irradiated concepts, where the reactor functions as both a solar receiver and a thermochemical converter, concentrated solar flux drives endothermic decomposition reactions such as

$$MOx \rightleftharpoons MOx - 1 + 1/2O_2$$

Thermal resistance and energy losses are decreased by this configuration's reduction of heat transfer interfaces. However, it places stringent requirements on reactor window materials, absorber coatings, and flux homogenization to avoid localized overheating and material failure. In indirectly heated systems, the solar receiver and the TCES reactor are separated by high-temperature heat exchangers. Although this architecture increases operational flexibility and protects reactive materials from abrupt changes in solar flux, it also increases thermal resistances and complicates the system, particularly when it comes to handling thermal expansion mismatch between the receiver, piping, and reactor vessels.

The primary problem with CSP–TCES integration is the mismatch between the temperature-dependent kinetics of thermochemical reactions and the extremely transient solar input. For maximum conversion, redox and hydration systems have limited ideal temperature windows. Incomplete conversion, uneven reaction fronts, and decreased storage efficiency are caused by variations in solar irradiance. Therefore, in order to maintain thermodynamic equilibrium conditions, sophisticated thermal control algorithms are needed to control receiver output, gas partial pressures, and reactor inlet temperatures in real time (Iaquaniello et al., 2014).

High-grade thermal energy is released by the exothermic reverse reaction during the discharge phase and needs to be effectively transferred to the CSP power block, which is usually a superheated steam Rankine or supercritical CO_2 Brayton cycle. Temperature stratification and lower turbine inlet temperatures result from packed-bed TCES reactors' low effective thermal conductivity, which severely restricts heat extraction rates. In order to get around this restriction, more research is being done on reactor designs that use metal foams, finned structures, or fluidized beds to increase effective heat transfer coefficients without compromising storage density.

System components experience severe thermomechanical fatigue when TCES reactors and CSP receivers are cyclically coupled. Internal heat exchangers, reactor shells, and absorber tubes all experience high thermal stresses as a result of rapid temperature ramps during startup and shutdown. Reactive gas environments also hasten the oxidation, corrosion, and embrittlement of metallic alloys, necessitating the use of diffusion barrier coatings, high-performance superalloys, and ceramic linings, all of which have a substantial impact on plant economics (Malagueta et al., 2014).

Heliostat field control, receiver flux distribution, reactor dynamics, and power block operation are just a few of the length and time scales that must be optimized in order to integrate TCES into CSP plants. The lack of large-scale TCES demonstrations that are commercially mature draws attention to the unsolved difficulties in transferring laboratory concepts to multi-MW plants while maintaining round-trip efficiency, safety, and economic viability.

5.6 Industrial and residential applications

5.6.1 Industrial applications

5.6.1.1 Waste heat recovery and process heat management

Many processes industry like steel, glass, food, tobacco, pulp and paper, chemical industry, and cement manufacturing industries release a large amount of waste heat. SHS materials are used to recover this waste heat (Brückner et al., 2015). The heat which is recovered is reused for combustion air preheating or feed water heating improving the plant efficiency (Miró et al., 2016).

Waste heat from foundry ovens is recovered by borehole TES. An estimation of 3800 MWh of heat can be stored at a temperature of 60–65 °C mitigating 1500 t/y of CO_2 (Fernández et al., 2021). Discontinuous waste heat from the foundry furnace can also be stored in a dual-TES media with packed bed which is in direct contact with a HTF at temperature up to 300°C. The stored heat is again used for various furnace processes up to 100°C. Mold cavity surrounded by sand molds with aluminum shots can be used as thermal energy system to recover the waste heat from solidification process of metal castings (Selvaraj et al., 2015). The shots which are near to the mold absorbs the heat during solidification, and later are separated from the mold cavity. The heated shots heat the scrap material, which is again used as input to the furnace thereby reducing the energy consumption. The waste heat from steelmaking industry from the electric furnace can be stored using PCM, molten salts and concrete. However, only a fraction of the exhaust energy can be recovered, but significant reduction of 45 kg of CO_2 is reported per ton of liquid steel (Steinparzer et al., 2012). LHS using high-temperature PCMs such as nitrate salts or eutectic metallic alloys is increasingly adopted where temperature stabilization is required, e.g., in batch furnaces and annealing lines. These systems smooth out thermal fluctuations and reduce fuel consumption during load variations.

5.6.1.2 Industrial solar thermal integration

Thermal energy system integrated with solar thermal systems enables supply of continuous process heat during shady days and a key strategy for reducing fossil fuel consumption. The solar industrial process heat systems collect the solar radiation from sun and store the extra energy in TES media. The heat from TES is supplied to industrial processes during shady days and helps in synchronizing the nonconsistent solar supply (Koçak et al., 2020). In mineral and mining operations where low temperature heat storage is required, large array of flat plate solar collectors are installed besides the hot water storage tanks (Tasmin et al., 2022). The low temperature heat is required for cleaning, electrolyte heating, or drying. In food processing sector, solar heated water is used in cleaning and sanitation purpose and supplied year-round reducing the dependency on fossil fuel (Tasmin et al., 2022). In dairy industry, parabolic trough integrated with TES module like thermal oil tank or hot water buffers supply heat for steam generation (Masera et al., 2022). Without TES solar heat cannot meet continuous industrial heat demands specifically on shady days. However, solar thermal integration with TES faces barriers like high capital investment, marketability, and limited design practice.

5.6.1.3 Power generation and load shifting

In cogeneration plants and CSP facilities, TES systems allow decoupling of electricity generation from real-time solar or fuel input. Two-tank molten salt sensible storage systems are widely used for 6–15 h of dispatchability, enabling power delivery during evening peak hours and improving plant capacity factor by more than 40%. Apart from electricity generation, load shifting from high grid demand to low grid demand is possible with TES, thus supporting grid stability. The surplus grid demands from intermittent sources like wind and solar can be converted into thermal energy and stored. It enhances power generation flexibility and grid reliability.

5.6.2 Residential applications of TES

5.6.2.1 Space heating and cooling

TES is mostly used for space heating and cooling load shifting in residential buildings. When incorporated into wallboards, ceilings, and floors, low-temperature PCMs (melting range 18–30 °C) store indoor heat during the day and release it at night, stabilizing indoor temperatures and lowering HVAC electricity consumption by up to 25%–40%. In hot climates, ice storage systems are used to store electricity for use at night when rates are lower. In order to provide cooling, water is frozen at night and melted during the day, greatly lowering the grid's peak electrical load.

5.6.2.2 Domestic hot water systems

In solar domestic hot water systems, water-based sensible TES tanks are widely used. Hot water is supplied at night and in the early morning by flat plate and evacuated tube solar collectors that charge the thermal storage tank during the day. Temperatures between 60 and 90 °C for storage guarantee hygienic operation while reducing the need for additional electric heating.

5.6.2.3 *Integration with heat pumps and smart grids*

By allowing residential heat pumps to operate during times of low electricity rates or high renewable generation, TES increases their efficacy. Flexible grid-responsive building energy systems are made possible by thermal batteries that store heat at moderate temperatures (40–80 °C) using PCMs or thermochemical sorption materials.

5.7 Conclusion

TES has emerged as a cornerstone technology for enabling large-scale renewable energy integration, improving industrial energy efficiency, and reducing peak electricity demand in residential and commercial sectors. SHS remains the most technologically mature and economically viable option for high-temperature applications such as CSP plants and industrial waste heat recovery, whereas LHS using PCMs has demonstrated substantial potential for medium- and low-temperature thermal management, particularly in building energy systems. Despite being in the precommercial stage, TCES is a desirable option for long-term and seasonal heat storage due to its unparalleled energy density and minimal standby losses. Despite these advancements, fundamental material, reactor, and system-integration issues prevent TES from being widely used. Long-term dependability in high-temperature applications is limited by corrosion, complex heat exchanger designs, limited thermal conductivity, and material deterioration under cyclic operation. Problems like PCM leakage, supercooling, low charging and discharging rates, and a lack of standardized design practices restrict market penetration in residential and building-integrated TES. Additionally, capital costs, system lifetime, and integration complexity continue to have an impact on TES's technoeconomic competitiveness.

References

Abddaim, E., Braik, R., Sakami, S., Boukhattem, L., 2025. Experimental investigation of intermediate rocks as sensible thermal energy storage materials: durability and characterization. J. Energy Storage 140. Article 119071.

Abrha, A.K., et al., 2024. Charging of an air–rock bed thermal energy storage system: experimental and CFD analysis for solar TES applications. Energies 17 (19). Article 4952.

Afflerbach, S., Trettin, R., 2019. A systematic screening approach for new materials for thermochemical energy storage and conversion based on the Strunz mineral classification system. Thermochim. Acta 674, 82–94.

Alkan, C., Alakara, E.H., 2025. Property development in *n*-alkane and *n*-alkane eutectic phase change materials for thermal energy storage applications. Solar Energy Advances 5. Article 100089.

Amiri, L., et al., 2024. Progress on Rock Thermal Energy Storage (RTES): A State-of-the-Art Review. Energy Science and Engineering. Advance online publication.

Andic, O., Copuroglu, O., Ramyar, K., 2008. Effect of high temperature on mechanical and microstructural properties of cement mortar. In: Proceedings of the 11th International Conference on Durability of Building Materials and Components, Istanbul, Turkey, pp. 527–533.

André, L., Abanades, S., 2020. Recent advances in thermochemical energy storage via solid–gas reversible reactions at high temperature. Energies 13 (22). Article 5859.

Banerjee, B., Chatterjee, S., Datta, J., Saha, S.D., 2022. Application of phase changing materials in a CSP plant for thermal energy storage: a review on recent developments. Mater. Today Proc. 62, 3502–3507.

Behzadi, S., Farid, M.M., 2014. Long-term thermal stability of organic PCMs. Appl. Energy 122, 11–16.

Boquera, L., Castro, J.R., Pisello, A.L., Cabeza, L.F., 2021a. Research progress and trends on the use of concrete as thermal energy storage material through bibliometric analysis. J. Energy Storage 38. https://doi.org/10.1016/j.est.2021.102562. Article 102562.

Boquera, L., Pons, D., Fernández, A.I., Cabeza, L.F., 2021b. Characterization of supplementary cementitious materials and fibers to be implemented in high-temperature concretes for thermal energy storage (TES) application. Energies 14 (16). Article 5190.

Brückner, S., Liu, S., Miró, L., Radspieler, M., Cabeza, L.F., Lävemann, E., 2015. Industrial waste heat recovery technologies: an economic analysis of heat transformation technologies. Appl. Energy 151, 157–167.

Cabeza, L.F., 2009. Advances in thermal energy storage systems. Energy Convers. Manag. 50, 275–278.

Cárdenas, B., León, N., 2013. High temperature latent heat thermal energy storage: phase change materials, design considerations and performance enhancement techniques. Renew. Sustain. Energy Rev. 27, 724–737.

Cengel, Y.A., Ghajar, A.J., 2015. Heat and Mass Transfer: Fundamentals and Applications, fifth ed. McGraw-Hill Education, New York.

Chung, O., Jeong, S.G., Yu, S., Kim, S., 2014. Thermal performance of organic PCMs/ micronized silica composite for latent heat thermal energy storage. Energy Build. 70, 180–185.

Codd, D.S., Gil, A., Manzoor, M.T., Tetreault-Friend, M., 2020. Concentrating solar power (CSP)—thermal energy storage (TES) advanced concept development and demonstrations. Curr. Sustain. Renew. Energy Rep. 7 (2), 17–27.

Cultrone, G., Sebastián, E., Elert, K., et al., 2004. Influence of mineralogy on the physical properties of fired clays. J. Eur. Ceram. Soc. 24, 547–564.

Deng, X., Li, C., Sun, X., Wang, C., Liu, B., Li, Y., Yang, H., 2024. Flame-retardant wood-based composite phase change materials based on PDMS/expanded graphite coating for efficient solar-to-thermal energy storage. Appl. Energy 368. Article 123454.

Dincer, I., Rosen, M.A., 2011. Thermal Energy Storage: Systems and Applications, second ed. Wiley, Chichester.

Dinker, A., Agarwal, M., Agarwal, G.D., 2017. Heat storage materials, geometry and applications: a review. J. Energy Inst. 90 (1), 1–17.

Duquesne, M., et al., 2021. Characterization of fatty acids as bio-based organic phase change materials for thermal energy storage. Energies 14 (17). Article 5467.

Fernández, A.I., Barreneche, C., Miró, L., Brückner, S., Cabeza, L.F., 2021. Waste heat recovery using thermal energy storage. In: Advances in Thermal Energy Storage Systems. Woodhead Publishing, Cambridge, pp. 639–653.

Fu, T., Wang, W., Fang, G., 2024. Thermal properties and applications of form-stable phase change materials for thermal energy storage and thermal management: a review. Energy Storage 6 (1). Article e533.

Goud, V.M., Satyanarayana, G., Ramesh, J., Pathanjali, G.A., Sudhakar, D.R., 2023. An experimental investigation and hybrid neural network modelling of thermal management of lithium-ion batteries using a non-paraffinic organic phase change material, myristyl alcohol. J. Energy Storage 72. Article 108395.

Han, R., Xing, S., Wu, X., Pang, C., Lu, S., Su, Y., Liu, Q., Song, C., Gao, J., 2022. Relevant influence of alkali carbonate doping on the thermochemical energy storage of Ca-based natural minerals during $CaO/CaCO_3$ cycles. Renew. Energy 181, 267–277.

Himran, S., Suwono, A., Mansoori, G.A., 1994. Characterization of alkanes and paraffin waxes for application as phase change energy storage medium. Energy Sources 16 (1), 117–128.

Höhlein, S., König-Haagen, A., Brüggemann, D., 2018. Macro-encapsulation of inorganic phase-change materials (PCM) in metal capsules. Materials 11 (9). Article 1752.

Hoivik, N., Greiner, C., Barragan, J., Iniesta, A.C., Skeie, G., Bergan, P., Blanco-Rodriguez, P., Calvet, N., 2019. Long-term performance results of concrete-based modular thermal energy storage system. J. Energy Storage 24. Article 100735.

Iaquaniello, G., Salladini, A., Mari, A., Mabrouk, A.A., Fath, H.E., 2014. Concentrating solar power (CSP) system integrated with MED–RO hybrid desalination. Desalination 336, 121–128.

Incropera, F.P., DeWitt, D.P., 2007. Fundamentals of Heat and Mass Transfer, sixth ed. Wiley, Hoboken, NJ.

Islam, M.M., Yu, T., Giannoccaro, G., Mi, Y., La Scala, M., Nasab, M.R., Wang, J., 2024. Improving reliability and stability of the power systems: a comprehensive review on the role of energy storage systems to enhance flexibility. IEEE Access 12, 152738–152765.

Jaguemont, J., Omar, N., Van den Bossche, P., Mierlo, J., 2018. Phase-change materials (PCM) for automotive applications: a review. Appl. Therm. Eng. 132, 308–320.

John, E.E., Hale, W.M., Selvam, R.P., 2010. Effect of high temperatures and heating rates on high strength concrete for use as thermal energy storage. In: Proceedings of ASME 2010 4th International Conference on Energy Sustainability, vol 43956, pp. 709–713.

Jotshi, C.K., Goswami, D.Y., Klausner, J., et al., 1996. Heat Transfer Characteristics of a High-Temperature Sensible Heat Storage Water Heater Using Cast Iron as a Storage Material [online].

Jurczyk, M., Spietz, T., Czardybon, A., Dobras, S., Ignasiak, K., Bartela, Ł., Uchman, W., Ochmann, J., 2024. Review of thermal energy storage materials for application in large-scale integrated energy systems—methodology for matching heat storage solutions for given applications. Energies 17 (14). Article 3544.

Kahwaji, S., White, M.A., 2021. Organic phase change materials for thermal energy storage: molecular structure, properties, and applications. Molecules 26 (21). Article 6635.

Khare, S., Dell'Amico, M., Knight, C., McGarry, S., 2013. Selection of materials for high-temperature sensible energy storage. Sol. Energy Mater. Sol. Cell. 115, 114–122.

Koçak, B., Fernandez, A.I., Paksoy, H., 2020. Review on sensible thermal energy storage for industrial solar applications and sustainability aspects. Sol. Energy 209, 135–169.

Kumar, R., Lee, D., Ağbulut, Ü., Kumar, S., Thapa, S., Thakur, A., Jilte, R.D., Saleel, C., Shaik, S., 2024. Different energy storage techniques: recent advancements, applications, limitations, and efficient utilization of sustainable energy. J. Therm. Anal. Calorim. 149 (5).

Kundu, R., Kar, S.P., Sarangi, R., 2022. Performance enhancement with inorganic phase change materials for the application of thermal energy storage: a critical review. Energy Storage 4 (5). Article e320.

Li, G., 2016. Sensible heat thermal storage energy and exergy performance evaluations. Renew. Sustain. Energy Rev. 53, 897–923.

Li, W., Gao, C., Hou, A., Qiu, J., Wang, S., 2024. One-pot in situ synthesis of expandable graphite-encapsulated paraffin composites for thermal energy storage. Chem. Eng. J. 481. Article 148541.

Liu, M., Saman, W., Bruno, F., 2012. Review on storage materials and thermal performance enhancement techniques for high temperature phase change thermal storage systems. Renew. Sustain. Energy Rev. 16 (4), 2118–2132.

Liu, J., Baeyens, J., Deng, Y., Tan, T., Zhang, H., 2020. The chemical CO2 capture by carbonation-decarbonation cycles. J. Environ. Manag. 260. Article 110054.

Lovegrove, K., Stein, W. (Eds.), 2012. Concentrating Solar Power Technology: Principles, Developments and Applications. Woodhead Publishing, Cambridge.

Lugolole, R., Mawire, A., Okello, D., Lentswe, K.A., Nyeinga, K., Shobo, A.B., 2019. Experimental analyses of sensible heat thermal energy storage systems during discharging. Sustain. Energy Technol. Assessments 35, 117–130.

Mahdi, J.M., Lohrasbi, S., Nsofor, E.C., 2019. Hybrid heat transfer enhancement for latent-heat thermal energy storage systems: a review. Int. J. Heat Mass Tran. 137, 630–649.

Malagueta, D., Szklo, A., Soria, R., Dutra, R., Schaeffer, R., Borba, B.S., 2014. Potential and impacts of concentrated solar power (CSP) integration in the Brazilian electric power system. Renew. Energy 68, 223–235.

Malinowski, M., Bodziacki, S., Famielec, S., Huptyś, D., Kurpaska, S., Latała, H., Basak, Z., 2025. Environmental impact assessment of heat storage system in rock-bed accumulator. Energies 18 (13). Article 3360.

Manar, G., Shalaby, M., Bakar, M.S., Parveez, B., Najeeb, M.I., Hassan, M.K., Al-Sowayan, S., Alawad, M.A., 2025. Bio-based composites with encapsulated phase change materials for sustainable thermal energy storage: a review. Polymers 17 (21). Article 2925.

Masera, K., Tannous, H., Stojceska, V., Tassou, S., 2022. Application of solar thermal heating and cooling energy to dairy processes: a case study. In: Proceedings of the 17th UK Heat Transfer Conference (UKHTC 2022).

Mika, Ł., Radomska, E., Sztekler, K., Gołdasz, A., Zima, W., 2025. Review of selected PCMs and their applications in the industry and energy sector. Energies 18 (5). Article 1233.

Miliozzi, A., et al., 2025. Experimental evaluation of a combined sensible and latent thermal energy storage system. Energies 18 (21). Article 5808.

Miró, L., Gasia, J., Cabeza, L.F., 2016. Thermal energy storage (TES) for industrial waste heat (IWH) recovery: a review. Appl. Energy 179, 284–301.

Mohamed, S.A., Al-Sulaiman, F.A., Ibrahim, N.I., Zahir, M.H., Al-Ahmed, A., Saidur, R., Yılbaş, B.S., Sahin, A.Z., 2017. A review on current status and challenges of inorganic phase change materials for thermal energy storage systems. Renew. Sustain. Energy Rev. 70, 1072–1089.

Nazari Sam, M., Caggiano, A., Mankel, C., Koenders, E., 2020. A comparative study on the thermal energy storage performance of bio-based and paraffin-based PCMs using DSC procedures. Materials 13 (7). Article 1705.

Neville, A.M., 2011. Properties of Concrete, fifth ed. Pearson Education, Harlow.

Nguyen, T.P., Nguyen, T.N., Ramadan, Z., Park, C.W., 2024. Comparative study on thermal performance of two graphite fin thermal energy storages based on experiment, simulation, and artificial neural network. Int. Commun. Heat Mass Tran. 156. Article 107645.

Nkhonjera, L., Bello-Ochende, T., John, G., King'ondu, C.K., 2017. A review of thermal energy storage designs, heat storage materials and cooking performance of solar cookers with heat storage. Renew. Sustain. Energy Rev. 75, 157–167.

Oliveski, R.D., Becker, F., Rocha, L.A., Biserni, C., Eberhardt, G.E., 2021. Design of fin structures for phase change material (PCM) melting process in rectangular cavities. J. Energy Storage 35. Article 102337.

Olivkar, P.R., Katekar, V.P., Deshmukh, S.S., Palatkar, S.V., 2022. Effect of sensible heat storage materials on thermal performance of solar air heaters: state-of-the-art review. Renew. Sustain. Energy Rev. 159. Article 112085.

Paul, A., Holy, F., Textor, M., Lechner, S., 2022. High-temperature sensible thermal energy storage as a crucial element of Carnot batteries: overall classification and technical review based on parameters and key figures. J. Energy Storage 56. Article 106015.

Peng, H., Zhang, D., Ling, X., Li, Y., Wang, Y., Yu, Q., She, X., Li, Y., Ding, Y., 2018. *n*-Alkanes phase change materials and their microencapsulation for thermal energy storage: a critical review. Energy Fuel. 32 (7), 7262–7293.

Purohit, B.K., Sistla, V.S., 2021. Inorganic salt hydrate for thermal energy storage application: a review. Energy Storage 3 (2). Article e212.

Qiu, S., Galbraith, R., 2013. Material compatibility study for thermal energy storage containment structure with phase change material. In: ASME International Mechanical Engineering Congress and Exposition, vol 56291. American Society of Mechanical Engineers, New York. Article V06BT07A075.

Rahjoo, M., Goracci, G., Gaitero, J.J., Martauz, P., Rojas, E., Dolado, J.S., 2022a. Thermal energy storage (TES) prototype based on geopolymer concrete for high-temperature applications. Materials 15 (20). https://doi.org/10.3390/ma15207086. Article 7086.

Rahjoo, M., Goracci, G., Martauz, P., Rojas, E., Dolado, J.S., 2022b. Geopolymer concrete performance study for high-temperature thermal energy storage (TES) applications. Sustainability 14 (3). https://doi.org/10.3390/su14031937. Article 1937.

Rahjoo, M., Rojas, E., Goracci, G., Gaitero, J.J., Martauz, P., Dolado, J.S., 2023. A numerical study of geopolymer concrete thermal energy storage: benchmarking TES module design and optimizing thermal performance. J. Energy Storage 74. Article 109389.

Rathod, M.K., Banerjee, J., 2013. Thermal stability of phase change materials used in latent heat energy storage systems: a review. Renew. Sustain. Energy Rev. 18, 246–258.

Rodriguez-Pastor, D.A., Garcia-Guzman, A., Marqués-Valderrama, I., Ortiz, C., Carvajal, E., Becerra, J.A., Soltero, V.M., Chacartegui, R., 2024. A flexible methanol-to-methane thermochemical energy storage system (TCES) for gas turbine (GT) power production. Appl. Energy 356. Article 122398.

Salehudress, Z.M., 2024. Performance Study of Low Temperature Air Heated Rock Bed Thermal Energy Storage System. Energy Storage. Advance Online Publication.

Salomoni, V.A., Majorana, C.E., Giannuzzi, G.M., Miliozzi, A., Di Maggio, R., Girardi, F., Mele, D., Lucentini, M., 2014. Thermal storage of sensible heat using concrete modules in solar power plants. Sol. Energy 103, 303–315.

Sarbu, I., Dorca, A., 2019. Review on heat transfer analysis in thermal energy storage using latent heat storage systems and phase change materials. Int. J. Energy Res. 43 (1), 29–64.

Sathyamurthy, R., 2024. Combined sensible heat and nano-enhanced latent heat energy storage for solar thermal evaporation: an experimental approach. J. Therm. Anal. Calorim. 149 (16), 8771–8785.

Selvaraj, J., Thenarasu, M., Aravind, S., Ashok, P., 2015. Waste heat recovery from castings and scrap preheating by recovered heat using an intermediate heat transfer medium. Appl. Mech. Mater. 813, 776–781.

Sharma, A., Tyagi, V.V., Chen, C.R., Buddhi, D., 2009. Review on thermal energy storage with phase change materials and applications. Renew. Sustain. Energy Rev. 13 (2), 318–345.

Singh, M.R., Giri, A., Lingfa, P., 2022. Enhancing heat transfer in phase change material-based cooling: a review. In: AIP Conference Proceedings, vol 2421. AIP Publishing, Melville, NY. No. 1, Article 060008.

Steinparzer, T., Haider, M., Fleischanderl, A., Hampel, A., Enickl, G., Zauner, F., 2012. Heat exchangers and thermal energy storage concepts for the off-gas heat of steelmaking devices. In: Journal of Physics: Conference Series, vol 395. IOP Publishing, Bristol (1), Article 012158.

Tasmin, N., Farjana, S.H., Hossain, M.R., Golder, S., Mahmud, M.P., 2022. Integration of solar process heat in industries: a review. Cleanroom Technol. 4 (1), 97–131.

Tripathi, B.M., Rathore, P.K.S., 2025. A Comprehensive Review of Sensible Heat Thermal Storage Materials for High-Temperature Applications. Energy Storage. Advance Online Publication.

Ushak, S., Marin, P., Galazutdinova, Y., Cabeza, L.F., Farid, M.M., Grágeda, M., 2016. Compatibility of materials for macro encapsulation of inorganic phase change materials: experimental corrosion study. Appl. Therm. Eng. 107, 410–419.

Vigneshwaran, K., Sodhi, G.S., Muthukumar, P., Arvind, V.K., Balamurugan, G., Sriram, S., Senthilmurugan, S., 2019a. Experimental investigation of a cast-steel based thermal energy storage system. Energy Proc. 158, 4664–4670.

Vigneshwaran, K., Sodhi, G.S., Muthukumar, P., Guha, A., Senthilmurugan, S., 2019b. Experimental and numerical investigations on high-temperature cast steel based sensible heat storage system. Appl. Energy 251. Article 113322.

Voronin, D., Ivanov, E., Gushchin, P., Fakhrullin, R., Vinokurov, V., 2020. Clay composites for thermal energy storage: a review. Molecules 25 (7). Article 1504.

Wang, S., Abdulridha, A., Bravo, J., Naito, C., Quiel, S., Suleiman, M., Romero, C., Neti, S., Oztekin, A., 2023. Thermal energy storage in concrete: review, testing, and simulation of thermal properties at relevant ranges of elevated temperature. Cement Concr. Res. 166. Article 107096.

Wokon, M., 2020. Reversible Redox Reactions for High Temperature Thermochemical Energy Storage. Universität Stuttgart, Stuttgart. Doctoral dissertation.

Wu, W., Wang, X., Xia, M., Dou, Y., Yin, Z., Wang, J., Lu, P., 2020. A novel composite PCM for seasonal thermal energy storage of solar water heating system. Renew. Energy 161, 457–469.

Yang, R., Howland, C.J., Liu, H.R., Verzicco, R., Lohse, D., 2024. Enhanced efficiency of latent heat energy storage by inclination. PRX Energy 3 (4). Article 043006.

Zhang, H., Baeyens, J., Cáceres, G., et al., 2013. Thermal energy storage: recent developments. Energy 63, 1–10.

Zhao, C.Y., Wu, Z.G., 2011. Heat transfer enhancement of high temperature thermal energy storage using metal foams and expanded graphite. Sol. Energy Mater. Sol. Cell. 95 (2), 636–643.

Zhao, Y., Zhao, C., Wen, T. and Markides, C.N., High-temperature Sensible Storage—Industrial Applications.

Zuo, J., Luo, H., Ling, Z., Zhang, Z., Fang, X., Zhang, W., 2024. Preparation of inorganic molten salt composite phase change materials and study on their electrothermal conversion properties. Indust. Chem. Mater. 2 (4), 571–586.

Further reading

Paroutoglou, E., Fojan, P., Gurevich, L., Hultmark, G., Afshari, A., 2021. Thermal analysis of organic and nano encapsulated electrospun phase change materials. Energies 14 (4). Article 995.

Saeed, R.M., Shigrekar, A., Mikkelson, D.M., George Rigby, A.C., Yang Hui Otani, C.M., Garrouste, M., Frick, K.L., Bragg-Sitton, S.M., 2022. Multilevel Analysis, Design, and Modeling of Coupling Advanced Nuclear Reactors and Thermal Energy Storage in an Integrated Energy System. Idaho National Laboratory (INL), Idaho Falls, ID.

U.S. Department of Energy, 2004. Energy Use, Loss and Opportunities Analysis: U.S. Manufacturing and Mining. U.S. Department of Energy, Industrial Technologies Program (DOE/ITP), Washington, DC.

Mechanical and electrical energy storage methods

6

Arun Teja Doppalapudi[1], *Abul Kalam Azad*[2] *and Amanullah Maung Than Oo*[3]
[1]School of Engineering and Technology, Central Queensland University, Mackay, QLD, Australia; [2]School of Engineering and Technology, Central Queensland University, Melbourne, VIC, Australia; [3]School of Science, Technology and Engineering, University of Sunshine Coast, Moreton Bay, QLD, Australia

6.1 Introduction

Energy storage systems (ESSs) are emerging as the vital technology for modern grids. Historically, electrical distribution systems were centralized, and power plants were operated according to demand. With the new renewable infrastructure in the energy market, accommodating this additional energy to the current grid confronts a major paradigm shift. In addition, the renewable energy sources such as solar, wind, and hydroelectricity are unpredictable and weather-dependent. The intermittent nature of energy from renewables poses a major challenge for traditional grid architectures, affecting grid stability and reliability. For instance, in 2023, Germany curtailed approximately 19 TWh of renewable electricity due to grid bottlenecks and sudden, weather-driven inflows of wind and solar power. That curtailed energy was 4% of Germany's total annual electricity generation (Wehrmann, 2024). For such cases, ESSs provide better support by filling the gap between surplus energy and consumption.

ESSs became the fundamental infrastructure, storing excess energy from renewables during surplus production periods and discharging it during peak demand periods. At Conference of the Parties (COP)28, several countries agreed to commit to tripling global installed renewable power capacity to at least 11,000 GW (GW) and doubling the global average annual rate of energy efficiency improvements from around 2% to over 4% by 2030 (COP28, 2024). To achieve this, substantial upgrades to grid flexibility and storage capacity are required. According to the International Energy Agency, approximately 1500 GW of energy storage capacity should be added to meet the goal of tripling renewables by 2030 (IEA, 2025; Jones, 2024; Aguado Molina et al., 2025). From the Global Energy Storage and Grids Pledge taken at COP29, countries are committed to jointly deploy 1500 GW of energy storage capacity and modernize or build 25 million kilometer of electricity grids by 2030 (IEA, 2024a). At COP30, energy storage is no longer an option but a necessary technology for achieving near-term renewable expansion, enhancing grid flexibility, and meeting global decarbonization objectives by 2030 (IRENA, 2025a). Several advanced ESS

Energy Storage Technologies. https://doi.org/10.1016/B978-0-443-45858-3.00015-2

technologies are deployed worldwide to address the energy crisis and have become necessary cornerstones for sustainable development.

Advancements in critical ESS technologies are now widely recognized as essential for achieving commercial maturity and enabling ESS deployment as a key component of global energy infrastructure. A broad range of ESS technologies is currently being deployed worldwide, with many systems already in commercial operation, others under active construction or scale-up, and several still in early research or conceptual development. In general, ESS technologies can be classified based on their energy conversion and storage mechanisms into mechanical, electrochemical (battery-based), electrical, thermal, and chemical storage categories (Rahman et al., 2020). Among these, mechanical and electrical/electrochemical storage systems have received significant attention due to their strong relevance to grid-scale renewable integration, fast response capability, and growing market adoption. Accordingly, this chapter focuses on mechanical and electrical ESSs, presenting their working principles, recent technological advancements, current operating projects, and key technical and economic challenges (Amirante et al., 2017). The following sections provide an in-depth discussion of these technologies, highlighting performance trends, deployment barriers, and future development directions.

6.2 Mechanical energy storage methods

6.2.1 Pumped hydro energy storage

Pumped hydro energy storage (PHES) is similar to the EES of a conventional hydrothermal power plant. In PHES, two or more water bodies at different elevations are connected and managed by either pumps or turbines (Rehman et al., 2015). During periods of low electricity demand, water is transferred from a lower-elevation water body to a higher-elevation one using pumps. To meet the demand for electricity, water will be diverted from higher to lower elevations, passing through turbines to generate electricity. In this case, the amount of energy storage depends on the elevation difference between the two water bodies and the volume of each. Despite its high capital costs, PHES is widely used as an energy storage technique (Zhao et al., 2025a).

Pumped storage hydropower is the world's largest energy storage technology, with a global installed capacity of nearly 200 GW, accounting for over 94% of the world's long-duration energy storage capacity. Conventional PHES systems provide convenient EESs. For instance, in the United States, the 3 GW Bath County PSH has 11 h of energy storage, providing power to 750,000 homes (IHA, 2025b). Fig. 6.1 depicts energy generation from PHES across North and Central America, South America, Europe, Africa, South and Central Asia, and East Asia and the Pacific. In 2024, a total of 189 GW of energy was produced from PHES across all regions, with the East Asia and Pacific region alone accounting for 52% of total PHES energy generation (IHA, 2025a). Countries in East Asia and the Pacific, such as China, Japan, and South Korea, produced approximately 90.8 GW of energy. Countries in the North and Central American region produced around 22.4 GW; of this, the United States alone

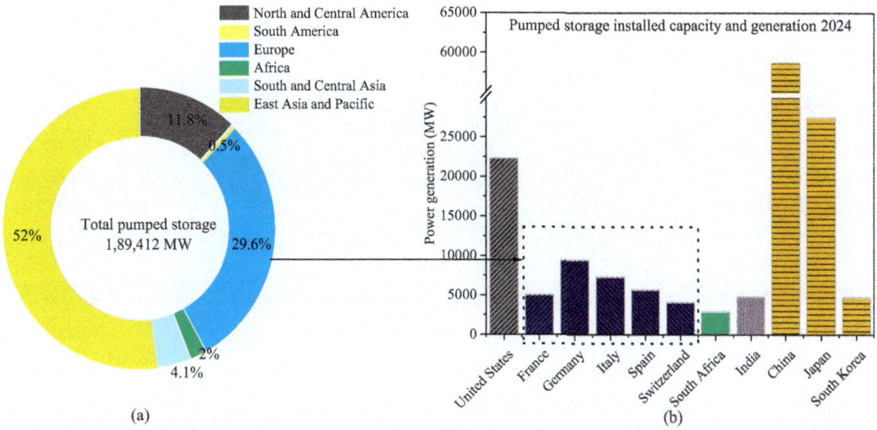

Figure 6.1 Worldwide pumped hydro power generation (IHA, 2025a).

accounted for 22.2 GW. In addition, countries in the East Asia and Pacific region, such as China, Japan, and South Korea, account for 90.8 GW of energy. Globally, 105 GW of new PHES systems are under construction, with over 90 GW in China alone (IHA, 2025a).

At present, PHES systems play a significant role in energy storage and electricity transition systems, supporting the renewable energy sector. PHES operational advantages, such as proven reliability, bulk energy storage, rapid response, controlled energy storage, and high efficiency, are well-suited to variable renewable energy sources. Several studies and numerical analyses suggest that PHES linked to wind and solar plants can significantly increase the utilization of renewables, thereby reducing reliance on conventional plants.

Although PHES was the best-suited alternative for energy storage, it cannot be relied upon for all aspects. For instance, the conventional PHES is being built on rivers that are already formed and connected to other rivers or reservoirs (Donalek, 2020). It is quite convenient to create a PHES system with such availabilities. However, in the case of off-river PHES (a closed-loop system), numerous area, water, geologic, and geographic factors are involved. The capital budget is expected to be very high. Moreover, to improve energy storage, the storage capacity must be increased, further increasing the plant's cost.

6.2.2 Compressed air energy storage

Compressed air energy storage (CAES) is another form of mechanical ESS, where energy is stored as pressurized air. Fig. 6.2 presents the general working principle of the CAES. It consists mainly of three stages: compression, storage, and expansion. The input energy is the excess electricity used to power the compressor, pressurizing the air during off-peak demand (charging mode). The pressurized air is stored in storage tanks/sites (adiabatic storage condition for Adiabatic-CAES), and during the

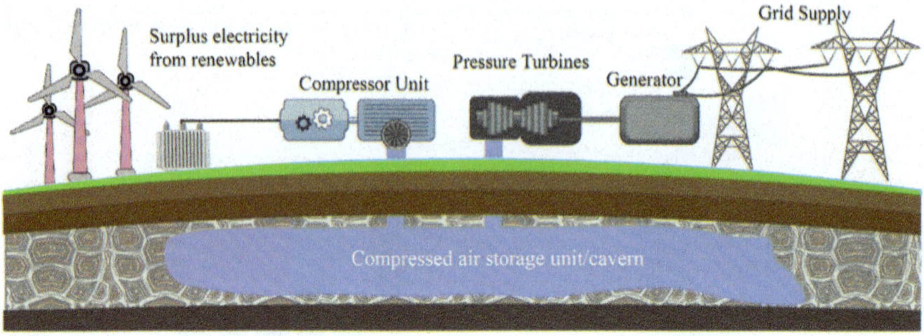

Figure 6.2 Compressed air energy storage system.

expansion process, it is sent to turbines, thereby generating electricity during peak demand periods.

The CAES technology has been under consideration since the 1950s. The first CAES technology was commercially started in the 1970s at the Huntorf plant in Germany. At first, the plant produced at 290 MW for a 2-hour discharge, and after modifications in 2007, the energy generation increased to 320 MW with the round-trip efficiency of 42% (Jafarizadeh et al., 2020). The plant's storage capacity was about 310,000 m^3, built in an abandoned underground mining site. The compressed air will be stored at a pressure of 43–70 bar at a depth of 600m. Another commercial plant operating in the United States, the McIntosh plant, has 110 MW of power generation capacity. The compressed air is stored in a 500,000 m^3 cavern at a depth of 450m, under pressures of 45–74 bar. The McIntosh CAES plant uses a waste-heat recovery process, which can reduce fossil fuel use by 25% compared to the Huntorf CAES plant (Nakhamkin et al., 1990). In 2025, China began operating the world's largest 300 MW CAES plant, Nengchu-1 (Yingcheng), in Hubei, with an integrated thermal storage system achieving 70% efficiency. The Yingcheng plant uses Advanced CAES (A-CAES) technology, avoiding natural gas combustion during the expansion stage and utilizing two deep salt caverns for storage, with a capacity of 700,000 m^3 at a depth of 600 m. There was ongoing interest in building additional commercial A-CAES plants. For instance, China, Australia, and Canada are building more CAES plants with A-CAES technology with the operating capacities of 100–300 MW (Tong et al., 2021; Wei et al., 2023). Although CAES technology is operating worldwide, several factors need attention to improve the system's overall efficiency.

First and foremost, mechanical components such as compressors, expanders, heat exchangers, and auxiliary components play a major role in the overall efficiency of the operating plant. In addition, a large-scale storage system is required to store large volumes of compressed air. Currently, the scope for such spaces is mostly salt caverns (Wu et al., 2023), abandoned mines (Wu et al., 2024), aquifers (Bi et al., 2025), and rock caverns (Ji et al., 2024). However, the risk matrix remains considerable due to factors such as creep deformation, irregular geometry, leakage, and collapse. Furthermore, additional integrated systems are needed to improve overall system efficiency and make CAES systems more economically and environmentally sustainable. For example, systems such as liquefied CAES (Olympios et al., 2021), compressed CO_2 ESS (Zhang et al., 2026), supercritical CAES (Guo et al., 2016), and advanced adiabatic CAES (Tola et al., 2017) have been explored to increase net output power relative to traditional CAES systems. Through ongoing research, the potential of the CAES system can be further investigated to improve energy storage capacity.

6.2.3 Flywheel energy storage systems

The flywheel energy storage systems (FESSs) system has recently attracted significant attention for its ability to charge/discharge large amounts of power in a short time. As shown in Fig. 6.3, it consists mainly of a rotating mass (flywheel) supported by a magnetic bearing, coupled to an electric machine, which is typically a permanent-magnet synchronous machine. The FESS system stores energy through the high-speed rotation of its flywheel rotor. During charging mode, electrical energy drives the motor to accelerate the flywheel, converting electrical energy into mechanical energy. In discharging mode, the rotating flywheel drives the motor to generate electrical power. The stored energy depends on the flywheel's rotational speed and moment of inertia, as presented in Eq. (6.1) (Li and Palazzolo, 2022).

$$E = \frac{1}{2} I_p \omega^2 \tag{6.1}$$

where E is the rotational kinetic energy, ω is the rotational speed, and I_p is the moment of inertia.

Compared to other mechanical ESSs, FESSs have high power density (2 to >5000 kWh_{out}), extremely fast charge/discharge capability, rapid response (seconds–minutes), long service life (>15 years), and high energy conversion efficiency

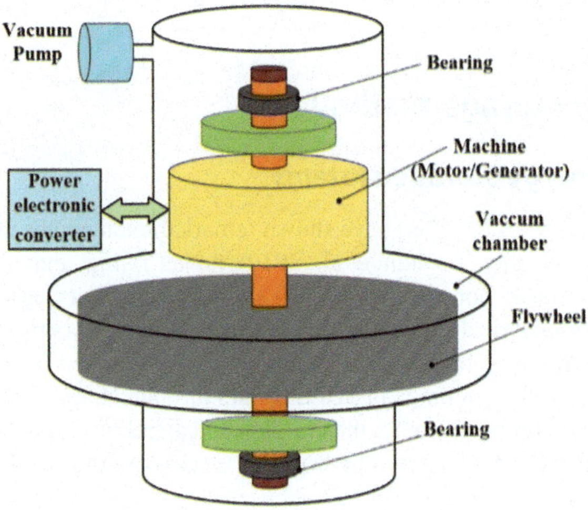

Figure 6.3 Structural layout of the flywheel energy storage system (Zhang et al., 2022).

(80%–95%) (IEA, 2024b). FESSs are slowly gaining attention for commercialization due to their environmental friendliness and high efficiency. For example, in 2011, USA Beacon Power started a 20 MW plant with 200 flywheels, connected to New York State Electric and Gas at 115 kV. In 2024, China built the largest 30 MW Dinglun FESS with only 120 flywheels, and connected to the grid at 110 kV, which can power up to 2000 households for a year (Rogers, 2024). An advanced high-speed magnetic levitation system was employed in the Dinglun FESS, resulting in a higher and more effective energy-storage capacity.

Despite several benefits, FESS faces significant technical and operational challenges. The major disadvantages include high idling loss, high cost, and electromechanical control complexity (Amiryar and Pullen, 2020). Mainly, losses occur across mechanical (windage and bearing friction), electromagnetic (magnetic bearing loss, motor loss, inverter loss), and auxiliary subsystems (vacuum equipment, cooling equipment, monitoring) (Amiryar and Pullen, 2020). Current research activities, as reflected in the provided material, focus strongly on improving support/suspension mechanisms, control robustness, and reliability. Traditional mechanical bearings have high friction and operating losses. In contrast, magnetic levitation bearings are complex and costly, and they require stable control, motivating new suspension concepts such as 5-DOF active magnetic bearing designs [30], high-temperature superconducting suspensions [31], and zero-flux coil guidance structures (including cross-connection designs that show improved performance) (Huang et al., 2024). However, motor losses remain dominant, converting to thermal energy and causing high temperatures that accelerate material aging and lead to irreversible demagnetization of permanent magnets (Pullen, 2019). On the control side, research emphasizes high-performance motor control and sensorless operation because installing physical sensors in vacuum-encapsulated systems is difficult, thereby increasing costs and maintenance burden. Methods such as model-reference adaptive system (MRAS)-based sensorless estimation, improved MRAS with online permanent-magnet flux identification, and sparrow speech algorithm-based parameter tuning are investigated to improve speed estimation accuracy and stability (Song et al., 2024).

6.3 Electrical energy storage methods

6.3.1 Supercapacitor energy storage systems

Supercapacitor energy storage systems (SCESs) have shown remarkable outcomes in terms of cycling lifetime, power density (Beaudin et al., 2010), power rating, and efficiency (Vazquez et al., 2010; Díaz-González et al., 2012). Supercapacitors, also referred to as ultracapacitors or electrical double-layer capacitors (EDLCs), consist of a pair of carbon electrodes separated by a porous membrane and immersed in an electrolyte (Kebede et al., 2022). Energy storage in SCES occurs through charge separation at the electrode–electrolyte interface, rather than through chemical reactions as in batteries. When a voltage is applied, charged ions from the electrolyte physically adsorb and desorb on both sides of the high-surface-area electrodes, forming an electric double layer (Prasad et al., 2019; Conway and Pell, 2003; Sathyan et al., 2022). This physical charge storage mechanism enables extremely fast charge and discharge processes and results in a significantly increased cycle lifetime (Saini et al., 2024). As a result, supercapacitors can withstand frequent, rapid cycling without degradation, making them well-suited for applications requiring short-term, high-power operation.

Supercapacitors have attracted significant interest due to their ability to bridge the performance gap between conventional capacitors and batteries or fuel cells in terms of power and energy storage (Saini et al., 2024). While lithium-ion (Li-ion) batteries possess 10–20 times higher energy density, supercapacitors offer 10–50 times higher power density (Engelbrecht et al., 2023). The substantially lower internal resistance of supercapacitors makes them particularly suitable for short-duration, high-power pulses, such as those required in fast frequency response and inertial response applications. The operational difference between batteries and supercapacitors can be illustrated using a water analogy: a battery system is represented by a large tank with a high total energy content, whereas a supercapacitor system is represented by a smaller bucket capable of delivering a very high instantaneous flow rate. Although both systems may be capable of supplying sufficient power for certain applications, the battery system is often overdimensioned for short-duration power delivery. In contrast, the supercapacitor system offers a more efficient solution.

Supercapacitors can be charged hundreds of times faster than Li-ion batteries, with discharge durations ranging from seconds to minutes, compared to hours for batteries (Rashid Khan and Latif Ahmad, 2025). Since the cost per kilowatt increases with higher energy-to-power (E/P) ratios, supercapacitors provide a more economical solution for systems with low E/P ratios (Ariyarathna et al., 2021; Engelbrecht et al., 2023). High cycling capability is essential for applications with frequent, short-term power injection, as batteries degrade with repeated cycling, while supercapacitors maintain stable performance (Pameté et al., 2023; Gee et al., 2013). The typical lifetime of Li-ion batteries is approximately 5–10 years or 500–800 cycles, whereas supercapacitors can be charged and discharged thousands of times, with projected service lifetimes of up to 20 years depending on the application.

From a safety perspective, supercapacitors are generally safe, less toxic, and pose a lower fire hazard. Unlike batteries, they can be fully discharged before maintenance, significantly reducing safety risks for maintenance personnel (Gopi and Ramesh, 2024). Due to their hybrid nature, supercapacitors exhibit characteristics of both electrochemical cells and conventional capacitors. Typical performance characteristics include power ratings of several megawatts, energy capacities of a few kilowatt-hours, discharge times of several minutes, cycle life exceeding 10^6 cycles, efficiency of 95%–98%, energy density of 4–7 $Wh\cdot kg^{-1}$, specific energy of 2.5–15 $Wh\cdot kg^{-1}$, specific power ranging from 500 to 10^4 $Wh\cdot kg^{-1}$, and self-discharge rates of 20%–40% (Luo et al., 2015; Siami et al., 2001; Kowal et al., 2011; Kebede et al., 2022).

Due to their relatively high self-discharge rates and costs, supercapacitors are generally not used for large-scale, long-duration energy storage, but rather for short-term storage and power-support applications (Iqbal et al., 2025; Admase et al., 2025). They are widely applied in automotive start–stop systems, uninterruptible power supplies, short-term backup during peak demand, voltage smoothing in renewable energy systems, and protection of electronic devices (Ranjan et al., 2025). The main drawbacks are low energy density, high self-discharge, and cost (Dissanayake and Kularatna-Abeywardana, 2024; Ike et al., 2016; Riaz et al., 2021). However, increasing demand for supercapacitors is contributing to cost reductions, and ongoing development of highly porous electrode materials and advanced electrolytes capable of operating at higher voltages is expected to enhance further capacitance and specific energy (Mandal et al., 2025; Kim et al., 2011).

6.3.2 Superconducting magnetic energy storage

Superconducting magnetic energy storage (SMES) operate on the basis of superconductivity, in which superconducting materials are cooled below their critical temperature to exhibit zero electrical resistance and zero magnetic-field dissipation. SEMS stores surplus energy in the form of direct current (DC) and stores it as a DC magnetic field. The DC in the superconductor continues to flow even after the voltage across it is removed. The DC will be converted back to AC during peak demand periods. Because of the near-zero resistance of these superconducting coils, the SMES's efficiency is very high (95%–98%), and its power density is high, due to the high current density in superconductivity (Adetokun et al., 2022). Hence, these have ultra-fast response times, with discharge occurring in milliseconds. Moreover, the superconducting coils are highly reliable, have a longer lifespan, and require minimal maintenance. However, several complexities are involved in the design and operation of SMES for commercial applications.

SMES mainly consists of a magnetized superconducting coil (MSC), a power conditioning system (PCS), a cryogenic system, and a control system, and each subsystem introduces technical and operational challenges (Xu et al., 2018). In the MSC, low-temperature superconducting coils (LTS) require near absolute zero temperatures (4.2 K) to enter the superconducting state, with dedicated liquid helium cryogenic refrigeration and continuous cooling (Hirano et al., 2016). Hence, high-temperature superconducting coils (HTS) were introduced using materials such as bismuth strontium calcium copper oxide, yttrium barium copper oxide, and magnesium diboride for scalable applications, which can operate at higher liquid nitrogen temperatures (77 K), making construction and operation more convenient and cost-effective (Friedman et al., 2003).

In addition, the PCS is a crucial component of SMES because it determines dynamic control performance and overall system capability and manages power transfer between the superconducting coil and the grid. Different PCS configurations have been developed, including thyristor-based PCS, voltage-source converter-based (VSC-PCS), and current-source converter-based (CSC-PCS), each with distinct control and harmonic performance characteristics (Guo et al., 2018). On the other hand, the cryogenic system remains a major source of complexity, as the coil must be maintained at sufficiently low temperatures to preserve superconductivity. This requires a dedicated cryogenic refrigerator using helium for LTS systems or liquid nitrogen for HTS systems, and system losses such as radiation, conduction, cold-to-hot current leads, and alternating current effects must be minimized to improve efficiency and reduce overall cost. Finally, the control system must continuously coordinate grid demand signals with power flow to and from the SMES coil while monitoring safety and system status, which becomes increasingly challenging as SMES are expected to actively regulate grid power across all four quadrants and support power system stability and power quality.

6.3.3 Battery energy storage technologies

BESS technology converts electrical energy into storable electrochemical potential energy. The power conversion system is an important component, and control and monitoring systems are the key components. These ensure safe, optimal performance, preventing overcharging and overdraining. There are various types of technology today, and among them, the most advanced and viable options are lithium-ion and sodium-ion batteries (SIBs). The following sections discuss these further.

6.3.3.1 Lithium-ion battery energy storage system

Li-ion batteries are emerging as a key part of the global clean energy transition and are dominating the energy storage market. One prominent advantage of this technology is its very high energy density (up to 200 Wh/kg), which enables a wide range of applications. Additionally, lithium-ion batteries have high energy efficiency (greater than 95%) and a long cycle life of approximately 3000 cycles at a deep discharge of 80% (Chen et al., 2020). Recent trends also suggest that Li-BESS is gaining popularity, and large commercial plants are currently operating worldwide, providing greater energy storage capacity. For example, the Edwards and Sanborn Solar project in California, USA, uses approximately 120,720 lithium-ion batteries for a 3287 MWh capacity, which is used to store and support the renewable energy from the 1.9 million solar panels (Himmelstein, 2024; Murray, 2024). Countries like Australia, China, and the United Kingdom are deploying Li-ion storage for grid support and renewable integration. Also, several innovative modifications have been made since the beginning to improve the size, quality, and efficiency of the batteries.

In the last decade, several advancements have been made in Li-ion batteries, with novel materials that offer a distinctive approach to energy storage. For instance, conventional Li-ion batteries use graphite as the anode (Speirs et al., 2014). To improve the performance and meet the future requirements, several novel anode materials have been introduced with alloy materials (Al, M_g, A_g, and S_b) (Kim et al., 2005; Zhong et al., 2015), silicon (Lin et al., 2023), carbon-based compounds (graphitic carbon and non-graphitic carbon) (Wu et al., 2003), and transition metals (Fe_3O_4, FeS_2, and MnO_2) (Ye et al., 2021). Besides several advanced cathode materials, such as lithium iron phosphate ($LiFePO_4$), lithium nickel cobalt aluminum oxide, lithium titanate ($Li_4Ti_5O_{12}$), and lithium nickel manganese cobalt oxide ($LiNiMnCoO_2$), were extensively studied. Among LIPs ($LiFePO_4$), which are gaining popularity, its market share has grown from 48% in 2021 to 85% in 2024, owing to lower costs and improved safety and cycle life (IRENA, 2025b). Beyond electrodes, trends in cell electrolytes and architecture are important. New liquid electrolytes and additives improve safety and low-temperature performance (Brijesh et al., 2025). Solid-state electrolytes (polymer or ceramic) are being developed for even higher energy density and intrinsic safety (Hasan et al., 2025). However, there are several concerns with Li-ion batteries, including safety, recycling requirements, battery degradation, and responsible mining. Additionally, there are several challenges for LIBs in aligning with the United Nations Sustainable Development Goals (Han et al., 2019).

6.3.3.2 Sodium-ion batteries energy storage systems

SIBs are emerging as the most popular batteries, overcoming the challenges posed by LiBs. Unlike lithium, sodium is abundantly available (423 times more plentiful than lithium), but has a larger atomic radius (Yabuuchi et al., 2014). SIBs are gaining popularity as next-generation BES systems due to their abundant resources, economic viability, and low environmental impact. The raw material is much cheaper for SIBs. For instance, the sodium costs around \$150/ton and lithium costs around \$5000/ton and this will be a great advantage to build large stationary BES systems (Wanison et al., 2024). Also, SIBs can charge to 80% in 15 min and maintain 90% capacity efficiency even in extreme cold conditions (−20 °C) (Yang et al., 2023; Tang et al., 2015). However, sodium has a lower energy capacity, and sodium ions are heavier than Lithium (Keiner et al., 2026). The energy density of the SIBs is around 140 Wh/kg, whereas that of the LIBs was recorded around 250 Wh/kg (Jang et al., 2026). In addition, the active materials in the SIBs are also abundant. For instance, SIBs use hard carbon as the anode material, whereas LIBs use graphite. According to the European Commission, natural graphite is a critical raw material (Union, 2014; Jasper et al., 2025), whereas synthetic graphite requires an energy-intensive process (Zhao et al., 2022).

The operating principles and battery cell configurations are similar to those of LIBs, and here the charge carriers are sodium ions (Na^+). During charging, an external power source drives Na^+ ions to de-intercalate from the cathode and migrate through the electrolyte toward the anode, while electrons flow through the external circuit. The Na^+ ions are stored within the anode material, converting electrical energy into chemical potential energy. During discharge, this process reverses: Na^+ ions migrate back to the cathode through the electrolyte, and electrons flow through the external circuit, delivering electrical energy to the load. The larger atomic radius of the Na ion (1.02 Å) compared to the Li ion (0.76 Å) makes the system complex and heavy, and it is difficult to select a relative electrode material for SIBESS's (Cao et al., 2012). Na^+ ions exhibit lower desolvation energies than Li^+ ions (by approximately 40–70 $kJmol^{-1}$), attributed to the weaker Lewis acidity of sodium. This leads to faster charge transfer kinetics and improved low-temperature performance. Also, hard carbon is used as an anode, with curved graphene-like layers with larger interlayer spacing (0.37–0.40 nm) and abundant nanopores. These structural features enable Na^+ storage via intercalation, pore filling, and adsorption mechanisms, yielding reversible capacities of 400–1000 $mAhg^{-1}$ at the material level (Cui et al., 2025).

On the cathode side, layered transition metal oxides ($NaMO_2$, where M = Fe, M_n, Co, Ni), polyanionic compounds such as $Na_3V_2(PO_4)_3$ and $NaFePO_4$ (Chliyah et al., 2026; Raghavan et al., 2024), and Prussian blue analogs (Goel et al., 2025) have emerged as promising candidates due to their open frameworks and stable Na^+ diffusion pathways. These materials offer competitive operating voltages (2.5–3.8 V) and good structural stability over extended cycling (Ma et al., 2011). From a system perspective, SIBs benefit from the high natural abundance and low cost of sodium resources, enhanced safety due to reduced dendrite formation, and improved thermal tolerance compared to LIBs. However, challenges remain in terms of lower gravimetric energy density (Slater et al., 2013), volumetric expansion of electrode materials (Klein et al., 2013), and long-term interfacial stability (Zhao et al., 2025b). Ongoing advancements in electrolyte formulations, surface coatings, and electrode architecture optimization are progressively closing the performance gap between sodium-ion and lithium-ion technologies (Morais et al., 2026).

6.3.4 Hydrogen fuel cell energy storage systems

In recent years, the hydrogen energy storage system (HFESS) has attracted significant attention due to its high energy density, long-term storage capability, controllable capacity, and flexible operation (Liang et al., 2026). The typical components of a HESS are an electrolyzer, a hydrogen storage tank, and a fuel cell. Excess energy from renewables will be converted into hydrogen (green hydrogen) via electrolysis, splitting water into hydrogen and oxygen. The hydrogen is then stored in storage tanks (salt caverns/other storage media) and later sent to the fuel cell to produce electricity during peak times (Liang et al., 2026). The fuel cell converts chemical energy into electrical energy through electrochemical reactions, producing water and heat as byproducts.

Research trends in HFESS primarily focus on improving hydrogen production, fuel cell performance, and grid-scale feasibility. Electrolysis research is advancing across multiple routes, including alkaline water electrolysis (Tüysüz, 2024), polymer electrolyte membrane water electrolysis (Longo et al., 2023), solid oxide electrolysis cells, and microbial electrolysis cells (Aravindan and Praveen Kumar, 2023). On the other hand, exploration of photocatalytic hydrogen production, where silicon-based photocatalysts are gaining attention due to their low toxicity, cost-effectiveness, high conductivity, and abundance (Islam et al., 2025). Here as well, performance improvements are linked to catalyst crystallinity, oxygen concentration, calcination temperature, surface modification, and synthesis pathways.

Fuel cell research is also focusing on efficiency optimization, durability, and cost reduction across multiple cell types. Proton exchange membrane fuel cells (PEMFCs) remain attractive for fast-response and mobile applications because they operate at 80 °C–90 °C (El Idi et al., 2026). PEMFCs offer quick start-up, high power density, and a compact design, but they face challenges due to the high cost of platinum catalysts and their sensitivity to hydrogen impurities. On the other hand, solid-oxide fuel cells (SOFCs) offer high efficiency and fuel flexibility and are well suited to stationary generation. However, SOFCs require high-temperature operation (650 °C–850 °C) (Quach et al., 2026), which raises concerns about cost and materials degradation, as well as slow start-up behavior. Across these systems, research is targeting improved catalyst utilization, advanced membrane-electrode assemblies, optimized flow-field design, and system integration strategies, such as combined heat and power, to increase overall energy utilization.

However, hydrogen fuel cell storage systems also face major disadvantages that currently limit wide deployment. The most persistent challenge is low efficiency (35%–70%) compared with other storage options such as SMES, BESS, and PHES (Alam Rimon et al., 2025). Additionally, high capital costs and limited infrastructure remain the biggest challenge for the commercialization of HFESS. Hydrogen fuel cells also require high-purity hydrogen (≥99.999% for PEM and ≥98.8 for alkaline) to improve performance and lifetime (Rahim Malik et al., 2023). Many technological advancements are needed for the HFESS to provide economically viable options.

6.4 Recent trends and future direction of energy storage systems

Future trends in mechanical and electrical/electrochemical storage systems are shifting toward hybridization rather than extending the current individual storage systems. As discussed above, in terms of capacity, storage duration, and response time, it is highly beneficial to coordinate them as multiple ESSs. For instance, hybrid energy storage systems (HESS) that combine high-energy-density batteries (lithium-ion with typical gravimetric energy densities of 150–3000 Wh/kg (Taabodi et al., 2025)) with high-power-density devices such as FESS or ultracapacitors (supercapacitors with power densities in the order of 5–10 kW/kg and cycle life often exceeding 10^5 cycles) can simultaneously meet fast power-discharge requirements and longer-duration energy needs (Hu et al., 2021). In such HESS architectures, batteries supply medium- to long-duration energy, while supercapacitors or flywheels compensate for fast power transients and peak demands. This coordinated operation improves round-trip efficiency (typically above 85% for modern battery-based systems) and extends battery lifetime by minimizing high-rate and frequent cycling.

The long-term storage is another important factor. Among the technologies discussed above, PHES remains the most mature technology for long-duration applications. But CAES designed with advanced adiabatic concepts offers a long-duration option: recent analyses show that adiabatic CAES can become economically viable for storage durations of about 10–100 h at current energy costs of roughly 200–400 USD/kWh. These can reach 500–1000 h if storage costs can be reduced to 1 USD/kWh through improved cavern engineering and scale-up (Yang et al., 2025). Hybrid mechanical configurations such as CAES combined with high-speed flywheel energy storage systems (CAES-HFESS) are particularly suitable for large-scale, combined-cycle power plants, where CAES provides bulk, long-duration capacity while flywheels deliver subsecond to second-level response and high cycling capability for frequency regulation and ramp-rate control. In addition, providing mechanical support to batteries with high-power flywheel or ultracapacitor modules can significantly reduce battery current peaks, thereby mitigating degradation mechanisms and enabling long-term, high-reliability applications. Overall, well-designed HESS architectures improve voltage stability and power quality by mitigating rapid voltage fluctuations during peak-demand periods.

6.5 Conclusion and recommendations

This chapter has reviewed mechanical, electrical, and electrochemical ESSs and their role in supporting large-scale renewable energy integration. PHES remains the dominant long-duration technology, accounting for more than 94% of global long-duration storage capacity, with nearly 200 GW installed worldwide. CAES has demonstrated commercial-scale operation at 110–320 MW, with recent advanced systems achieving efficiencies of up to 70%. FESSs provide very fast response with efficiencies of 80%–95% and power ratings up to 30 MW, making them suitable for frequency regulation and short-term grid support.

On the other hand, supercapacitors exhibit power densities of 500–10,000 $Wh{\cdot}kg^{-1}$, efficiencies of 95%–98%, and cycle lives exceeding 10^6 cycles, making them highly suitable for high-frequency, short-duration applications. Besides, SMES systems offer ultra-fast millisecond-level response and efficiencies of 95%–98%, but their deployment is limited by cryogenic requirements and high capital costs. Battery ESSs, particularly lithium-ion, dominate the market due to their high energy density (up to 200–250 $Wh{\cdot}kg^{-1}$), round-trip efficiencies above 95%, and cycle life of approximately 3000 cycles at 80% depth of discharge. SIBs provide lower energy density (~140 $Wh{\cdot}kg^{-1}$) but offer cost advantages and improved low-temperature performance. Hydrogen fuel cell systems enable long-term storage with energy densities exceeding 30,000 $Wh{\cdot}kg^{-1}$ (hydrogen basis), but their overall round-trip efficiency remains limited to 35%–70%.

Future energy storage deployment should prioritize expanding long-duration technologies such as PHES and advanced CAES to achieve storage durations above 10 h at low cost. Hybrid energy storage architectures should be adopted to integrate high-energy batteries with high-power devices (supercapacitors or flywheels), thereby reducing peak battery current stress and improving the overall lifetime. At the same time, material research should aim to increase energy density, improve thermal stability, and enhance cycle life. Recycling and recovery of critical materials should exceed 90% to support long-term sustainability. Artificial intelligence-based control should be used for multi-timescale operation, such as coordinating fast flywheel response with battery and PHES dispatch, to improve grid stability and reduce operating costs. In addition, AI-based digital twins should support system design, performance monitoring, and optimal sizing of hybrid ESSs.

References

Adetokun, B.B., Oghorada, O., Abubakar, S.J.A., 2022. Superconducting magnetic energy storage systems: prospects and challenges for renewable energy applications. J. Energy Storage 55, 105663.

Admase, A.T., Asrade, E.D., Fanta, S.W., 2025. A comprehensive review on energy storage materials and technologies: applications of nanofabrication techniques for enhanced performance and efficiency. Mater. Renew. Sustain. Energy 14, 56.

Aguado Molina, R., Cartelle Barros, J.J., De La Cruz López, M.D.P., Lara Coira, M., Del Caño Gochi, A., 2025. A comparative sustainability assessment of several grid energy storage technologies. Appl. Energy 396, 126248.

Alam Rimon, S.T., Mourshed, M., Kibria, M.G., 2025. Proton exchange membrane fuel cells: advances in materials development, performance optimization, and future outlook. Energy Convers. Manag. X 27, 101102.

Amirante, R., Cassone, E., Distaso, E., Tamburrano, P., 2017. Overview on recent developments in energy storage: mechanical, electrochemical and hydrogen technologies. Energy Convers. Manag. 132, 372–387.

Amiryar, M.E., Pullen, K.R., 2020. Analysis of standby losses and charging cycles in flywheel energy storage systems. Energies 13, 4441.

Aravindan, M., Praveen Kumar, G., 2023. Hydrogen towards sustainable transition: a review of production, economic, environmental impact and scaling factors. Results Eng. 20, 101456.

Ariyarathna, T., Kularatna, N., Gunawardane, K., Jayananda, D., Steyn-Ross, D.A., 2021. Development of supercapacitor technology and its potential impact on new power converter techniques for renewable energy. IEEE J. Emerg. Selected Topics Industr. Electr. 2, 267–276.

Beaudin, M., Zareipour, H., Schellenberglabe, A., Rosehart, W., 2010. Energy storage for mitigating the variability of renewable electricity sources: an updated review. Energy Sustain. Develop. 14, 302–314.

Bi, Z., Guo, Y., Yang, C., Yang, H., Wang, L., He, Y., Guo, W., 2025. Numerical investigation of fluid dynamics in aquifers for seasonal large-scale hydrogen storage using compositional simulations. Renew. Energy 239, 122101.

Brijesh, K., Jareer, M., Lakshmi Sagar, G., Mukesh, P., Amudha, A., Mandal, D., Nagaraja, H. S., Shahgaldi, S., 2025. Advanced electrolyte additives for lithium-ion batteries: classification, function, and future directions. J. Phys. Chem. C 129, 11221–11251.

Cao, Y., Xiao, L., Sushko, M.L., Wang, W., Schwenzer, B., Xiao, J., Nie, Z., Saraf, L.V., Yang, Z., Liu, J., 2012. Sodium ion insertion in hollow carbon nanowires for battery applications. Nano Lett. 12, 3783–3787.

Chen, T., Jin, Y., Lv, H., Yang, A., Liu, M., Chen, B., Xie, Y., Chen, Q., 2020. Applications of lithium-ion batteries in grid-scale energy storage systems. Trans. Tianjin Univ. 26, 208–217.

Chliyah, S., Rabii, S., Lahmidi, A., Chtita, S., Kouali, M.E.L., Errougui, A., 2026. Temperature influence on the solvation, transport, and dielectric properties of the {$NaPF_6$–Dimethyl sulfoxide} electrolyte system used in sodium-ion battery technology: a molecular dynamics approach and DFT study. In: Chemical Thermodynamics and Thermal Analysis, p. 100271.

Conway, B., Pell, W., 2003. Double-layer and pseudocapacitance types of electrochemical capacitors and their applications to the development of hybrid devices. J. Solid State Electrochem. 7, 637–644.

COP28, 2024. Global Renewables and Energy Efficiency Pledge.

Cui, Z., Li, B.-Y., Xiong, H., Li, T., Xie, M.-X., Hu, J.-Y., Qiu, X., Gui, Z.-Q., Zhou, R., Shi, L.-L., Ju, Z.-C., Chen, Y.-X., 2025. A review of ways to improve the performance of hard carbon anodes in low-temperature sodium-ion batteries. N. Carbon Mater. 40, 1246–1263.

Díaz-González, F., Sumper, A., Gomis-Bellmunt, O., Villafáfila-Robles, R., 2012. A review of energy storage technologies for wind power applications. Renew. Sustain. Energy Rev. 16, 2154–2171.

Dissanayake, K., Kularatna-Abeywardana, D., 2024. A review of supercapacitors: materials, technology, challenges, and renewable energy applications. J. Energy Storage 96, 112563.

Donalek, P.J., 2020. Pumped storage hydro: then and now. IEEE Power Energy Mag. 18, 49–57.

El Idi, M.M., Ankouni, M., Ben Haj Slama, R., Affonso Nobrega, P., Hajjar, A., Atli, A., 2026. Critical review of the latest advances in thermal management techniques for PEM fuel cells. Renew. Sustain. Energy Rev. 226, 116199.

Engelbrecht, T., Isaacs, A., Kynev, S., Matevosyan, J., Niemann, B., Owens, A.J., Singh, B., Grondona, A., 2023. Statcom technology evolution for tomorrow's grid: E-statcom, statcom with supercapacitor-based active power capability. IEEE Power Energy Mag. 21, 30–39.

Friedman, A., Shaked, N., Perel, E., Gartzman, F., Sinvani, M., Wolfus, Y., Kottick, D., Furman, J., Yeshurun, Y., 2003. HT-SMES operating at liquid nitrogen temperatures for electric power quality improvement demonstrating. IEEE Trans. Appl. Supercond. 13, 1875–1878.

Gee, A.M., Robinson, F.V., Dunn, R.W., 2013. Analysis of battery lifetime extension in a small-scale wind-energy system using supercapacitors. IEEE Trans. Energy Convers. 28, 24–33.

Goel, G., Sharma, M., Tripathi, S.K., 2025. Prussian blue analogue as cathode materials in sodium ion batteries: a review. J. Energy Storage 126, 116995.

Gopi, C.V.V.M., Ramesh, R., 2024. Review of battery-supercapacitor hybrid energy storage systems for electric vehicles. Results Eng. 24, 103598.

Guo, H., Xu, Y., Chen, H., Zhou, X., 2016. Thermodynamic characteristics of a novel supercritical compressed air energy storage system. Energy Convers. Manag. 115, 167–177.

Guo, W., Li, D., Cai, F., Zhao, C., Xiao, L., 2018. Z-source-converter-based power conditioning system for superconducting magnetic energy storage system. IEEE Trans. Power Electron. 34, 7863–7877.

Han, X., Lu, L., Zheng, Y., Feng, X., Li, Z., Li, J., Ouyang, M., 2019. A review on the key issues of the lithium ion battery degradation among the whole life cycle. eTransportation 1, 100005.

Hasan, M.M., Haque, R., Jahirul, M.I., Rasul, M.G., Fattah, I.M.R., Hassan, N.M.S., Mofijur, M., 2025. Advancing energy storage: the future trajectory of lithium-ion battery technologies. J. Energy Storage 120, 116511.

Himmelstein, S., 2024. Largest US Solar-Plus-Storage System Now Complete. GlobalSpec.

Hirano, N., Watanabe, T., Nagaya, S., 2016. Development of cooling technologies for SMES. Cryogenics 80, 210–214.

Hu, W., Shang, Q., Bian, X., Zhu, R., 2021. Energy management strategy of hybrid energy storage system based on fuzzy control for ships. Int. J. Low-Carbon Technol. 17, 169–175.

Huang, D., Jiao, C., Fang, J., 2024. Design and research of a new type of flywheel energy storage system equipped with permanent magnet synchronous motor and zero-flux coils. In: International Conference of Electrical, Electronic and Networked Energy Systems. Springer, pp. 509–520.

IEA, 2024a. Chairs' Summary and Call to Action: COP29-IEA High-Level Energy Transition Dialogues.

IEA, 2024b. Technology: Flywheel Energy Storage.

IEA, 2025. COP28 Tripling Renewable Capacity Pledge 2025: Update. Paris.

IHA, 2025a. 2025 World Hydropower Outlook.

IHA, 2025b. Pumped Storage Hydropower.

Ike, I.S., Sigalas, I., Iyuke, S., 2016. Understanding performance limitation and suppression of leakage current or self-discharge in electrochemical capacitors: a review. Phys. Chem. Chem. Phys. 18, 661–680.

Iqbal, M.F., Nasir, F., Shabbir, F., Babar, Z.U.D., Saleem, M.F., Ullah, K., Sun, N., Ali, F., 2025. Supercapacitors: an emerging energy storage system. Adv. Energy Sustain. Res. 6, 2400412.

IRENA, 2025a. Global Utilities Set Out USD 1 Trillion Investment Plans at COP30 as Grid Spend Grows.

IRENA, 2025b. Renewable Power Generation Costs in 2024. International Renewable Energy Agency, Abu Dhabi.

Islam, A., Islam, M.T., Teo, S.H., Mahmud, H., Swaraz, A.M., Rehan, A.I., Rasee, A.I., Kubra, K.T., Hasan, M.M., Salman, M.S., Waliullah, R.M., Hasan, M.N., Sheikh, M.C., Uchida, T., Awual, M.E., Hossain, M.S., Znad, H., Awual, M.R., 2025. Progress in silicon-based materials for emerging solar-powered green hydrogen (H_2) production. Adv. Colloid Interface Sci. 343, 103558.

Jafarizadeh, H., Soltani, M., Nathwani, J., 2020. Assessment of the Huntorf compressed air energy storage plant performance under enhanced modifications. Energy Convers. Manag. 209, 112662.

Jang, H., Youn, B., Han, Y., Lee, D., 2026. Bridging the gap from basic principles to advanced applications in sodium-ion batteries. J. Energy Storage 142, 119547.

Jasper, F.B., Baumann, M.J., Ersoy, H., Smith, A., Büchele, S., Bohn, N., Binder, J.R., Neuhaus, D.H., Weil, M., 2025. Comparative life cycle assessment of Prussian white and NVP/C-based sodium-ion batteries based on primary laboratory data. ChemSusChem 18 e202500268.

Ji, W., Wang, S., Wan, J., Cheng, S., He, J., Shi, S., 2024. Stability analysis of surrounding rock of multi-cavern for compressed air energy storage. Adv. Geo-Energy Res. 13, 169–175.

Jones, D., 2024. The Clean Flexibility COP?.

Kebede, A.A., Kalogiannis, T., Van Mierlo, J., Berecibar, M., 2022. A comprehensive review of stationary energy storage devices for large scale renewable energy sources grid integration. Renew. Sustain. Energy Rev. 159, 112213.

Keiner, D., Jasper, F., Bogdanov, D., Lopez, G., Peters, J., Baumann, M., Breyer, C., Weil, M., 2026. Sodium-ion battery cost projections and their impact on the global energy system transition until 2050. J. Energy Storage 146, 119861.

Kim, J.W., Ryu, J.H., Lee, K.T., Oh, S.M., 2005. Improvement of silicon powder negative electrodes by copper electroless deposition for lithium secondary batteries. J. Power Sources 147, 227–233.

Kim, T.Y., Lee, H.W., Stoller, M., Dreyer, D.R., Bielawski, C.W., Ruoff, R.S., Suh, K.S., 2011. High-performance supercapacitors based on poly (ionic liquid)-modified graphene electrodes. ACS Nano 5, 436–442.

Klein, F., Jache, B., Bhide, A., Adelhelm, P., 2013. Conversion reactions for sodium-ion batteries. Phys. Chem. Chem. Phys. 15.

Kowal, J., Avaroglu, E., Chamekh, F., Šenfelds, A., Thien, T., Wijaya, D., Sauer, D.U., 2011. Detailed analysis of the self-discharge of supercapacitors. J. Power Sources 196, 573–579.

Li, X., Palazzolo, A., 2022. A review of flywheel energy storage systems: state of the art and opportunities. J. Energy Storage 46, 103576.

Liang, J., Wu, S., Lu, T., 2026. Portfolio selection and optimal planning for hydrogen energy storage systems composed of heterogeneous electrolyzer and fuel cell technologies in industrial park multi-energy systems. Appl. Energy 403, 127001.

Lin, J., Wang, L., Xie, Q., Luo, Q., Peng, D.L., Buddie Mullins, C., Heller, A., 2023. Stainless steel-like passivation inspires persistent silicon anodes for lithium-ion batteries. Angew. Chem. 135 e202216557.

Longo, S., Cellura, M., Guarino, F., Ferraro, M., Antonucci, V., Squadrito, G., 2023. Chapter 7 - life cycle assessment of solid oxide fuel cells and polymer electrolyte membrane fuel cells: a review. In: Scipioni, A., Manzardo, A., Ren, J. (Eds.), Hydrogen Economy, second ed. Academic Press.

Luo, X., Wang, J., Dooner, M., Clarke, J., 2015. Overview of current development in electrical energy storage technologies and the application potential in power system operation. Appl. Energy 137, 511–536.

Ma, X., Chen, H., Ceder, G., 2011. Electrochemical properties of monoclinic $NaMnO_2$. J. Electrochem. Soc. 158, A1307.

Mandal, S., Mendhe, A.B., Rakhade, H.M., Barse, N.S., Roy, M., Rosaiah, P., Park, T., Lee, H.-S., Mendhe, A.C., Kim, D., 2025. Recent advancement and design in supercapacitor hybrid electrode materials: bridging the gap between energy and power density. Chem. Eng. J. Adv. 21, 100690.

Morais, D.F.D.S., Tirado, J.L., Pérez-Vicente, C., Da Motta, F.V., Lavela, P., Bomio, M., Lavela, S., 2026. Unlocking the performance of sodium-ion batteries by coating $Na_3V_2(PO_4)_3$ with Nb_2O_5. Acta Phys. Chim. Sin. 42, 100180.

Murray, C., 2024. California solar-plus-storage project with world's largest BESS fully online. In: Energy Storage News.

Nakhamkin, M., Andersson, L., Swensen, E., Howard, J., Meyer, R., Schainker, R., Pollak, R., Mehta, B., 1990. AEC 110 mw CAES plant; Status of project. In: Proceedings of the American Power Conference. United States).

Olympios, A.V., Mctigue, J.D., Farres-Antunez, P., Tafone, A., Romagnoli, A., Li, Y., Ding, Y., Steinmann, W.-D., Wang, L., Chen, H., 2021. Progress and prospects of thermo-mechanical energy storage—a critical review. Prog. Energy 3, 022001.

Pameté, E., Köps, L., Kreth, F.A., Pohlmann, S., Varzi, A., Brousse, T., Balducci, A., Presser, V., 2023. The many deaths of supercapacitors: degradation, aging, and performance fading. Adv. Energy Mater. 13, 2301008.

Prasad, G., Shetty, N., Thakur, S., 2019. Supercapacitor technology and its applications: a review. IOP Conf. Ser. Mater. Sci. Eng. 561, 012105.

Pullen, K.R., 2019. The status and future of flywheel energy storage. Joule 3, 1394–1399.

Quach, T.-Q., Lee, D., Park, J.Y., Kim, Y.S., Lee, S., Jang, H.J., Lee, S.Y., Bae, Y., 2026. Comprehensive analysis of a full-scale solid oxide fuel cell stack fueled by ammonia. Energy Convers. Manag. 348, 120618.

Raghavan, P., Das, A., Mj, J.F., 2024. Advanced Technologies for Rechargeable Batteries: Metal Ion, Hybrid, and Metal-Air Batteries. CRC Press.

Rahim Malik, F., Yuan, H.-B., Moran, J.C., Tippayawong, N., 2023. Overview of hydrogen production technologies for fuel cell utilization. Eng. Sci. Technol. Int. J. 43, 101452.

Rahman, M.M., Oni, A.O., Gemechu, E., Kumar, A., 2020. Assessment of energy storage technologies: a review. Energy Convers. Manag. 223, 113295.

Ranjan, A., Srivastava, I., Kumar, A., 2025. Supercapacitors aiding new and renewable energy generation for integrated power harvesting and storage systems. In: Verma, D.K., Aslam, J. (Eds.), Supercapacitors: Fundamentals, Advances and Future Applications. Royal Society of Chemistry.

Rashid Khan, H., Latif Ahmad, A., 2025. Supercapacitors: overcoming current limitations and charting the course for next-generation energy storage. J. Ind. Eng. Chem. 141, 46–66.

Rehman, S., Al-Hadhrami, L.M., Alam, M.M., 2015. Pumped hydro energy storage system: a technological review. Renew. Sustain. Energy Rev. 44, 586–598.

Riaz, A., Sarker, M.R., Saad, M.H.M., Mohamed, R., 2021. Review on comparison of different energy storage technologies used in micro-energy harvesting, WSNs, low-cost microelectronic devices: challenges and recommendations. Sensors 21, 5041.

Rogers, D., 2024. China connects world's biggest flywheel energy storage system to grid. In: Global Construction Review.

Saini, P., Yadav, J.K., Rani, B., Pandey, A.P., Dixit, A., 2024. High-performance symmetric supercapacitor using cost-efficient iron oxide (Fe_3O_4) nanoparticles. Energy Storage 6, e608.

Sathyan, M., Jandas, P., Venkatesan, M., Pillai, S.C., John, H., 2022. Electrode material for high performance symmetric supercapacitors based on superparamagnetic Fe_3O_4 nanoparticles modified with cetyltrimetylammonium bromide. Synth. Met. 287, 117080.

Siami, S., Joubert, C., Glaize, C., 2001. High frequency model for power electronics capacitors. IEEE Trans. Power Electron. 16, 157–166.

Slater, M.D., Kim, D., Lee, E., Johnson, C.S., 2013. Sodium-ion batteries. Adv. Funct. Mater. 23, 947–958.

Song, G., Wu, Z., Zheng, X., Zhang, J., Yang, P., Zhang, Z., 2024. Control strategy of flywheel energy storage system for improved model reference adaptive system based on tent-sparrow search algorithm. Electronics 13.

Speirs, J., Contestabile, M., Houari, Y., Gross, R., 2014. The future of lithium availability for electric vehicle batteries. Renew. Sustain. Energy Rev. 35, 183–193.

Taabodi, M.H., Niknam, T., Sharifhosseini, S.M., Asadi Aghajari, H., Shojaeiyan, S., 2025. Electrochemical storage systems for renewable energy integration: a comprehensive review of battery technologies and grid-scale applications. J. Power Sources 641, 236832.

Tang, J., Dysart, A.D., Pol, V.G., 2015. Advancement in sodium-ion rechargeable batteries. Curr. Opin. Chem. Eng. 9, 34–41.

Tola, V., Meloni, V., Spadaccini, F., Cau, G., 2017. Performance assessment of adiabatic compressed air energy storage (A-CAES) power plants integrated with packed-bed thermocline storage systems. Energy Convers. Manag. 151, 343–356.

Tong, Z., Cheng, Z., Tong, S., 2021. A review on the development of compressed air energy storage in China: technical and economic challenges to commercialization. Renew. Sustain. Energy Rev. 135, 110178.

Tüysüz, H., 2024. Alkaline water electrolysis for green hydrogen production. Acc. Chem. Res. 57, 558–567.

Union, I., 2014. Communication from the Commission to the European Parliament, the Council, the European Economic and Social Committee and the Committee of the Regions. Brussels, 202020, 20. http://www.xploit-eu.com/pdfs/Europe.

Vazquez, S., Lukic, S.M., Galvan, E., Franquelo, L.G., Carrasco, J.M., 2010. Energy storage systems for transport and grid applications. IEEE Trans. Ind. Electron. 57, 3881–3895.

Wanison, R., Syahputra, W.N.H., Kammuang-Lue, N., Sakulchangsatjatai, P., Chaichana, C., Shankar, V.U., Suttakul, P., Mona, Y., 2024. Engineering aspects of sodium-ion battery: an alternative energy device for Lithium-ion batteries. J. Energy Storage 100, 113497.

Wehrmann, B., 2024. Curtailing of renewable power increases in Germany in 2023 as re-dispatch costs recede. In: Journalism for the Energy Transition.

Wei, X., Ban, S., Shi, X., Li, P., Li, Y., Zhu, S., Yang, K., Bai, W., Yang, C., 2023. Carbon and energy storage in salt caverns under the background of carbon neutralization in China. Energy 272, 127120.

Wu, Y., Rahm, E., Holze, R., 2003. Carbon anode materials for lithium ion batteries. J. Power Sources 114, 228–236.

Wu, F., Gao, R., Li, C., Liu, J., 2023. A comprehensive evaluation of wind-PV-salt cavern-hydrogen energy storage and utilization system: a case study in Qianjiang salt cavern, China. Energy Convers. Manag. 277, 116633.

Wu, F., Liu, Y., Gao, R., 2024. Challenges and opportunities of energy storage technology in abandoned coal mines: a systematic review. J. Energy Storage 83, 110613.

Xu, Y., Ren, L., Zhang, Z., Tang, Y., Shi, J., Xu, C., Li, J., Pu, D., Wang, Z., Liu, H., Chen, L., 2018. Analysis of the loss and thermal characteristics of a SMES (superconducting magnetic energy storage) magnet with three practical operating conditions. Energy 143, 372–384.

Yabuuchi, N., Kubota, K., Dahbi, M., Komaba, S., 2014. Research development on sodium-ion batteries. Chem. Rev. 114, 11636–11682.

Yang, T., Luo, D., Liu, Y., Yu, A., Chen, Z., 2023. Anode-free sodium metal batteries as rising stars for lithium-ion alternatives. iscience 26.

Yang, D., Wang, Y., Wang, J., Rui, Z., He, W., 2025. Cost-reducing adiabatic compressed air energy storage for long duration energy-storage applications. iScience 28, 113967.

Ye, H., Zheng, G., Yang, X., Zhang, D., Zhang, Y., Yan, S., You, L., Hou, S., Huang, Z., 2021. Application of different carbon-based transition metal oxide composite materials in lithium-ion batteries. J. Electroanal. Chem. 898, 115652.

Zhang, J.W., Wang, Y.H., Liu, G.C., Tian, G.Z., 2022. A review of control strategies for flywheel energy storage system and a case study with matrix converter. Energy Rep. 8, 3948–3963.

Zhang, T., Gao, J., Du, Q., Xie, M., Sun, Q., 2026. Study on the dynamic characteristics of adsorption-type compressed CO_2 energy storage system: based on the cycle self-consistent regulation mechanism of high-pressure storage tank. Renew. Energy 260, 125177.

Zhao, L., Ding, B., Qin, X.Y., Wang, Z., Lv, W., He, Y.B., Yang, Q.H., Kang, F., 2022. Revisiting the roles of natural graphite in ongoing lithium-ion batteries. Adv. Mater. 34, 2106704.

Zhao, H., Zhu, B., Jiang, B., 2025a. Comprehensive assessment and analysis of cavitation scale effects on energy conversion and stability in pumped hydro energy storage units. Energy Convers. Manag. 325, 119370.

Zhao, L., Costa, S.I.R., Chen, Y., Fitzpatrick, J.R., Naylor, A.J., Kolosov, O., Tapia-Ruiz, N., 2025b. A comparative study of solid electrolyte interphase evolution in ether and ester-based electrolytes for {Na}-ion batteries. PRX Energy 4, 033002.

Zhong, Y., Yang, M., Zhou, X., Zhou, Z., 2015. Structural design for anodes of lithium-ion batteries: emerging horizons from materials to electrodes. Mater. Horiz. 2, 553–566.

Energy storage: Fundamentals, methods, and renewable energy applications

7

Taimoor Ali Khan [1, 2], *Tanzeela Asghar* [3, 4], *Yuwalee Unpaprom* [5] *and Rameshprabu Ramaraj* [6]

[1]College of Electrical Engineering and Computer Science, National Taiwan University (NTU), Taipei, Taiwan; [2]Research Center for Information Technology Innovation, Academia Sinica, Taipei, Taiwan; [3]College of Science, National Taiwan University (NTU), Taipei, Taiwan; [4]Institute of Physics, Academia Sinica, Taipei, Taiwan; [5]Program in Biotechnology, Maejo University, Chiang Mai, Thailand; [6]School of Renewable Energy, Maejo University, Chiang Mai, Thailand

7.1 Introduction

The energy shift in the world today is also characterized by high-scale implementation of variable renewable energy, especially solar photovoltaic energy and wind power. Though these are the key resources to the decarbonization process, they are the cause of variability, uncertainty in forecasting, and nondispatchable profile of generation that complicates traditional power-system operations. Due to the increased use of renewables, the system operators need flexibility that is in new kinds, to handle ramping events, reserve demand, and temporary balancing, but without jeopardizing reliability and resiliency (Ai et al., 2024; Lappalainen et al., 2024; Shahzad and Jasińska, 2024).

The storage of energy has become one of the most diverse flexibility tools that can be offered to contemporary energy systems. Storage (based on technology and scale) can offer rapid frequency response, voltage regulation, peak shaving, renewable firming, smooth reserve, congestion decongestion, and cross-timescale energy shifting of seconds-long views up to seasons-long views. Storage in the grids which are rich in renewable means that the curtailment is minimized, the portion of the clean energy can be used, and the operational value of the distributed and utility-scale generation resources is enhanced (Liu et al., 2023; Lund et al., 2025; Reveles-Miranda et al., 2024).

The performance of storage is determined by the hardware performance as well as the effectiveness of asset monitoring, management, and connection to the broader energy networks. The physics-based models and rule-based controllers are still significant; however, they might fail in nonlinear degradation, partial observability, manufacturing variability, and dynamically changing operating conditions. Machine learning and artificial intelligence have complementary prediction, diagnostics, optimization, and adaptive decision support capabilities, especially in cases where the

Energy Storage Technologies. https://doi.org/10.1016/B978-0-443-45858-3.00013-9

amount of operational data is a large stream (Fasogbon et al., 2025; Ukoba et al., 2024).

In this chapter, the multiscale perspective of energy storage is discussed with reference to AI and machine learning. It will discuss the storage basics, AI/ML basics, the intelligent storage data ecosystem, pack-level applications, and device-level applications of renewable energy applications. The batteries are the most actively highlighted as they prevail in recent research on the topic of AI in storage, though supercapacitors, flow batteries, pumped hydro, thermal storage, and hydrogen-based directions are also noted where they are applicable (Ajibade et al., 2023; Kiasari et al., 2024; Yao et al., 2023).

7.1.1 Chapter organization

This chapter proceeds from fundamentals to applications. Section 7.1 reviews storage technologies, performance metrics, modeling foundations, and uncertainty sources. Section 7.2 introduces the AI/ML concepts most relevant to storage engineering. Section 7.3 discusses the data ecosystem and deployment architecture. Section 7.4 examines device-level applications such as SOC/SOH estimation, fault diagnosis, thermal management, and charging control. Section 7.5 addresses system-level renewable energy applications, including forecasting, scheduling, microgrid energy management, and hybrid storage coordination. Sections 7.6–7.8 discuss representative applications, critical challenges, and future research directions, respectively, before Section 7.9 concludes this chapter.

7.2 Fundamentals of energy storage systems

7.2.1 Definition and classification of energy storage

Energy storage refers to the temporary conversion of energy into a retained form so that it can be released when needed. Storage technologies are commonly classified into electrochemical, mechanical, thermal, chemical, and electrical categories. This classification is useful because each class differs in energy density, power density, duration, response time, efficiency, cost, and deployment context (Fig. 7.1), all of which influence both engineering design and the suitability of AI/ML methods (Ajibade et al., 2023; Kiasari et al., 2024; Yao et al., 2023).

7.2.2 Major storage technologies relevant to renewable energy

Electrochemical storage includes lithium-ion, sodium-ion, lead-acid, and flow batteries. Mechanical storage includes pumped hydro, compressed air, and flywheels. Thermal storage encompasses sensible, latent, and thermochemical systems, while chemical storage includes hydrogen and power-to-gas pathways. Supercapacitors occupy a distinct niche due to their very high power density and long cycle life. In renewable energy systems, these technologies are selected according to duration,

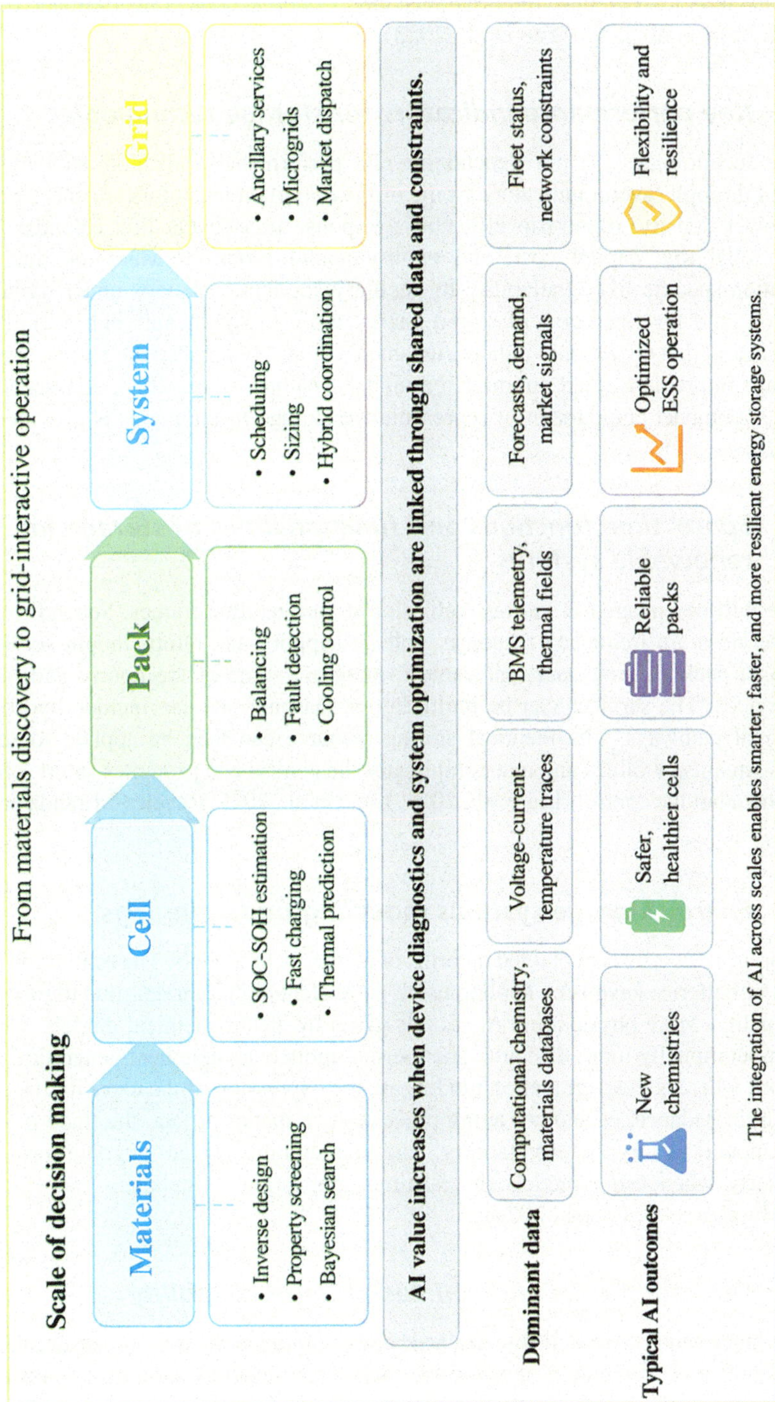

Figure 7.1 Multiscale map of AI applications in energy storage: Materials, cell, pack, system, and grid levels.

site characteristics, response requirements, and economic constraints (Ajibade et al., 2023; Kiasari et al., 2024; Yao et al., 2023).

7.2.3 Key performance indicators for storage technologies

Storage technologies do not depend on one performance indicator, as they are assessed through several indicators. Some of the important indicators are energy density, power density, round-trip efficiency, response time, cycle life, calendar life, safety, reliability, capital cost, and environmental footprint. Measures that are application-specific like availability to ancillary services, behavior under dynamic operation, and levelized cost of stored energy are also significant (Fig. 7.2). They are indicators that do not only indicate technical suitability but also the representation of storage in predictive and optimization models (Ajibade et al., 2023; Kiasari et al., 2024). A compact comparison of representative storage technologies is provided in Table 7.1.

7.2.4 Space–time functions and functionalities of storage in renewable systems

Storage will be engaged in various activities at various time frames. Second to minutes services are frequency response, voltage support, and ramp smooth services. Arbitrage, peak shaving, and renewable firming are some of the hourly and intraday services. The services can be multi-day or seasonal that can include hydrogen or thermal pathways. The identical storing resource can then be applied to serve various streams of values and that complicates the control and promotes smart scheduling and adaptive control (Liu et al., 2023; Lund et al., 2025; Reveles-Miranda et al., 2024).

7.2.5 Energy storage systems modeling underpinnings

The storage of energy is explained in terms of hierarchy of models. The estimation and control of batteries have been applied using equivalent circuit models due to its computationality. More physical information is given by electrochemical models, which are computationally more difficult. The thermal models include heat generation and transport, whereas the degradation models are the expression of the capacity loss, increase in impedance, or another aging process. As a matter of fact, the trade-off between model fidelity, interpretability, and deplorability is an ongoing process, particularly when using models in real-time applications (Ghalkhani and Habibi, 2023; Hussain and Musilek, 2024).

7.2.6 Operation risks and sources of unpredictability

Internal uncertainty, like cell-to-cell uncertainty, degradation, and immeasurable internal state, and exogenous uncertainty, including ambient conditions, renewable

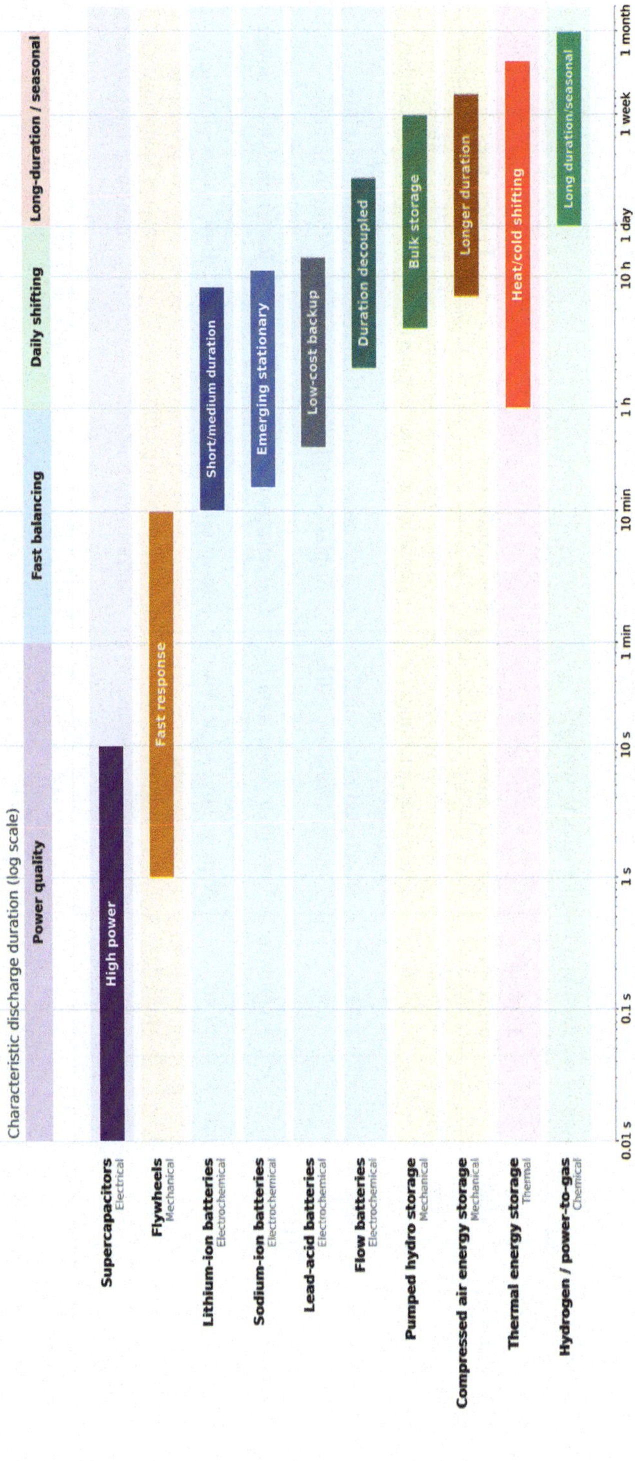

Figure 7.2 Taxonomy of energy storage technologies and characteristic discharge durations relevant to renewable energy systems.

Table 7.1 Succinct comparison of the exemplar technologies in storage.

Technology	Class	Typical duration	Key strengths	Key limitations	Typical uses
Lithium-ion battery	Electrochemical	Minutes to 4+ h	High efficiency, fast response, modular design	Thermal management needs, degradation, critical materials	PV plus storage, microgrids, ancillary services
Flow battery	Electrochemical	2–12 h	Decoupled power and energy, long cycle life	Lower energy density, balance-of-plant complexity	Renewable shifting, longer-duration microgrids
Supercapacitor	Electrical	Seconds to minutes	Very high power, rapid cycling, long life	Low energy density	Ramp smoothing, power quality, hybrid ESS
Pumped hydro	Mechanical	4 h to multiday	Mature, large scale, low specific cost	Site-specific, long lead time	Bulk energy shifting, reserve provision
Hydrogen/ P2G	Chemical	Multi-day to seasonal	Long duration, sector coupling potential	Lower round-trip efficiency, infrastructure needs	Seasonal balancing, industrial integration

variability, load variations, and price volatility, are factors that influence the operation of storage. These uncertainties interrelate with each other in a manner that is hard to describe using rules or parameters that are fixed in place. Subsequently, even an effective type of storage can become inefficient in case its monitoring and decision-making layers are unable to adjust to the changing circumstances (Cordeiro-Costas et al., 2023; Liu et al., 2023; Ghalkhani and Habibi, 2023).

7.2.7 Why are the mentioned traits present AI/ML opportunities

The dynamics of the energy storage are nonlinear, multiscale, time-varying, and multidimensional data streams of sensors and supervisory systems. These characteristics specifically ensure that storage is especially appropriate to the AI/ML applications in estimation, anomaly detection, forecasting, and optimization. Simultaneously, deployment in safety-critical applications must have physically plausible, computationally tractable, and domain shift and sensor noise resistant methods (Fasogbon et al., 2025; Hussain and Musilek, 2024; Ukoba et al., 2024).

7.3 Fundamentals of artificial intelligence and machine learning

7.3.1 Basic concepts and terminology

Artificial intelligence can be described generally as the computational techniques that may simulate the activities related to learning, thinking, or making decisions. Machine learning is a branch of AI that is learned to understand the relationships based on data, whereas deep learning is the neuron architecture of multilayer networks that are able to learn hierarchical representations. The data-driven, model-driven, and hybrid methods are frequently useful in energy storage, as numerous applied systems tend to utilize attained relationships with structure (Maurya et al., 2024; Selvaraj and Vairavasundaram, 2023).

7.3.2 Types of learning problems in energy storage

The storage applications can be projected onto problem classes of standard machine learning. SOC, SOH, RUL, temperature, or power prediction ones are regressed. The fault modes or safety states are classified. Pattern discovery or battery grouping support has been assisted by clustering. As much as normal operation is detected through anomaly detection, anomaly detection occurs in situations where there are few labeled fault data. Time-series forecasting aids in making predictions of renewable, load, and price predictions, whereas sequential decision-making aids in dispatch and charging control (Maurya et al., 2024; Selvaraj and Vairavasundaram, 2023).

7.3.3 Main learning paradigms

The most common paradigm that is used in energy storage is supervised learning, particularly to estimate the state and for predictive maintenance. The unsupervised and semi-supervised approaches prove to be helpful in cases of limited and uncertain labels. Reinforcement learning is appealing in the operation of a storage system since it is able to acquire policies over time in unpredictability. The new paradigms include transfer learning, federated learning, and physics-informed learning that are made to enhance the generalization, maintain privacy, or inject domain constraints in the learning process (Maurya et al., 2024; Selvaraj and Vairavasundaram, 2023; Zhao et al., 2024).

7.3.4 Common algorithm families

Linear and nonlinear regression, support vector, decision tree, random forest, gradient boosting, feedforward neural network, CNNs, RNNs, LSTM, GRU, autoencoders, and transformers were selected as common algorithms when working with storage-related AI/ML. Dependent on the characteristics of the input data, the quality of labels, interpretability needs, tolerance to uncertainty, and computing error, this choice of method will be made (Borah et al., 2024; Maurya et al., 2024; Selvaraj and Vairavasundaram, 2023; Zhao et al., 2024).

7.3.5 Model development workflow

A good model development process would consist of well-defined problem formulation, data collection process, preprocessing, feature construction, training-validation-testing, and hyperparameter optimization process. In time-series applications, a chronological split is typically much better than a random split, in that it is more realistic in terms of the deployment situation. The other aspects that should be evaluated in storage systems include the aging, seasonal drift, change in operating regime, and the cost of prediction errors when operating within actual constraints (Maurya et al., 2024; Selvaraj and Vairavasundaram, 2023).

7.3.6 Performance evaluation metrics

Mean absolute error, root square error, mean absolute percentage error, or coefficient of determination are widely used in order to estimate the regression tasks. Such metrics needed to classify include precision, recall, F1-score, confusion matrices, and ROC-AUC. In the case of probabilistic forecasting, probabilistic and sharpness are required. Statistical accuracy is not enough in the storage applications and inference latency, computer cost, robustness, and engineering implications should be put into consideration (Borah et al., 2024; Zhao et al., 2024).

7.3.7 Explainability, robustness, and trustworthiness

Black-box behavior may be an issue in the safety-critical storage applications like thermal management, fault diagnostics, or protection support. The tools of explainability, feature-importance tests, quantification of uncertainty, and the combination of physical constraints should hence be significant toward credible deployment. One of the practically useful AI models is the one that can solve and perform under such circumstances as dirty inputs, variable operating conditions, and partial sensor degradation, in addition to being human-understandable and auditable (Borah et al., 2024; Zhao et al., 2024). In this chapter, Fig. 7.3 and Table 7.2 give a brief conceptual overview of the workflow and method-task mapping that was used.

7.3.8 Data acquisition and sensing infrastructure

AI/ML is only useful to a certain extent based on the manner in which data are gathered. The strategies of sensor placement, sensor calibration, sampling rate, time synchronization, and communication architecture have an effect on the model quality. In a multicomponent system of renewable energy, the data will frequently be received in various subsystems and with varying time resolutions and this poses a challenge in integrating the data. The model performance with respect to the high-fidelity sensing can be considerably impaired in terms of poor synchronization or low-fidelity sensing conditions, even though state-of-the-art algorithms are applied (Naseri et al., 2023; Ward et al., 2022; Zhang et al., 2023).

7.3.9 Preprocessing and quality control of data

Raw storage data normally need denoising, outlier removal, missing-value treatment, normalization, and resampling prior to their application in machine learning. These

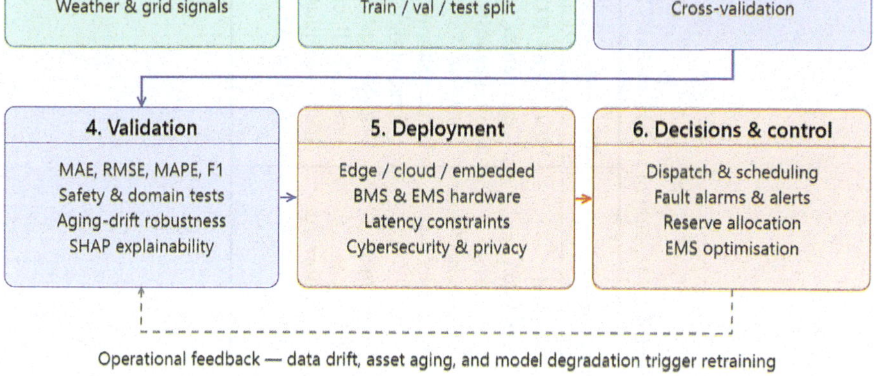

Figure 7.3 Generic AI/ML workflow for energy storage analytics and control, including data acquisition, preprocessing, model training, validation, deployment, and feedback.

Table 7.2 Mapping of AI/ML methods to storage problems, representative inputs, outputs, and evaluation metrics.

Method family	Example storage tasks	Typical inputs	Typical outputs	Common metrics
Supervised regression	SOC, SOH, RUL estimation	Voltage, current, temperature, cycle history	Continuous state or life estimate	MAE, RMSE, MAPE
Classification/ anomaly detection	Fault diagnosis, safety alarms	Residuals, thermal signatures, impedance features	Fault class or anomaly score	Precision, recall, F1-score, ROC-AUC
Sequence models	Load, renewable, price forecasting	Historical time series, weather, calendar data	Multi-step forecast	RMSE, MAPE, calibration
Reinforcement learning	Dispatch, fast charging, reserve allocation	System state, forecast summary, tariff state	Control action or policy	Reward, cost, constraint satisfaction
Physics-informed/ hybrid	State estimation, thermal prediction	Sensor data plus model states	Physically constrained prediction	Error metrics, robustness, plausibility

are steps that should be informed physically. As an example, signatures of impending faults can be flattened by aggressive smoothing, and every indication of a transient aspect can be produced by naive interpolation. Good preprocessing is not just a statistical task; it is really an engineering step that is domain-specific (Ward et al., 2022; Zhang et al., 2023).

7.3.10 Learning features and features engineering

The concept of feature engineering has become very useful with respect to storage applications. They can be time-domain statistics, electrochemical signatures, temperature gradient, relaxation features, cumulative throughput, and frequency-domain descriptors. All of them can enhance the prediction performance and interpretation. Meanwhile, some deep learning techniques are capable of automatic features extraction in raw sequence or images more often. Expert-based hybrid strategies using learned and expert based strategies are usually effective particularly when labeled datasets are scarce (Ward et al., 2022; Zhang et al., 2023).

7.3.11 Data set limitations and applicable constraints

The storage datasets cannot satisfy the requirements of storage system despite their rapid development due to the lack of public data, completion labels, biased classes, limited chemistries, and unfavorable representation of unusual faults or degradation over time. Laboratory databases may be clean and unrealistic, whereas field databases may be realistic, noisy, and weakly labeled. Such an incongruity provides great difficulty with generalization and benchmarking (Ward et al., 2022; Zhang et al., 2023).

7.3.12 Digital twins, synthetics, and simulation

Artificial data and simulation are also becoming popular in order to supplement meager measurements, investigate rare events, and aid in the analysis of scenarios. Digital twins are further developed to give constantly updated virtual models of tangible storage materials. Such tools are particularly well used in size predictive diagnostics and control, but must be as effective at collecting degradation, thermal behavior, and system-level interactions as they can reach. The biases may also be brought about due to poorly calibrated synthetic environments (Naseri et al., 2023).

7.3.13 Deployment architecture

Each of the aforementioned Software deployment architectures provides other tradeoffs. Knowledge computing and fleet-level learning are supported on the cloud. Edge solutions offer a lower latency rate and communication requirements. Embedded systems are also mandatory in BMS or EMS equipment (hardware), but would demand strict memory and processing constraints. Model complexity, retraining strategy, and cybersecurity exposure, as well as maintenance requirements, are consequently determined through deployment architecture (Naseri et al., 2023).

7.3.14 Security data administration and cyber-physical matters

With the increase in the level of storage being digitized, the questions of privacy, interoperability, cybersecurity, data ownership, and standardization gain heightened significance. The interconnected storage resources can be exposed to the threats of data manipulation, networking malfunctions, or physical attacks due to cyber-attacks. Reliable AI implementation is thus a concern not only based on model output but also secure and connector-ubiquitous, information infrastructure (Hu et al., 2020; Naseri et al., 2023; Xu et al., 2023).

7.4 AI/ML at the component, cell, and pack levels

7.4.1 Materials and component discovery using AI

The materials discovery and component design is also increasingly applicable to steel-based, especially electrodes, electrolytes, and separators, as well as solid-state structures, with the usage of AI. The screening of design space that is high dimensional can be made faster with the help of surrogate models, Bayesian optimization, and generative methods and can be lessened through exhaustive experimental trials. AI at this level helps in reducing design phases and more focus on experiments instead of laboratories being used (Cao et al., 2025; Shang et al., 2024).

7.4.2 Battery and other storage establishment state estimation

One of the best developed AI applications is used in storage through state estimation. SOC, SOH, SOP, and SOP cannot be directly measured using simple sensors, but they are key to efficient operation having a safe environment. Nonlinear vectors between voltage, current, temperature, and history of operations can be learned with regardless of the data with the use of ML-based estimators which is complemented with either equivalent-circuit structure or electrochemical structure to enhance robustness (Hu et al., 2020; Xu et al., 2023).

7.4.3 Use case degradation modeling and prognostics

The objectives of degradation prognostics are to predict the loss of capacity, increment in resistance or the remaining useful life at varying usage conditions and environmental conditions. By analyzing the cycling data, impedance CS severity indicators of long-term degradation can be identified using AI/ML, where reduced-order physics models fail to capture such long-term phenomena. These predictions are particularly useful when they are incorporated in the control strategies where the wear of assets is clearly taken into consideration (Xu et al., 2023).

7.4.4 Fault diagnosis and anomaly detection

Storage defects can be sensor failure, internal short circuit, loose interconnections, cell imbalance, abnormal self-discharge, and coolant problems. Due to the rather small number of labeled fault data, one-class classification, autoencoders, and hybrid diagnostic models are also popular. The AI-based diagnostics will be able to enhance the ability to warn in good time, yet it is necessary to implement them as a component of the layered protection systems instead of considering them as the safety functions (Cao et al., 2025).

7.4.5 Health and safety, thermal regulation

Temperature has a high impact on the performance, degradation, and safety within the electrochemical storage. AI/ML may be used to aid in temperature estimation, prediction of hotpowers, optimization of the cooling system, as well as in the detection of dangerous thermal behavior at an early stage. Thermal-based models, sensor fusion, uncertainty awareness, and conservative supervisory logic have the most plausible methods of combining them, particularly in situations where thermal runaway risk would need to be controlled in real time (Cao et al., 2025; Shang et al., 2024).

7.4.6 Optimizing policy fast charging and charging

The high charge rate enhances accessibility but heightens the chances of high degradation rate, lithium deposition, and heat. The AI-based charging policies such as adaptive control and the reinforcement learning aim to stabilize the time charging and the limitations of aging and safety. Ideally, the most appropriate cases of such methods lie within the limits of physically acceptable current, voltage, and temperatures (Doan et al., 2024).

7.4.7 Resolving also involves cell balancing and cell pack management

Cell heterogeneity, thermal variations, as well as mismatch, which builds up with time, are limiting factors to pack-level performance. The AI/ML may be useful in predicting the evolution of the imbalance, detection of limiting cells, active balancing coordination, and optimization of pack-level functioning during nonuniform aging. This is especially necessary in high battery capacity without any external access and old battery packs where preliminary heterogeneity is an important parameter (Singh and Gupta, 2023).

7.4.8 Comparative evaluation of device-level AI

A device level does not have a universal best method of AI. Rich networks can also do well with big data and be challenging to implement on limited devices. The classical approach to machine learning can be more comprehensible and computationally runnable but feature engineering is frequently critically needed. Hybrid methods usually present a reasonable tradeoff as they take advantage of data usage without violating physical constraints and reasoning of safety (Cao et al., 2025; Doan et al., 2024; Shang et al., 2024; Singh and Gupta, 2023). The key areas of deployment and trade-off of deployment that are recurrent and used across many device-level applications are summarized in Table 7.3. Also, Fig. 7.4 demonstrates the overall pipeline.

Table 7.3 Device-level applications of AI/ML in batteries and related storage technologies.

Application	Representative targets	Typical methods	Deployment focus	Main cautions
State estimation	SOC, SOH, SOP, SOE	ANN, LSTM, Kalman-ML hybrids	BMS accuracy and usable energy	Domain shift with aging
Degradation prognostics	Capacity fade, resistance growth, RUL	Gradient boosting, LSTM, Bayesian models	Lifetime-aware control	Need long-horizon labels
Fault diagnosis	Sensor faults, shorts, abnormal self-discharge	Autoencoders, one-class models, classifiers	Early warning and root-cause screening	Rare-fault class imbalance
Safety and thermal management	Hotspots, runaway risk, cooling demand	PINN, surrogate thermal models, anomaly detection	Real-time supervisory support	False negatives are costly
Fast charging	Charging current profile, target charge time	RL, adaptive control, surrogate optimization	Speed-lifetime-safety trade-off	Guardrails required
Pack management and balancing	Cell mismatch, balancing priority	Clustering, optimization, predictive control	Heterogeneity-aware pack operation	Embedded compute limits

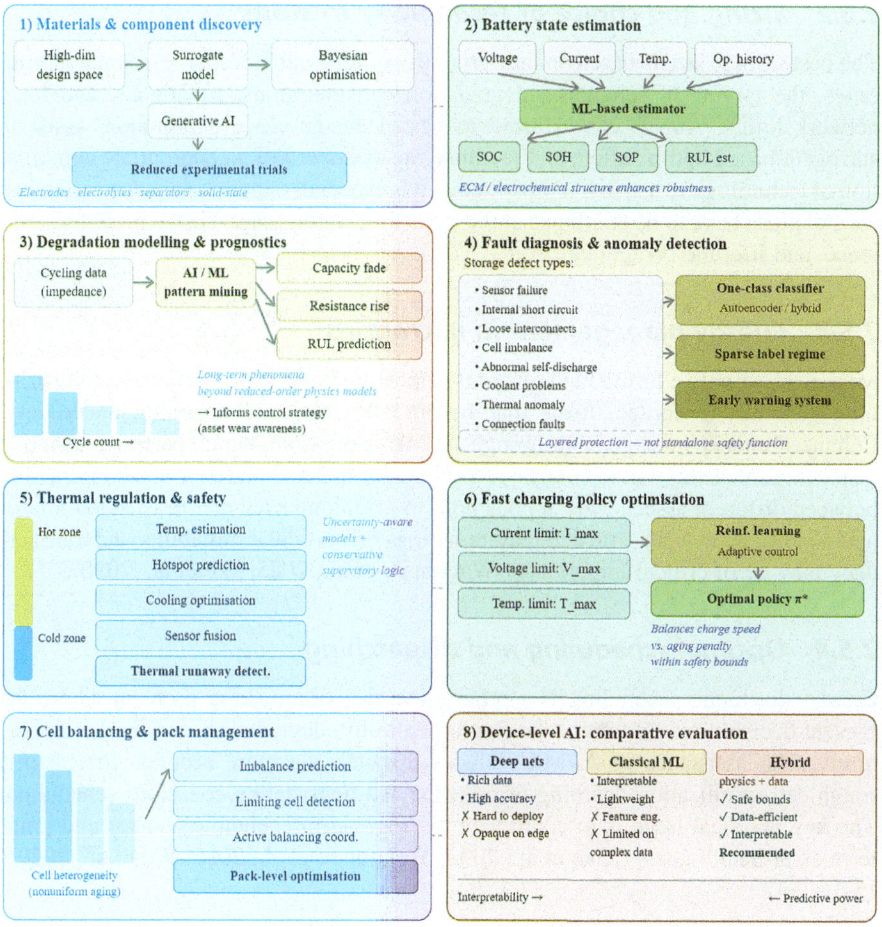

Figure 7.4 Overview of an AI/ML-enabled battery management and optimization pipeline across the battery lifecycle.

7.5 AI/ML for system-level energy storage in renewable energy contexts

7.5.1 Forecasting tasks that support storage operation

Solar generation, wind power, demand, price, and network forecasts are all heavily relied upon to make decisions at the system level, that is, the storage decisions. Nonlinear temporal relationships and heterogeneous predictors in the AI/ML model are able to include weather forecasts, calendar variables, historical loads, and market signals (Fig. 7.4). The quality of the forecast has a direct impact on storage, reserve, and economic performance, particularly in systems that are highly renewable (Xiao et al., 2023).

7.5.2 Sizing and choice of technology to store

The best storage size is based on the variability of renewable resources, demand properties, the targets of reliability, degradation considerations, economics, and local network limits. AI/ML can be used to hasten design-space exploration, assist in surrogate-based optimization, and facilitate the discovery of an appropriate combination of technology to use in hybrid systems. It is especially applicable in the case when several goals have to be combined at the same time, in this case, cost, emissions, resilience, and lifetime (Waghmare et al., 2025).

7.5.3 Energy management in microgrids

Microgrids supply a fertile field of smart storage in that it is a combination of distributed generation, storage, loads that become flexible, and operating uncertainty. Enhanced energy management through AI/ML is possible, which performs adaptive dispatch, forecasting-based control, recognition of demand patterns, and coordination between different assets. This applies both to grid-connected and to islanded microgrids, although in weak-grid and remote settings, in particular, reliability and transparency may be of critical importance (Waghmare et al., 2025; Wu et al., 2020).

7.5.4 Optimal scheduling and dispatching

Charge–discharge scheduling is a progressive decision-making problem where the present decision would impact the future flexibility, degradation, and market opportunity. The assistance of AI/ML includes dispatch via more accurate forecasting, rough-cut optimization, learning of policies, and degradation-conscious scheduling. The key practical issue is to coordinate the short-term economic choices and long-term asset-health goals (Xiao et al., 2023; Waghmare et al., 2025; Wu et al., 2020).

7.5.5 Hybrid renewable-storage systems

The hybrid battery-supercapacitor, battery-hydrogen, or battery-thermal are also some of the renewable applications of hybrid storage. These systems can be coordinated with AI/ML that assigns tasks in regard to the timescale, efficiency, state, and wear features. This enables the high-power briefs, daily balancing, and long-term storage to be assigned to the most appropriate component in a coordinated system (Tang et al., 2023).

7.5.6 Control and reinforcement learning

The importance of reinforcement learning is due to storage control as it inherently involves sequential decision-making in the face of uncertainty. Over time, RL and deep reinforcement learning can learn policies of charging, discharging, reserve allocation, and dispatch. But in reality, operational deployment, there must be good reward design, a good training environment, and solid safety guardrails. In most instances,

RL can be regarded as supplementary to MPC and other constrained control strategies, in addition to not being a full substitute (Wu et al., 2020).

7.5.7 Storage supporting grid and ancillary services

Storage on a grid-scale is becoming more and more useful as a source of frequency regulation, voltage support, spinning reserve, black start, and congestion management. Artificial intelligence/machine learning can be used to enhance these services with the prediction of availability, headroom estimation, fleet coordination, and service stacking. These applications require an accurate state of assets and a high level of integration between both the diagnostics that occur at the device-level and the control that occurs on the system-level (Liu et al., 2023; Lund et al., 2025; Reveles-Miranda et al., 2024).

7.5.8 EVs and second-life battery

New opportunities for AI-enabled storage arise with the electric vehicle charging and second-life stationary battery. AI/ML may assist in the coordination of chargers, prediction of arrival, load leveling in the sites, possible vehicle-to-grid engagement, and health evaluation of used batteries. Such applications are particularly critical to the distributed energy systems, where the uncertainties in behavior and the varying status of batteries are key factors (Reveles-Miranda et al., 2024).

7.5.9 The economic and environmental assessment at the system level

Technological outcomes, instead of accuracy in prediction, should be used in determining the value of AI-enabled storage. With more efficient scheduling, prediction and health-conscious control of revenue, curtailment, lifetime, and utilization of assets will be enhanced. Carbon-conscious operation and lifecycle-sensitive dispatch offer other opportunities to balance the storage regulation and the objectives of sustainability (Ma et al., 2024). Fig. 7.5 and Table 7.4 summarize the system architecture and application contexts discussed in this section.

7.6 Representative applications and case studies

7.6.1 Solar PV plus battery energy storage in residential and commercial systems

PV-battery systems are widely deployed in the residential and commercial environment to achieve the self-consumption maximization, decrease the demand charge, and enhance resilience. The AI/ML is usually used in load prediction, photovoltaic forecasting, and adaptive scheduling (Fig. 7.5). Remarkably small improvements in prediction can render behind-the-meter storage very useful in situations where decisions are activated every day with regard to long terms of operation (Ma et al., 2024).

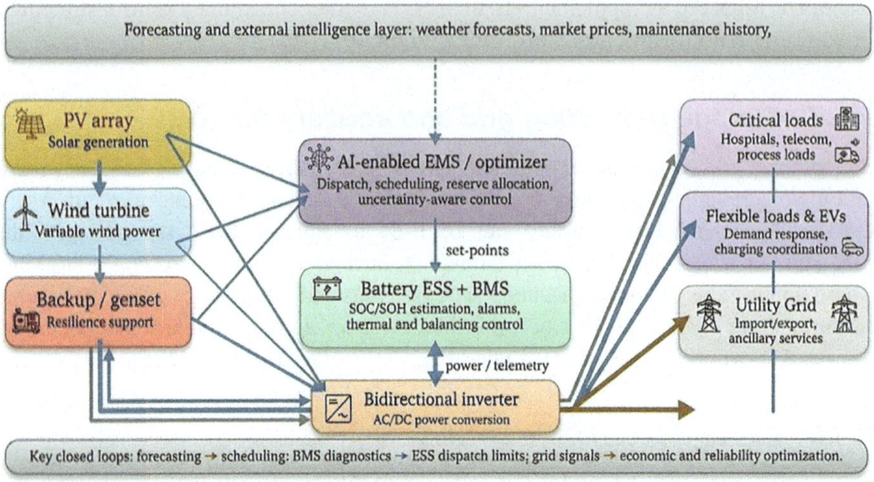

Figure 7.5 Architecture of an AI-enabled renewable microgrid with storage, forecasting inputs, supervisory control, BMS/EMS layers, and grid interaction.

7.6.2 Wind-storage hybrid systems

The wind-storage systems are reliant on storage to even out the variation and curtailment and give higher dispatchability. The AI/ML makes contributions based on wind prediction, ramp-event prediction, and trade-off between the smoothing aims of short-term and over a longer horizon, Dispatch goals. The paper is an indicator of the significance of the multiscale decisions that are made during the period of meteorological uncertainty (Ma et al., 2024; Xiao et al., 2023).

7.6.3 Microgrids intended to be renewable in remote and islanded communities

Solar, wind, diesel generators, and batteries are frequently installed together on remote and islanded microgrids with high reliability criteria. AI/ML has the potential to assist in demand prediction, predictive maintenance, fault capture and robust energy control. Under such conditions, simplicity, maintainability, and robustness might have a larger value in comparison with the optimal application of algorithm sophistication (Ma et al., 2024; Waghmare et al., 2025).

7.6.4 Integration of renewable energy storage in utility scale

Storage can be used at utility scale at firming, relieving the congestion, and providing reserves as well as market arbitrage. With the assistance of AI/ML, price forecasts, asset diagnostics, and dispatch optimization are united in a single operational strategy. The largest dilemma is how to weigh between the short-term market value and the long-term deterioration and risk (Ma et al., 2024; Tang et al., 2023; Wu et al., 2020).

Table 7.4 System-level renewable energy applications of AI-enabled storage.

Context	AI/ML role	Primary objective	Typical storage assets	Example benefits
PV plus battery	Load and PV forecasting; schedule optimization	Self-consumption and demand-charge reduction	Lithium-ion batteries	Better utilization and cost savings
Wind-storage hybrid	Ramp forecasting and smoothing dispatch	Curtailment reduction and firming	Li-ion or hybrid ESS	Improved dispatchability
Renewable microgrid	Demand forecasting and resilient EMS	Reliability and fuel displacement	Batteries, diesel hybrid, supercapacitor	Lower diesel use and better resilience
Utility-scale storage	Market forecasting and degradation-aware bidding	Arbitrage and ancillary services	Grid-scale battery systems	Revenue stacking with asset protection
EV charging hub	Arrival prediction and coordinated charging	Peak shaving and power quality	Battery or battery-supercapacitor	Reduced grid stress
Battery-hydrogen hybrid	Multi-timescale coordination	Daily plus long-duration balancing	Battery plus electrolyzer/hydrogen	Greater flexibility across sectors

7.6.5 EV charging hub, with renewable generation and storage

The charging stations that have local renewables and storage are characterized by high-power peaks, doubtful arrivals, and reduced connection capacities to the grid. AI/ML could be used to assist in charging coordination, demand forecasting, queue-conscious power management, and setting up of hybrid storage. This app explains why user behavior can become a significant factor in technical optimization (Ma et al., 2024).

7.6.6 Hydrogen-battery hybrid designs of long-term storage

The hydrogen battery systems balance short periods of electrochemical activity with longer periods of storage. AI/ML will be able to organize the timing of batteries to achieve faster balancing and the moments of directing the excess electricity to hydrogen or subsequent conversion. These systems apply to the sector coupling and multi-energy optimization (Borah et al., 2024).

7.6.7 Cross-case synthesis

There are a number of themes that are shared across these cases. To begin with, the quality of the forecasting has a potent effect on the storage value. Second, the operation of the system should not be displayed separately because device health should be part of the system. Third, isolated assets do not always perform better than hybrid ones whenever several timescales are considered. Lastly, the success of the deployment is extremely reliant on the quality of data, interpretability, and operational realism as opposed to the accuracy of benchmarks themselves (Ma et al., 2024).

7.7 Challenges, limitations, and critical issues

7.7.1 Data-related limitations

Lack of quality with respect to quantity, public, and representative storage data, where little has been done yet, is one of the greatest setbacks toward progress. The rare faulting, long-horizon aging, cross-climatic variation, and realistic field environment are underrepresented, especially. This limits the development of an algorithm as well as fair benchmarking (Ward et al., 2022; Zhang et al., 2023).

7.7.2 The transferability and model generalization

Models that have been trained on a single chemistry, manufacturer, duty cycle, or site can easily fail to achieve performance when this is transferred to another site. This inability to generalize is particularly troublesome in storage since the systems change as they grow and work in different conditions of environments and market rules. It has high potential, and transfer learning and domain adaptation are two methods to address the problem, and a strong cross-domain performance is a significant unresolved issue (Borah et al., 2024; Fasogbon et al., 2025).

7.7.3 The interpretability and transparency of decisions are considered in section

There might be reluctance among the operators and asset owners to have black-box models that change the charging limits and alarms or maintenance actions. The ability to explain and audit is consequently required, particularly in industrial and utility environments where accountability of operations is important. The issue of interpretability has to be a requirement of design and not an afterthought (Fasogbon et al., 2025; Yao et al., 2023).

7.7.4 Real-time implementation constraints

Numerous models of high-performing research are constructed offline with generous assumptions of computations. Actual implementation has to work within a very narrow band on memory, latency, bandwidth, and embedded processing capacity. Storage-related practical AI needs thus to have efficient architectures, model compression, and deployment-considerable design (Fasogbon et al., 2025; Zhou et al., 2024).

7.7.5 Safety, reliance, and fail-safe design

Storage systems also have safety-critical issues, especially when used in high-energy battery packs and grid-support applications. When a false negative happens in the detection of an anomaly or in the recommendation of an unsafe charge, the consequences can be disproportionate. The supervisory logic as well as the uncertainty awareness and traditional protection layers must therefore be coupled to AI models (Fasogbon et al., 2025; Shang et al., 2024).

7.7.6 Benchmarking and reproducibility

The comparison analysis is not an easy task since various studies tend to analyze varied datasets, preprocessing pipelines, metrics, and definitions. In the absence of standard benchmarks and reportability, one can hardly differentiate genuine improvement in methodology-related issues from tuning of the dataset at hand. Dependable advancements consequently entail the existence of open data and reusable processes (Zhang et al., 2023).

7.7.7 Cybersecurity and privacy

Cyber risks are more consequential as storage is increasing in connection. Incorrect information injection, model pollution, and interrupting communications might result in a poor operation or dangerous behaviors. Privacy is also important when setting up behind-the-meter and fleet where the data generated during the choice of operations shows the behavior of the users or proprietary information of the system (Fasogbon et al., 2025; Zhou et al., 2024).

7.7.8 Sustainability of the AI itself

The use of AI is characterized as a facilitator of sustainability, whereas the training and implementation of complex models require the use of computational resources as well. The cost of having complex models, both in terms of the environment and the economy, must be counterbalanced by realistic benefits. Smoother and more powerful models, rather than a little more accurate and very resource-consuming ones, may suit many applications (Fasogbon et al., 2025).

7.7.9 Commercialization and regulatory barriers

Success in algorithms is not all that is required to make an industrial deployment. The storage solutions offered by the AI solutions should meet criteria in the areas of standards, certification, interoperability, warranty, cybersecurity, and bankability. Such barriers are biased toward the approaches that are open, tested, and can be scheduled in terms of long functional periods of operation (Fasogbon et al., 2025). Table 7.5 accentuates the main challenges and a number of indicative mitigation directions set.

Table 7.5 Challenges, limitations, and indicative research priorities for AI/ML in energy storage.

Challenge	Why it matters	Indicative mitigation paths
Data availability and quality	Weak labels and narrow datasets limit trustworthy training	Shared benchmarks, better sensing, and metadata standards
Generalization and transferability	Models fail across sites, chemistries, and climates	Transfer learning, domain adaptation, and hybrid physics
Interpretability and governance	Operators need auditable decisions in safety-critical systems	Explainable AI, human-in-the-loop review, and model documentation
Real-time deployment	Embedded hardware and latency constrain usable models	Edge optimization, model compression, and hierarchical control
Safety and cybersecurity	Data tampering or unsafe actions can propagate physically	Defense-in-depth, anomaly monitoring, and fail-safe logic
Benchmarking and standards	Inconsistent protocols slow comparison and adoption	Open testbeds, standardized metrics, and interoperable workflows

7.8 Emerging research directions

7.8.1 Physics-informed and hybrid AI models

The possibility of physics-informed and hybrid models should be taken as one of the brightest perspectives of storage AI. This can be done by using submethods to enhance generalization, interpretability, and safety through embedding conservation laws or trends in degradation, or attainable operating constraints in learning algorithms. They particularly are appealing in storage since data-driven models are purely data-based and are not always resistant when leaving the training field (Borah et al., 2024).

7.8.2 Uncertainty-sensitive and risk-sensitive learning

It is also true that future storage intelligence will demand probabilistic intelligences more than unitary intelligences. Safety-conscious techniques could be used to charge, allocate the reserves, and plan the maintenance much more favorably. Control involving the consideration of risk is especially applicable in renewable systems when elusive meteorological and market uncertainty affect them (Borah et al., 2024).

7.8.3 Digital twins and closed-loop intelligent control

Digital twins are becoming real-time decision support systems, which integrate physical models, sensor streams, and AI-based analytics together. Digital twins can aid in predicting maintenance in storage systems, estimation of operating state, detection of faults, and optimization using a scenario. Their complete worth will determine how accurate they will be in the long run with the aging of systems and change of the operating conditions (Borah et al., 2024; Naseri et al., 2023).

7.8.4 Transfer learning, federated learning, and distributed intelligence

At a fleet level, a storage deployment will open prospects to cross-site learning and does not entail the complete centralization of data. Transfer learning will enable the decrease in the burden of retraining, whereas federated learning will be able to save privacy and data ownership. The strategies have great potential in the case of utility fleets, EV charging networks, and distributed residential storage portfolios (Borah et al., 2024).

7.8.5 Younger generations: Independent experimentation and materials discovery speed

Materials material in the materials level, there is a high chance that alterations via AI-guided experimentation and closed-loop optimization, explorations to safer, cheaper, and more sustainable storage chemistries can be faster. Robotic experimentation combined with surrogate modeling and adaptive search can have the advantage of reducing the development time and allow a company to conduct a faster cycle of materials design and device validation (Borah et al., 2024; Fasogbon et al., 2025).

7.8.6 Multienergy and sector-coupled systems

The storage of energy infrastructures will be activated in the multienergy blueprint, where electricity will be linked to heat, hydrogen, transport, and buildings. The AI/ML can be used to organize these infrastructures that are linked together in time spans and goals to take location smarts out of individual elements and combine that with information about systems at a large scale (Borah et al., 2024; Ma et al., 2024; Tang et al., 2023).

7.8.7 Second-life pathways and the circular economy: the second-life cycle

Using AI, the consumption of used batteries and other storage parts will probably undergo increased sorting, diagnosing, repurposing, and recycling decisions. The safe and cost-effective redeployment of data-driven residual capacity and risk evaluation is needed in the second-life applications. This makes AI lie within its operations or its lifecycle stewardship (Borah et al., 2024; Ma et al., 2024).

7.8.8 Toward scalable, standards-based, and trustworthy deployment

The success of AI in the long term in storage would be determined by the availability of open benchmarks, interoperable platforms, secure data channels, interfaces where operators are the primary users, and reputable validation practices. The field is no longer concerned with proof-of-concept models, but is increasingly concerned with issues of scalability, governance, and long-term real-world values (Borah et al., 2024; Fasogbon et al., 2025).

7.9 Conclusion

AI and ML are transforming energy storage through the entire value chain with solutions emerging in materials discovery and cell design, to pack-level diagnostics and system-level control. Their core value is to transform the complicated and dynamic operational information into operational insight that is actionable to promote estimation operations, diagnostics, as well as forecasting, scheduling and optimization. The importance of AI in storage in the case of renewable energy systems is more about optimization of storage as an adaptive, context-aware and economically optimal asset of flexibility as well as enhanced predictive accuracy. This is even more significant when power systems are decarbonizing, decentralizing, and digitizing. The further advancement will be determined by the combination of quality data, selected algorithms, physical reliability, cybersecurity, reproducible benchmarking, and deployable digital infrastructure. The best answers will probably be hybrid, credible, and sustainable long-term integration of operations as opposed to shallow benchmarks functioning.

References

Ai, Q., Xiang, J., Liu, Y., Qu, L., Cao, J., Li, X., Wang, Y., 2024. Multi-scenario flexibility assessment of power systems considering renewable energy output uncertainty. Front. Energy Res. 12, 1359233.

Ajibade, S.-S., Zaidi, A., Al Luhayb, A.S.M., Adediran, A.O., Voumik, L.C., Rabbi, F., 2023. New insights into the emerging trends research of machine and deep learning applications in energy storage: a bibliometric analysis and publication trends. Int. J. Energy Econ. Pol. 13 (5), 303–314.

Borah, M., Wang, Q., Li, W., Jiang, S., Shi, J., Moura, S., 2024. Synergizing physics and machine learning for advanced battery management. Commun. Eng. 3, 134.

Cao, R., Zhang, Z., Shi, R., Lu, J., Zheng, Y., Sun, Y., Liu, X., Yang, S., 2025. Model-constrained deep learning for online fault diagnosis in Li-ion batteries over stochastic conditions. Nat. Commun. 16, 56832.

Cordeiro-Costas, M., Villanueva, D., Eguía-Oller, P., Granada-Álvarez, E., 2023. Intelligent energy storage management trade-off system applied to deep learning predictions. J. Energy Storage 61, 106784.

Doan, N.Q., Shahid, S.M., Choi, S.J., Kwon, S., 2024. Deep reinforcement learning-based battery management algorithm for retired electric vehicle batteries with a heterogeneous state of health in BESSs. J. Energy Storage 84, 110881.

Fasogbon, S.K., Fetuga, I.A., Oyeniran, A.T., Shaibu, S.A., Afolabi, S., Ndokwu, T., 2025. Artificial intelligence applications for energy storage: a comprehensive review. Energies 18 (17), 4718.

Ghalkhani, M., Habibi, S., 2023. Review of the Li-ion battery, thermal management, and AI-based battery management system for EV application. Energies 16 (1), 185.

Hu, X., Xu, L., Lin, X., Pecht, M., 2020. Battery lifetime prognostics. Joule 4 (2), 310–346.

Hussain, A., Musilek, P., 2024. Energy management of buildings with energy storage and solar photovoltaic: a diversity in experience approach for deep reinforcement learning agents. Energy AI 15, 100313.

Kiasari, M., Ghaffari, M., Aly, H.H., 2024. A comprehensive review of the current status of smart grid technologies for renewable energies integration and future trends: the role of machine learning and energy storage systems. Energies 17 (16), 4032.

Lappalainen, K., Sherrill, R., Gevorgian, V., Krad, M., et al., 2024. Status of Power System Transformation: Leading Topics of 2024. National Renewable Energy Laboratory, Golden, CO. NREL/TP-6A20-91357.

Liu, X., Zhao, T., Deng, H., Wang, P., Liu, J., Blaabjerg, F., 2023. Microgrid energy management with energy storage systems: a review. CSEE J. Power Energy Syst. 9 (2), 483–504.

Lund, H., Thellufsen, J.Z., Østergaard, P.A., Sorknæs, P., Skov, I.R., Mathiesen, B.V., 2025. Role and trends of flexibility options in 100% renewable energy system analyses towards the Power-to-X Economy. Renew. Sustain. Energy Rev. 213, 115565.

Ma, R., Tao, S., Sun, X., Ren, Y., Sun, C., Ji, G., Xu, J., Wang, X., Zhang, X., Wu, Q., Zhou, G., 2024. Pathway decisions for reuse and recycling of retired lithium-ion batteries considering economic and environmental functions. Nat. Commun. 15, 9.

Maurya, V., Saini, D.K., Yadav, M., Pal, N., 2024. A study of different machine learning algorithms for state of charge estimation in lithium-ion battery pack. Energy Storage 6 (1), e658.

Naseri, F., Gil, S., Barberà, C., Schaltz, E., Pedersen, J.L., 2023. Digital twin of electric vehicle battery systems: comprehensive review of the use cases, requirements, and platforms. Renew. Sustain. Energy Rev. 179, 113280.

Reveles-Miranda, M., Ramirez-Rivera, V., Pacheco-Catalán, D., 2024. Hybrid energy storage: features, applications, and ancillary benefits. Renew. Sustain. Energy Rev. 192, 114196.

Selvaraj, A., Vairavasundaram, I., 2023. A comprehensive review of state of charge estimation in lithium-ion batteries used in electric vehicles. J. Energy Storage 72, 108777.

Shahzad, S., Jasińska, E., 2024. Renewable revolution: a review of strategic flexibility in future power systems. Sustainability 16 (13), 5454.

Shang, M., Wang, S., Fan, Y., Fernandez, C., Blaabjerg, F., 2024. Research progress in fault detection of battery systems: a review. J. Energy Storage 98, 113079.

Singh, R., Gupta, M., 2023. Neural network-based energy management system for smart microgrids using deep learning. Energy Convers. Manag. 268, 115931.

Tang, Q., Shi, X., Liu, L., 2023. Blockchain-based decentralized energy management system integrating machine learning for autonomous energy distribution. IEEE Trans. Smart Grid 14 (2), 981–993.

Ukoba, K., Olatunji, K.O., Adeoye, E., Jen, T.-C., Madyira, D.M., 2024. Optimizing renewable energy systems through artificial intelligence: review and future prospects. Energy Environ. 35 (8), 3833–3879.

Waghmare, A.V., Singh, V.P., Varshney, T., Sanjeevikumar, P., 2025. A systematic review of reinforcement learning-based control for microgrids: trends, challenges, and emerging algorithms. Discov. Appl. Sci. 7, 738.

Ward, L., Babinec, S., Dufek, E.J., Howey, D.A., Viswanathan, D.M., Viswanathan, V., Bloom, M.J., Bloom, I.D., Trask, B., Duong, K., Jain, A., Smith, K.A., 2022. Principles of the battery data genome. Joule 6 (10), 2253–2271.

Wu, B., Widanage, W.D., Yang, S., Liu, X., 2020. Battery digital twins: perspectives on the fusion of models, data and artificial intelligence for smart battery management systems. Energy AI 1, 100016.

Xiao, H., Pu, X., Pei, W., Ma, L., Ma, T., 2023. A novel energy management method for networked multi-energy microgrids based on improved DQN. IEEE Trans. Smart Grid 14 (6), 4912–4926.

Xu, X., Li, Z., Hong, S., Zheng, D., Ma, J., 2023. Fast capacity prediction of lithium-ion batteries using aging mechanism-informed bidirectional long short-term memory network. Reliab. Eng. Syst. Saf. 234, 109185.

Yao, Z., Lum, Y., Johnston, A., Mejia-Mendoza, L.M., Zhou, X., Luo, Y., Aspuru-Guzik, A., Sargent, E.H., Seh, Z.W., 2023. Machine learning for a sustainable energy future. Nat. Rev. Mater. 8, 202–215.

Zhang, J., Wang, Y., Jiang, B., He, H., Huang, S., Wang, C., Zhang, R., Han, T., Ouyang, M., 2023. Realistic fault detection of Li-ion battery via dynamical deep learning. Nat. Commun. 14, 5940.

Zhao, J., Feng, X., Pang, Q., Fowler, M., Lian, Y., Ouyang, M., Burke, A.F., 2024. Battery safety: machine learning-based prognostics. Prog. Energy Combust. Sci. 102, 101142.

Zhou, Y., Zhang, X., Khan, A., Liu, Y., 2024. AI, IoT, and big data-enabled energy management system for renewable microgrids: improving reliability and reducing costs. J. Energy Storage 58, 103225.

Optimization techniques for energy storage system design: Models, methods, and applications

8

Tajeddine Khalili[1], *Amal Kadri*[2], *Yosra M'zabiri*[2], *Yassine Benkirane*[3], *Abdelhadi Raihani*[2], *Mohammed Kissaoui*[2], *Bachir Benhala*[3] and *Mourad Fakhfakh*[4]
[1]ENSA of Beni Mellal, Sultan Moulay Slimane University, Beni-Mellal, Morocco; [2]ENSET of Mohammedia, Hassan II University of Casablanca, Mohammedia, Morocco; [3]Sidi Mohammed Ben Abdellah University, Fez, Morocco; [4]University of Sfax, Sfax, Tunisia

8.1 Introduction

The rapid transition toward low-carbon energy systems has led to an unprecedented increase in the penetration of variable renewable energy sources (RES), particularly solar photovoltaic (PV) and wind power (Ullah et al., 2024; Ali et al., 2025). While these technologies are essential to achieving global decarbonization targets, their inherent intermittency and limited predictability introduce significant operational and planning challenges for modern power systems. As a consequence, system flexibility has emerged as a key requirement to ensure reliability, economic efficiency, and secure operation under high renewable penetration levels (Khalili et al., 2025). Energy storage systems (ESS) play a central role in addressing these challenges by decoupling energy generation from consumption in time (Malik et al., 2025). By absorbing surplus energy during periods of high renewable production and releasing it when generation is insufficient or demand is high, storage systems enable renewable integration, mitigate curtailment, and support grid stability (Wu et al., 2019). Among the available storage technologies, battery energy storage systems (BESSs) have attracted particular attention due to their fast response, modularity, scalability, and rapidly declining costs (Chatzigeorgiou et al., 2024).

Despite their increasing deployment across generation, transmission, and distribution levels, the integration of storage systems remains a complex engineering task (Gholami et al., 2024). Storage systems are capital-intensive assets whose technical performance and economic viability strongly depend on design parameters such as power rating, energy capacity, operational strategy, and lifetime utilization (Shahbazbegian et al., 2023). Inadequate sizing or suboptimal operation may result in poor asset utilization, accelerated degradation, or insufficient economic returns, thereby limiting the effectiveness of storage-based solutions. As a result, storage system design has

Energy Storage Technologies. https://doi.org/10.1016/B978-0-443-45858-3.00016-4

increasingly been formulated as an optimization problem (Bamisile et al., 2024). Optimization-based approaches allow the explicit consideration of technical constraints, economic objectives, and system-level interactions, enabling systematic exploration of trade-offs between competing goals such as cost minimization, reliability enhancement, and emissions reduction (da Silva et al., 2023). Over the past decade, a wide range of optimization techniques—ranging from classical mathematical programming to metaheuristic and artificial intelligence (AI)-based methods—have been applied to storage-related design and operation problems (Gomes et al., 2024; Zhang and Strbac, 2025).

Recent literature has provided comprehensive reviews of optimization techniques for hybrid renewable energy systems integrating energy storage, with a strong emphasis on system-level planning and scheduling (Ansari et al., 2023). However, in many of these studies, storage design emerges as a secondary outcome of broader system optimization rather than as a primary object of investigation. Consequently, a unified perspective that explicitly focuses on storage system design, encompassing modeling choices, optimization formulations, and application-driven trade-offs, remains fragmented across the literature. The objective of this chapter is to address this gap by providing a structured overview of optimization techniques for storage system design, with emphasis on the underlying methods, models, and practical applications. Rather than focusing on a single energy system architecture, this chapter adopts a storage-centric perspective, highlighting how different modeling assumptions and optimization strategies influence design outcomes. In addition to reviewing established optimization approaches, this chapter discusses emerging trends and illustrates key concepts through a reproducible case study based on simulated data and Python-based optimization tools.

8.2 Energy storage systems: Technologies and design variables

8.2.1 Overview of energy storage technologies

ESSs cover a wide range of technologies that differ significantly in their physical principles, technical characteristics, and fields of application. From a power system perspective, these technologies are usually classified according to the form in which energy is stored, namely mechanical, electrical, electrochemical, thermal, and chemical storage (Sahoo and Timmann, 2023; Elalfy et al., 2024). This classification is useful because it highlights the fundamental trade-offs between power capability, energy capacity, response time, efficiency, lifetime, and cost, which are central to storage system design. Mechanical energy storage technologies represent some of the earliest large-scale storage solutions deployed in power systems. Pumped storage hydropower remains the most established technology in this category, benefiting from high efficiency, long operational lifetime, and large storage capacity (Killingtveit, 2019).

However, its deployment is strongly constrained by geographical conditions and long development times (Odoi-Yorke et al., 2024). Other mechanical storage options, such as compressed air energy storage (CAES) and flywheels, address different needs. CAES can provide large energy capacities but typically involves complex infrastructure and lower overall efficiency, while flywheel systems are mainly suited for short-duration applications requiring high power and fast response rather than bulk energy storage (AlJassmi et al., 2026). Electrical energy storage technologies, including supercapacitors and superconducting magnetic energy storage, are characterized by extremely fast response times and very high-power density (Shuja et al., 2024). These characteristics make them particularly attractive for power quality improvement and grid stabilization services. Nevertheless, their relatively low energy density and high investment cost limit their use to specific applications where energy duration is not a primary requirement.

Electrochemical ESSs, and in particular BESSs, have experienced the most rapid development in recent years (Dong et al., 2025). This evolution has been driven by technological improvements, economies of scale, and cost reductions, especially for lithium-ion battery technologies (Tasneem et al., 2025). Owing to their modularity and scalability, BESS can be deployed across a wide range of power and energy levels, from small residential systems to large utility-scale installations. In addition to lithium-ion batteries, other battery technologies such as sodium-ion batteries, flow batteries, and lead-acid batteries continue to be investigated for stationary storage applications where specific cost, safety, or lifetime requirements dominate (Piat et al., 2025). Thermal ESSs store energy in the form of heat and are commonly associated with applications such as concentrated solar power plants, district heating networks, and industrial processes (Vallese et al., 2026). Technologies based on sensible heat, latent heat, or thermochemical storage can offer relatively low-cost solutions for long-duration energy storage. However, their integration into electrical power systems is often indirect and application-dependent, which limits their flexibility compared to electrochemical storage solutions (Ong et al., 2024). Chemical ESSs, most notably hydrogen-based storage, provide a pathway for long-term and seasonal energy storage by converting electrical energy into chemical energy through electrolysis (Mulky et al., 2024). The stored energy can later be reconverted into electricity or used directly in other sectors. Despite their potential, these systems currently face challenges related to low round-trip efficiency, infrastructure requirements, and economic competitiveness.

Given the diversity of available storage technologies and their complementary characteristics, recent research has increasingly explored hybrid ESSs that combine multiple storage technologies within a single system (Leon et al., 2020). Such hybrid configurations aim to exploit the advantages of each technology, for example by combining high power capability with large energy capacity. This technological diversity further reinforces the need for systematic optimization approaches capable of identifying appropriate storage solutions tailored to specific system requirements and application contexts.

8.2.2 Key design variables

The design of an ESS is governed by a set of technical and economic variables that are strongly interdependent. Unlike conventional generation units, storage systems do not generate energy but rather shift it in time, which means that their design must be consistent with both the expected operating conditions and the intended application (Castro and Guillen, 2026). In practice, selecting storage design parameters is rarely a straightforward task, as improving one performance aspect often comes at the expense of another. One of the most fundamental design variables is the rated power of the storage system, usually expressed in megawatts (MW). This parameter defines the maximum charging and discharging capability and determines the system's ability to provide power-oriented services such as frequency regulation, peak shaving, or congestion relief (Zhao et al., 2023). A storage system with insufficient power rating may fail to respond effectively to fast system dynamics, whereas an oversized power rating can lead to unnecessary investment costs. Closely related to the rated power is the energy capacity, typically expressed in megawatt-hours (MWh). The relationship between power, energy, and storage duration is often expressed in a simplified form as

$$E = P \cdot T \tag{8.1}$$

where E is the nominal energy capacity, P is the rated power, and T is the storage duration in hours. Although this relation is simple, it provides a useful first-order insight into the trade-off between power-oriented and energy-oriented storage designs (Gervaz and Favre, 2024). Short-duration systems are generally suited for fast-response services, while longer durations are required for energy shifting and renewable integration.

In practical applications, the full nominal energy capacity of a storage system is rarely usable. State-of-charge (SOC) operating limits are imposed to ensure safe operation and to limit long-term degradation, particularly for battery-based storage systems (Hassan et al., 2022). As a result, the usable energy capacity can be expressed as

$$E_{\text{usable}} = E_{nom}(SOC_{max} - SOC_{min}) \tag{8.2}$$

where E_{nom} is the nominal energy capacity and SOC_{max} and SOC_{min} denote the upper and lower SOC limits, respectively. These constraints directly affect the effective flexibility of the storage system and must be accounted for in both design and optimization studies. Another important set of design variables is related to efficiency, including charging efficiency, discharging efficiency, and overall round-trip efficiency (Su et al., 2023). Energy losses reduce the amount of usable energy and have a direct impact on operational profitability, especially in applications involving frequent cycling. Although efficiency values are often treated as constants in simplified studies, they may vary with operating conditions such as power level and temperature, which introduces additional modeling complexity.

The lifetime and degradation behavior of the storage system represents a critical design consideration, particularly for BESSs (Yagci et al., 2025). Degradation is influenced by multiple factors, including depth of discharge (DoD), cycling frequency, charge and discharge rates, and operating temperature. Design choices that favor aggressive operation may improve short-term performance but accelerate degradation and reduce system lifetime. For this reason, recent studies increasingly incorporate degradation-aware models that explicitly link operational decisions to long-term asset health (Safavi et al., 2025). Finally, economic parameters play a decisive role in storage system design. Capital expenditure (CAPEX), operational expenditure (OPEX), replacement costs, and financial assumptions such as discount rates and project lifetime determine the economic feasibility of a given storage configuration (Koubar et al., 2025). The relative contribution of power-related and energy-related costs varies across technologies and must be reflected in the design process. These economic considerations are often integrated directly into optimization-based frameworks through cost functions or economic constraints (Foslie et al., 2023). Overall, the interaction between power rating, energy capacity, operational limits, efficiency, degradation, and economic parameters makes storage system design a highly coupled problem. This complexity explains why heuristic approaches are often insufficient and motivates the use of optimization techniques capable of systematically exploring trade-offs and identifying balanced design solutions.

8.3 Modeling of energy storage systems

8.3.1 Energy balance and state-of-charge dynamics

A meaningful optimization of ESSs requires mathematical models that are sufficiently accurate to capture the main physical and operational constraints, while remaining simple enough to be tractable in optimization frameworks (Razzhivin et al., 2023). Among the different modeling aspects, the representation of energy balance and SOC dynamics constitutes the foundation upon which most storage optimization problems are built. The SOC represents the amount of energy stored in the system at a given time, usually expressed as a fraction of the nominal energy capacity. For a discretized time horizon, the SOC evolution can be described through a simple energy balance equation that links consecutive time steps. Assuming a fixed time step Δt, the SOC at time step $k+1$ can be written as

$$SOC_{k+1} = SOC_k + \frac{\eta_c P_k^c \Delta t}{E_{nom}} - \frac{P_k^d \Delta t}{\eta_d E_{nom}} \tag{8.3}$$

where SOC_k is the SOC at time step k, P_k^c and P_k^d denote the charging and discharging powers, respectively, E_{nom} is the nominal energy capacity, and η_c and η_d are the charging and discharging efficiencies (Abbasi et al., 2025). This formulation assumes that charging and discharging do not occur simultaneously, an assumption commonly enforced through operational constraints in optimization models. This discrete-time SOC formulation captures the essential energy conservation principle governing storage operation. Despite its simplicity, it has been widely adopted in the literature due to its compatibility with linear and mixed-integer optimization methods (de Lima et al., 2025). More detailed electrochemical models exist, but they are often impractical for system-level optimization problems due to their computational complexity (Cassano and Sossan, 2025). Charging and discharging powers are themselves constrained by the rated power of the storage system. These constraints can be expressed as

$$0 \leq P_k^c \leq P_{max}, 0 \leq P_k^d \leq P_{max} \tag{8.4}$$

where P_{max} denotes the maximum allowable charging or discharging power (Jamroen, 2022). These limits reflect both converter ratings and battery safety considerations. In practical applications, power constraints play a decisive role in determining the ability of the storage system to respond to fast variations in demand or generation. In addition to power constraints, the SOC must remain within predefined operational limits to ensure safe operation and to limit long-term degradation. These limits are typically expressed as:

$$SOC_{min} \leq SOC_k \leq SOC_{max} \tag{8.5}$$

where SOC_{min} and SOC_{max} represent the minimum and maximum allowable SOC levels (Attia et al., 2024). These constraints directly restrict the usable energy capacity of the storage system and influence its effective flexibility. Energy capacity constraints are implicitly enforced through the SOC limits, but they can also be expressed explicitly in terms of stored energy. Let Ek denote the stored energy at time step k, then:

$$E_k = SOC_k \cdot E_{nom} \tag{8.6}$$

with the corresponding bounds:

$$E_{min} \leq E_k \leq E_{max} \tag{8.7}$$

where $E_{min} = SOC_{min}E_{nom}$ and $E_{max} = SOC_{max}E_{nom}$ (Zhou et al., 2025). This alternative formulation is sometimes preferred in energy-based optimization models, particularly when multiple storage technologies are considered. Together, the SOC dynamics, power limits, and energy constraints define the feasible operating region of the storage system. These equations form the mathematical backbone of most storage optimization problems and provide the necessary link between physical system behavior and higher-level optimization objectives. In the following sections, these core models will be extended to include efficiency variations, degradation effects, and economic considerations.

8.3.2 Efficiency and loss modeling

Efficiency modeling plays a central role in the mathematical representation of ESSs, as losses directly affect both the technical performance and the economic value of storage operation (Kampker et al., 2025). Even relatively small efficiency assumptions can significantly influence optimal operating strategies, especially in applications involving frequent charging and discharging cycles. At the most basic level, storage losses are commonly represented through charging and discharging efficiencies, denoted by η_c and η_d, respectively. These parameters account for losses in power electronics, internal battery resistance, and auxiliary components. In system-level models, they are often assumed to be constant and independent of operating conditions, an assumption that allows the SOC dynamics to remain linear and compatible with classical optimization techniques (Lucaferri et al., 2023). Charging efficiency reflects the fraction of electrical energy that is effectively stored in the battery during charging, while discharging efficiency represents the fraction of stored energy that can be delivered to the grid or load. As shown in the SOC evolution equation presented earlier, these efficiencies affect the energy balance asymmetrically, as losses occur at different stages of the charge and discharge processes. Although this representation is simplified, it captures the dominant loss mechanisms relevant for most planning and operational studies. An alternative and commonly reported metric is the round-trip efficiency, which represents the ratio between the energy delivered during discharge and the energy absorbed during charging over a complete cycle. It is typically expressed as

$$\eta_{rt} = \eta_c \cdot \eta_d \tag{8.8}$$

Round-trip efficiency provides a convenient aggregated indicator of storage performance and is often used for technology comparison and high-level assessment (Zhang and Yang, 2021). However, relying solely on this metric can obscure the distinct roles of charging and discharging losses, particularly in optimization problems where charge and discharge decisions are made independently at each time step. In practice, efficiency is not strictly constant. Battery efficiency can vary with power level, SOC, temperature, and aging state (Xie et al., 2025). For instance, operating a battery at high power rates or near its SOC limits may lead to increased losses. Detailed electrochemical and thermal models are capable of capturing these effects, but their complexity often makes them unsuitable for large-scale or long-horizon optimization problems (Kurz et al., 2024).

As a result, a trade-off must be made between model accuracy and computational tractability. Simplified efficiency models, based on constant charging and discharging efficiencies, are widely adopted in system-level optimization studies because they strike a reasonable balance between realism and solvability (Hong et al., 2024). More detailed models, incorporating efficiency variations or nonlinear loss terms, are typically reserved for short-term operational analysis or component-level studies. The choice of efficiency model should therefore be guided by the objectives of the study and the required level of detail. For strategic design and sizing problems, simplified efficiency representations are generally sufficient and allow clear insights into the impact of losses on optimal storage configurations. In contrast, detailed efficiency modeling becomes more relevant when analyzing real-time control strategies or thermal and degradation effects, which are beyond the scope of many planning-oriented optimization frameworks.

8.3.3 Degradation and lifetime modeling

Degradation and lifetime considerations play a critical role in the design and optimization of ESSs, particularly for battery-based technologies. Unlike conventional power system assets, batteries experience progressive performance loss as a function of time and usage, which directly affects their usable capacity, efficiency, and ultimately their economic value (Yagci et al., 2025). Ignoring degradation effects may lead to overly aggressive operating strategies and unrealistic estimates of system profitability. Battery degradation is commonly decomposed into two main components: calendar aging and cycling aging. Calendar aging refers to capacity loss that occurs as a function of time, even when the battery is not actively cycled. It is primarily influenced by factors such as average SOC, temperature, and storage duration (Guo et al., 2025). Cycling aging, on the other hand, is associated with charge–discharge activity and depends on the number of cycles, DoD, and operating conditions (Hu et al., 2024). In practical applications, both mechanisms act simultaneously and contribute to overall battery degradation.

Among the operational factors influencing cycling aging, the DoD plays a particularly important role. Deeper charge–discharge cycles generally accelerate degradation, while shallow cycling tends to extend battery lifetime (Kim and Shin, 2023). This effect introduces a direct coupling between operational strategy and long-term asset health. For instance, an optimization strategy that maximizes short-term economic gains by exploiting the full SOC range may significantly reduce battery lifetime, whereas more conservative operation can preserve capacity at the expense of reduced flexibility. From a modeling perspective, detailed electrochemical degradation models exist and can accurately capture the underlying physical mechanisms. However, their complexity and data requirements make them impractical for system-level optimization studies, especially when long time horizons or multiple scenarios are considered (Kuzhiyil et al., 2025). As a result, most planning and design-oriented studies rely on simplified degradation models that approximate the impact of usage on battery lifetime. A common approach is to represent degradation as a cumulative function of energy throughput or equivalent full cycles. In its simplest form, the incremental degradation over a time step k can be expressed as

$$\Delta D_k = \alpha E_k^{\text{throughput}} \tag{8.9}$$

where $E_k^{\text{throughput}}$ denotes the charged or discharged energy during the time step and α is a degradation coefficient derived from empirical data (Wang et al., 2025). Over the system lifetime, total degradation is obtained by summing the incremental contributions across all time steps. Although this formulation is approximate, it provides a convenient way to link operational decisions to long-term capacity loss. More refined simplified models incorporate the effect of DoD by weighting energy throughput according to cycle depth. This can be achieved by introducing DoD-dependent degradation coefficients or piecewise-linear approximations that penalize deep cycles more heavily than shallow ones. Such models remain compatible with optimization frameworks while capturing the main qualitative behavior observed in real batteries (Kim and Shin, 2023). In degradation-aware optimization, lifetime constraints are often introduced explicitly by limiting the allowable cumulative degradation or by associating a degradation cost with battery usage. In the latter case, degradation is treated as an internal cost that reflects future replacement or capacity loss (Hassan et al., 2025). This approach allows optimization algorithms to balance immediate economic benefits against long-term degradation effects, leading to more realistic and sustainable operating strategies. Overall, simplified degradation and lifetime models provide a practical compromise between physical accuracy and computational efficiency. While they do not capture all degradation mechanisms in detail, they are well suited for design and planning studies where the objective is to assess trade-offs and identify robust storage configurations rather than predict precise battery aging behavior.

8.3.4 Economic modeling

Economic modeling provides the link between the physical behavior of ESSs and the optimization objectives used in design and planning studies. Since storage systems are capital-intensive assets with limited lifetime, their economic performance depends not only on operational benefits but also on how investment, operating, and replacement costs are accounted for over the project horizon (Cruz-De-Jesús et al., 2025). A consistent economic representation is therefore essential for meaningful optimization results.

8.3.4.1 Investment and operating costs

The economic cost of a storage system is commonly decomposed into CAPEX and OPEX. CAPEX typically includes costs related to the energy storage medium, power conversion system, balance-of-plant, and installation. These costs are often expressed as a combination of power-related and energy-related components, reflecting the dual nature of storage systems (Kelly and Leahy, 2020). A simplified representation of the total investment cost can be written as

$$C_{inv} = C_P P_{max} + C_E C_{nom} \tag{8.10}$$

where C_P and C_E denote the unit costs associated with power and energy capacity, respectively. This formulation is widely used in design optimization studies because it allows the impact of power and energy sizing decisions to be evaluated explicitly. Operating costs generally include maintenance, auxiliary consumption, and, in some cases, variable costs related to system usage. For battery-based storage, *OPEX* is often relatively small compared to *CAPEX* and is frequently modeled as a fixed annual cost proportional to the installed capacity (Nicoli et al., 2024). In optimization models, operating costs are either included as a constant term or aggregated over the system lifetime.

8.3.4.2 Replacement modeling

Due to degradation, ESSs—particularly batteries—may require partial or full replacement during the project lifetime. Ignoring replacement effects can lead to overly optimistic economic assessments, especially for long planning horizons (Ochoa-Barragán et al., 2023). Replacement modeling is therefore an important aspect of economic representation. A common approach is to assume that the battery is replaced once a predefined degradation threshold is reached, for example when the usable capacity falls below a certain fraction of its nominal value. The corresponding replacement cost can then be included explicitly in the economic model as

$$C_{rep} = \sum_{r} \frac{C_{bat}^{(r)}}{(1+i)^{t_r}} \tag{8.11}$$

where $C_{bat}^{(r)}$ is the replacement cost at replacement time tr, and i is the discount rate (Oosterom and Hall, 2022). This formulation allows future replacement costs to be discounted and compared consistently with initial investment costs. In simplified studies, replacement costs are sometimes incorporated indirectly by associating a degradation-related cost with battery usage. In this case, degradation is treated as an internal economic penalty reflecting future replacement or loss of capacity. This approach is particularly convenient in optimization frameworks, as it avoids the need for explicit lifetime simulation while still discouraging excessive battery usage (Amini et al., 2023).

8.3.4.3 Financial assumptions

Economic modeling inevitably relies on a set of financial assumptions that influence the outcome of optimization studies. These assumptions typically include the project lifetime, discount rate, inflation, and, in some cases, residual value at the end of life. Among these parameters, the discount rate plays a central role, as it determines the relative importance of upfront investment versus long-term operational benefits (Oosterom and Hall, 2022). To enable fair comparison between different storage configurations, costs are often annualized using standard financial metrics such as the annualized cost or net present value (NPV). The total discounted cost over the project lifetime can be expressed as

$$C_{tot} = C_{inv} + \sum_{k} \frac{C_{op,k} + C_{deg,k}}{(1+i)^{k}} \tag{8.12}$$

where $C_{op,k}$ and $C_{deg,k}$ denote operational and degradation-related costs at time step k, respectively (Shabani and Yan, 2025). This formulation provides a unified economic objective that can be minimized within an optimization framework. Overall, the level of detail used in economic modeling should be consistent with the objectives of the study and the resolution of the physical models. While simplified economic representations may overlook certain real-world complexities, they are often sufficient to capture the dominant cost drivers and to support informed design and optimization decisions. In the next section, these economic models will be integrated with physical constraints to formulate complete optimization problems for storage system design.

8.4 Optimization techniques for storage system design

8.4.1 Optimization problem formulation

At the design stage, the storage system problem can be formulated as a constrained optimization problem in which sizing and operation decisions are determined simultaneously. The purpose of this formulation is to identify storage configurations and operating strategies that best satisfy predefined economic or technical objectives while respecting physical, operational, and financial constraints (Manassaldi et al., 2024). Let the optimization horizon be discretized into N time steps. The formulation relies on a vector of decision variables, an objective function, and a set of constraints derived from the models introduced in Section 8.3 (Lekhel et al., 2026).

8.4.1.1 Decision variables

The decision variables can be grouped into design variables and operational variables. Design variables are time-invariant and typically include:

$$x_d = \{P_{max}, E_{nom}\} \tag{8.13}$$

These variables determine the rated power and nominal energy capacity of the storage system and directly affect both investment costs and feasible operating ranges (Pan et al., 2025).

Operational variables are time-dependent and defined for each time step $k = 1, ..., N$:

$$x_o(k) = \left\{P_k^c, P_k^c, SOC_k\right\} \tag{8.14}$$

Together, the full decision vector is given by:

$$x = \{x_d, x_o(1), ..., x_o(N)\} \tag{8.15}$$

This joint treatment of sizing and operation is a defining characteristic of storage system design optimization (Yang et al., 2024).

8.4.1.2 Objective function

The objective function represents the criterion to be optimized. In most storage design problems, the objective is economic and formulated as the minimization of total discounted cost over the project lifetime (García-Miguel et al., 2024). A generic objective function can be written as:

$$\min_{x} J = C_{inv}(x_d) + \sum_{k=1}^{N} \frac{C_{op}(k) + C_{deg}(k)}{(1+i)^{t_k}} + C_{rep} \tag{8.16}$$

where $C_{inv}(x_d)$ depends on design variables, $C_{op}(k)$ represents operating costs at time step k, $C_{deg}(k)$ accounts for degradation-related costs, C_{rep} captures discounted replacement costs, and i is the discount rate (Yang et al., 2025).

This formulation allows operational decisions to influence long-term economic outcomes through degradation and replacement effects, without explicitly simulating the entire battery lifetime at a fine temporal resolution. In revenue-oriented applications, the objective may alternatively be formulated as the maximization of NPV or operational profit. From a mathematical standpoint, such formulations are equivalent after sign inversion and appropriate cost–revenue aggregation (Wang et al., 2024).

8.4.1.3 Constraints

The feasible set of the optimization problem is defined by constraints that enforce physical consistency and operational limits. These constraints are inherited directly from the models introduced in Section 8.3 and are not repeated here. They can be expressed compactly as: Energy balance and SOC dynamics, Power limits and SOC bounds, Degradation or lifetime constraints, Economic or policy-related constraints. Collectively, these constraints ensure that the solution corresponds to a physically realizable and economically meaningful storage system. In compact form, the optimization problem can be written as:

$$\min_{x} J(x) \quad s.t. \quad \begin{matrix} g(x) = 0 \\ h(x) \leq 0 \end{matrix} \tag{8.17}$$

where $g(x)$ represents equality constraints, such as energy balance and SOC dynamics, and $h(x)$ represents inequality constraints, including power limits, SOC bounds, degradation-related limits, and economic or policy constraints.

8.4.1.4 Single-objective versus multi-objective formulations

In single-objective optimization, the problem is formulated with a single scalar objective, typically economic. This approach yields one optimal solution and is well suited for planning studies where a clear performance metric is required. In multi-objective optimization, several objectives are optimized simultaneously, for example:

$$min\{J_1(x), J_2(x), ...\} \tag{8.18}$$

where J_1 may represent cost, J_2 degradation, or J_3 emissions. The solution of such problems is a set of Pareto-optimal points rather than a unique optimum. While multi-objective formulations provide valuable insight into trade-offs, they increase computational complexity and are typically used in exploratory design studies rather than large-scale operational optimization. Overall, this formulation provides a unified mathematical framework for storage system design optimization. The structure of the problem directly influences the choice of solution method, which motivates the review of classical, metaheuristic, and advanced optimization techniques presented in the following sections.

8.4.2 Classical optimization methods

Classical optimization methods form the foundation of many storage system design and planning studies due to their solid theoretical basis and the availability of efficient solvers (Minh et al., 2025). When storage models are expressed using linear or piecewise-linear relationships, these methods provide reliable and computationally efficient solutions.

8.4.2.1 Linear programming

Linear programming (LP) is applicable when both the objective function and all constraints are linear in the decision variables. A generic LP formulation can be written as:

$$\min_{x} c\, x \;\; s.t. \quad Ax = b, x \geq 0 \tag{8.19}$$

In storage system optimization, this formulation arises naturally when efficiencies are assumed constant and degradation effects are represented through linear cost terms (Hachez et al., 2025; Hassan et al., 2025). LP formulations are widely used for operational planning problems such as energy arbitrage and peak shaving, where continuous charging and discharging decisions are optimized over time. The main advantage of LP lies in its scalability and guaranteed global optimality. However, its modeling capability is limited to continuous variables and linear relationships.

8.4.2.2 Mixed-integer linear programming

Mixed-integer linear programming (MILP) extends LP by allowing a subset of decision variables to take integer or binary values. The general MILP formulation can be expressed as:

$$\min_{x} c\, x \;\; s.t. \quad Ax = b, x \in \mathbb{R}^n, x_j \in \mathbb{Z}\, \forall j \in I \tag{8.20}$$

where I denotes the set of indices corresponding to integer variables (de Lima et al., 2025). In storage optimization, MILP formulations are commonly used to model logical constraints such as preventing simultaneous charging and discharging, representing installation decisions, or enforcing piecewise-linear degradation models (Gómez-Quiles et al., 2025). Although MILP problems are computationally more demanding than LP problems, they remain tractable for many design and planning applications, thanks to advances in solver technology.

8.4.2.3 Optimal power flow

When storage systems are embedded within power networks, optimization problems must account for power flow constraints. A generic optimal power flow (OPF) problem can be expressed as:

$$\min_{x} J(x) \quad s.t. \quad \begin{matrix} g(x) = 0 \\ h(x) \leq 0 \end{matrix} \tag{8.21}$$

where $f(x)$ represents power balance equations and $h(x)$ includes operational limits such as line capacities and voltage bounds (Mirzaei et al., 2019). Storage systems are incorporated into OPF formulations as controllable injection or absorption devices at specific buses, allowing coordinated optimization of network operation and storage behavior. Depending on the level of modeling detail, OPF formulations may be linearized or remain nonlinear.

8.4.3 Metaheuristic optimization techniques

Metaheuristic optimization techniques are frequently employed in storage system design problems when classical optimization methods become difficult to apply due to nonlinearity, nonconvexity, or discrete decision spaces (Gusain et al., 2023). These methods search for high-quality solutions by iteratively improving a population of candidate solutions based on stochastic or bio-inspired mechanisms, without requiring gradient information or convex problem structure (Wang et al., 2022). In the following, the general optimization problem defined in Section 8.4.1 is assumed, with decision vector x and objective function $J(x)$.

8.4.3.1 Genetic algorithms

Genetic algorithms are population-based optimization methods inspired by natural evolution. At each generation g, a population of candidate solutions is represented as:

$$\mathscr{P}^{(g)} = \{x_1^{(g)}, x_2^{(g)}, \ldots, x_M^{(g)}\} \tag{8.22}$$

Each individual $x_i^{(g)}$ encodes a feasible storage system design and operation strategy. Its quality is evaluated using the objective function:

$$f_i^{(g)} = \mathcal{J}\left(x_i^{(g)}\right) \tag{8.23}$$

The evolution process relies on three main operators:

- selection, favoring individuals with lower objective values,
- crossover, generating offspring by combining decision variables of parent solutions,
- mutation, introducing random perturbations to maintain population diversity.

The population is updated iteratively until a stopping criterion is met, such as a maximum number of generations or convergence of the best fitness value (Saxena et al., 2025; Ricardo et al., 2023).

8.4.3.2 Particle swarm optimization

Particle swarm optimization (PSO) models the collective behavior of interacting agents, called particles, moving through the search space. Each particle iii is characterized at iteration k by a position vector $x_i^{(k)}$ and a velocity vector $v_i^{(k)}$. The velocity and position updates are governed by:

$$v_i^{(k+1)} = \omega v_i^{(k)} + c_1 r_1 \left(p_i^{best} - x_i^{(k)}\right) + c_2 r_2 \left(g^{best} - x_i^{(k)}\right) \tag{8.24}$$

$$x_i^{(k+1)} = x_i^{(k)} + v_i^{(k+1)} \tag{8.25}$$

where:

- ω is the inertia weight,
- c_1 and c_2 are acceleration coefficients,
- r_1, r_2 are random variables uniformly distributed in $[0, 1]$,
- p_i^{best} is the best position found by particle i,
- g^{best} is the best position found by the swarm (Indramaya and Suyanto, 2021).

PSO is particularly effective for continuous storage sizing problems and parameter tuning, but may require additional mechanisms to handle constraints and discrete variables (Assogna et al., 2025).

8.4.3.3 Differential evolution

Differential evolution (DE) is a population-based optimization method that generates new candidate solutions by combining existing ones through vector differences. For each target vector $x_i^{(k)}$, a mutant vector is generated as:

$$v_i^{(k)} = x_{r_1}^{(k)} + F\left(x_{r_2}^{(k)} - x_{r_3}^{(k)}\right) \tag{8.26}$$

where r_1, r_2, r_3 are distinct indices randomly selected from the population, and F is a scaling factor controlling mutation strength (Wang and Yu, 2024). The mutant vector is then combined with the target vector through crossover to form a trial vector $u_i^{(k)}$. Selection is performed according to:

$$x_i^{(k)} = \begin{cases} u_i^{(k)}, \textit{if } J\left(u_i^{(k)}\right) < J\left(x_i^{(k)}\right) \\ x_i^{(k)}, \textit{otherwise} \end{cases} \tag{8.27}$$

This greedy selection strategy allows DE to maintain strong global search capability while steadily improving solution quality.

Metaheuristic algorithms offer significant flexibility in storage system design optimization, as they can accommodate nonlinear degradation models, discrete sizing variables, and multi-objective formulations without major reformulation of the problem (Wang et al., 2022). However, their computational cost generally scales with the number of objective function evaluations, which may limit their applicability to large-scale or long-horizon problems. In practice, metaheuristics are often used at the design stage or in combination with classical optimization methods, for example to determine candidate storage sizes that are later refined using LP or MILP formulations (Heroual et al., 2024).

8.4.4 Advanced and hybrid optimization approaches

As storage system models increasingly account for uncertainty, long-term dynamics, and complex interactions with energy markets and power networks, classical and purely deterministic optimization methods may become insufficient. Advanced and hybrid optimization approaches have therefore gained attention as a means to explicitly address uncertainty, enhance robustness, and combine complementary strengths of different optimization paradigms (Zhang et al., 2025).

8.4.4.1 Stochastic optimization

Stochastic optimization explicitly accounts for uncertainty by modeling uncertain parameters as random variables with known or estimated probability distributions. In energy storage design, uncertainties typically arise from renewable generation, load demand, market prices, and, to a lesser extent, degradation behavior (Prasanthi et al., 2026). A generic stochastic optimization problem can be formulated as:

$$\min_{x} \mathbb{E}_{\xi}[J(x, \xi)] \quad s.t \quad g(x, \xi) \leq 0 \tag{8.28}$$

where ξ denotes a vector of random variables and $\mathbb{E}[.]$ represents the expectation operator. In practice, stochastic storage optimization problems are often solved using scenario-based formulations, where uncertainty is represented by a finite set of scenarios with associated probabilities (Fan et al., 2025). While this approach provides a more realistic representation of system behavior, it significantly increases problem size and computational complexity, particularly for long planning horizons.

8.4.4.2 Robust optimization

Robust optimization offers an alternative to stochastic approaches by seeking solutions that remain feasible and near-optimal under worst-case realizations of uncertainty, without requiring precise probability distributions (Zhai et al., 2020). A generic robust optimization problem can be written as:

$$\min_{x} \max_{\xi \in \mathcal{U}} J(x, \xi) \quad s.t \quad g(x, \xi) = 0, h(x, \xi) \leq 0 \tag{8.29}$$

where $\mathcal{U}$ denotes a predefined uncertainty set.

In the context of storage system design, robust optimization is particularly attractive when uncertainty is difficult to characterize statistically, such as long-term price volatility or policy changes. Robust formulations typically lead to more conservative solutions compared to stochastic optimization, trading optimality under nominal conditions for improved reliability and feasibility across a wide range of scenarios (Eingartner et al., 2024).

8.4.4.3 AI-assisted and hybrid methods

More recently, AI techniques have been increasingly combined with classical and metaheuristic optimization methods to address the growing complexity of storage system design problems. Rather than replacing optimization models, AI techniques are often used to augment them (Parvathareddy et al., 2025). AI-assisted approaches include:

- surrogate modeling to approximate expensive objective functions,
- learning-based prediction of prices, loads, or degradation behavior,
- reinforcement learning for operational policy discovery.

Hybrid optimization frameworks typically combine AI-based components with mathematical optimization. A general hybrid structure can be expressed as:

$$x^* = arg \min_{x} J(x, \xi) \tag{8.30}$$

where ξ denotes AI-generated forecasts or model parameters.

In storage design, such hybrid approaches have been used to reduce computational burden, improve solution quality under uncertainty, and enable adaptive optimization strategies. However, their performance strongly depends on data availability and model generalization, and their interpretability remains a concern in planning-oriented studies.

Advanced and hybrid optimization approaches provide powerful tools for addressing uncertainty and complexity in storage system design. Stochastic and robust optimization offer complementary perspectives on uncertainty handling, while

AI-assisted methods introduce new opportunities for scalability and adaptability. Nevertheless, these approaches often come at the cost of increased model complexity and reduced transparency. As a result, they are most commonly applied in advanced research studies or decision-support tools, rather than as replacements for classical optimization methods in large-scale planning. In practice, hybrid frameworks combining deterministic optimization with selective advanced components offer a promising compromise between realism and tractability.

8.5 Optimization under market and grid constraints

While Sections 8.3 and 8.4 focused on modeling and solution techniques, practical deployment of ESSs is fundamentally shaped by market rules and grid-level constraints. In real systems, storage does not operate in isolation: its value is determined by electricity prices, ancillary service markets, network limitations, and uncertainty in both supply and demand. Optimization under these constraints is therefore essential to translate theoretical benefits into realizable economic gains (McPherson et al., 2020).

8.5.1 Energy market participation

One of the primary value streams for ESSs is participation in electricity markets. Storage can exploit temporal price variations by charging during low-price periods and discharging when prices are high, a strategy commonly referred to as energy arbitrage (Liu et al., 2025). In market-oriented formulations, the operating cost or revenue term in the objective function becomes price-dependent:

$$C_{op}(k) = \lambda_k P_k \tag{8.31}$$

where λ_k denotes the market price at time step k, and P_k represents net charging or discharging power.

Market participation introduces additional constraints related to bidding rules, minimum bid sizes, and settlement periods. These constraints are often exogenous to the storage system itself but strongly influence optimal operating strategies (Ren et al., 2023). As a result, optimization models must align their temporal resolution and decision structure with market design.

8.5.2 Ancillary services

Beyond energy arbitrage, storage systems can provide ancillary services such as frequency regulation, spinning reserve, and voltage support. These services are typically compensated through separate markets or contractual arrangements and often impose stricter technical requirements on response time and availability (Naseri et al., 2023). From an optimization perspective, ancillary service provision introduces availability constraints, ensuring that sufficient capacity is reserved to meet service commitments. For example:

$$SOC_k \geq SOC_{min}^{res} \tag{8.32}$$

where SOC_{min}^{res} represents the minimum SOC required to deliver a contracted service.

Ancillary services often offer higher remuneration than energy arbitrage but reduce the flexibility available for other uses. Optimizing storage participation therefore requires balancing opportunity costs across competing services (Ferrucci et al., 2025).

8.5.3 Revenue stacking

In practice, storage systems rarely rely on a single revenue stream. Instead, their economic viability often depends on revenue stacking, that is, the simultaneous participation in multiple markets and services (Biggins et al., 2022). Revenue stacking introduces coupling between operating decisions and multiple revenue terms:

$$J = C_{inv} - \sum_k \left(R_k^{\text{energy}} + R_k^{anc}\right) + \ldots \tag{8.33}$$

where R_k^{energy} and R_k^{anc} denote revenues from energy and ancillary services, respectively.

This multiservice participation significantly increases optimization complexity, as actions beneficial for one service may limit participation in another. Proper constraint formulation is therefore critical to prevent infeasible or economically inconsistent solutions (Seward et al., 2022).

8.5.4 Grid constraints and congestion

When storage systems are connected to distribution or transmission networks, grid constraints become a dominant factor in optimization. Line capacity limits, voltage constraints, and congestion can restrict charging and discharging at specific locations and times (Van Someren et al., 2025). These constraints are typically incorporated through network-aware formulations, such as optimal power flow models:

$$h_{grid}(x) \leq 0 \tag{8.34}$$

Grid constraints can significantly alter the value of storage by:

- limiting access to low-cost energy,
- reducing achievable arbitrage margins,
- or creating localized value through congestion relief.

As a result, location-specific optimization is often required to accurately assess storage benefits (Gao et al., 2025).

8.5.5 Uncertainty in prices and renewable energy sources

Market prices and renewable generation are inherently uncertain, and this uncertainty directly affects storage operation and profitability. Forecast errors in prices, wind, or solar production can lead to suboptimal or even infeasible operating strategies if not properly accounted for (Lin et al., 2025). Uncertainty is commonly addressed using stochastic or robust optimization techniques, as discussed in Section 8.4.4. In market-oriented storage optimization, uncertainty-aware formulations are particularly important for avoiding excessive risk exposure and ensuring reliable service delivery (Rojas-Innocenti et al., 2025). In practice, uncertainty handling often relies on a combination of forecasting, scenario generation, and conservative constraint design. The level of sophistication adopted depends on the application context and the acceptable trade-off between expected profit and operational risk (Huo et al., 2021).

Optimization under market and grid constraints highlights the gap between idealized storage models and real-world deployment. Market rules, network limitations, and uncertainty introduce structural constraints that fundamentally shape optimal storage behavior. As a result, storage system design cannot be decoupled from its regulatory and grid context. This section sets the stage for application-oriented studies, where theoretical optimization frameworks are evaluated under realistic market conditions and network constraints. In the following sections, these concepts are illustrated through representative applications and case studies.

8.6 Applications of optimized storage system design

8.6.1 Renewable energy integration

One of the most prominent applications of optimized ESS design is the integration of variable RESs, particularly solar PV and wind power. The inherent intermittency and limited predictability of renewable generation introduce operational challenges that can be mitigated through appropriately sized and optimally operated storage systems (Jaradat and Khatib, 2025). At high penetration levels, storage becomes a key flexibility resource for maintaining system balance and operational reliability (Das et al., 2020). Storage systems enable temporal decoupling between renewable generation and consumption, thereby improving renewable utilization, reducing grid stress, and enhancing overall system flexibility. Optimization plays a central role in determining how storage should be sized and operated to achieve these objectives under realistic technical and economic constraints.

PV generation combined with BESSs (PV–BESS) has become a widely adopted configuration in both grid-connected and behind-the-meter applications. In such systems, storage can absorb excess PV generation during periods of high irradiance and release energy when generation is insufficient or demand is high (Gonçalves et al., 2023). From an optimization perspective, PV–BESS design involves jointly determining PV capacity, battery power and energy ratings, and operational strategies. The objective typically balances investment costs against operational benefits such as increased self-consumption or reduced energy purchase from the grid. Optimization models ensure that battery operation is coordinated with PV output while respecting SOC limits and degradation constraints.

In power systems with high renewable penetration, curtailment of renewable generation is often unavoidable due to network constraints or demand mismatch. Energy storage provides a means to reduce curtailment by capturing surplus renewable energy that would otherwise be wasted (Reibsch et al., 2023). Curtailment reduction can be expressed conceptually as minimizing the difference between available renewable generation and utilized energy:

$$E_{\text{curt}} = \sum_{k} max\left(0, P_k^{RES} - P_k^{\text{load}} - P_k^{\text{stor}}\right) \tag{8.35}$$

Optimization frameworks incorporate this mechanism either by explicitly penalizing curtailed energy or by maximizing renewable utilization subject to system constraints. Properly sized storage can significantly reduce curtailment levels, particularly in systems with pronounced diurnal or seasonal generation patterns.

Energy shifting refers to the transfer of energy from periods of high renewable generation to periods of higher demand or higher economic value. In PV–BESS systems, this typically involves charging the battery during midday solar peaks and discharging during evening demand peaks. Optimization ensures that energy shifting is performed in a manner that accounts for efficiency losses, degradation costs, and market prices. Rather than blindly maximizing energy transfer, optimized strategies balance short-term gains against long-term asset health. This is particularly important in systems where storage is cycled daily, as aggressive energy shifting may accelerate battery aging (Rajah et al., 2024).

Optimized storage design significantly enhances the value of renewable energy integration by increasing self-consumption, reducing curtailment, and enabling effective energy shifting. While the underlying mechanisms are conceptually simple, their economic and technical implications are highly dependent on storage sizing, operating strategy, and system context. Optimization frameworks provide the necessary tools to navigate these trade-offs and to identify storage configurations that deliver robust benefits under realistic operating conditions. This application area naturally motivates the use of case studies, where the interaction between PV generation, storage operation, and economic outcomes can be quantified and visualized.

8.6.2 Storage as a transmission asset

Beyond energy balancing and renewable integration, ESSs can also be deployed as network assets to support transmission and distribution grids. In this role, storage is not primarily valued for energy arbitrage but for its ability to manage power flows, relieve congestion, and defer costly grid reinforcements. Optimization is essential to identify when and where storage can effectively substitute or complement traditional transmission investments (Barbar et al., 2022).

Transmission congestion occurs when power flows exceed the thermal or stability limits of network components. In congested areas, energy storage can absorb power during periods of high flow and inject power during periods of scarcity, effectively reshaping local power flows (Tarashandeh and Karimi, 2021). From an optimization perspective, congestion relief is achieved by constraining storage operation in coordination with network limits:

$$P_k^{\text{line}} \leq P_{max}^{\text{line}} \tag{8.36}$$

Storage charging and discharging decisions are then optimized to reduce violations of these constraints while minimizing system cost. Properly located storage can significantly reduce congestion frequency and severity, particularly in networks with high renewable penetration or limited transfer capacity.

One of the most economically attractive applications of storage as a transmission asset is the deferral of grid investments. Transmission upgrades are capital-intensive and often designed to accommodate peak demand conditions that occur infrequently. Storage can be used to shave these peaks, allowing existing infrastructure to operate within limits for longer periods. In optimization frameworks, this effect is captured by comparing the cost of storage deployment against the avoided or postponed cost of grid reinforcement:

$$C_{\text{stor}} \leq C_{\text{upgrade}} \tag{8.37}$$

When this condition holds, storage becomes a cost-effective alternative to traditional grid expansion. Optimization models determine the minimum storage capacity and optimal operating strategy required to achieve the desired deferral, taking into account uncertainty in load growth and generation patterns (Martínez et al., 2024).

The concept of virtual transmission lines refers to the use of geographically distributed storage systems to emulate the functionality of physical transmission capacity. By coordinating charging and discharging across different network locations, storage can effectively transfer energy in time rather than space, reducing reliance on new transmission corridors. This concept is particularly relevant in systems facing siting constraints, long permitting timelines, or public opposition to new transmission infrastructure. Optimization plays a key role in determining how storage units should be dispatched collectively to provide equivalent network support while respecting local constraints (Wang and Li, 2025). Although virtual transmission cannot fully replace physical lines in all scenarios, it offers a flexible and scalable complement that can be deployed incrementally and adapted as system conditions evolve.

Viewing storage as a transmission asset fundamentally changes its valuation and optimization objectives. Instead of maximizing energy-based revenues, the focus shifts toward minimizing system costs, improving reliability, and enhancing network flexibility. These applications require network-aware optimization models that integrate storage operation with power flow constraints and long-term planning considerations. The use of storage for congestion management and investment deferral illustrates the growing convergence between generation, storage, and transmission planning. As power systems evolve toward higher levels of decentralization and renewable integration, such hybrid planning approaches are expected to play an increasingly important role.

8.6.3 Utility-scale and system-level applications

At the utility and system level, ESSs are deployed not as isolated assets, but as integral components of large-scale power systems. In this context, optimization focuses on system-wide performance metrics such as reliability, operational cost, flexibility, and resilience (Li and Deusen, 2025). Storage design and operation decisions are therefore closely coupled with generation portfolios, network constraints, and long-term planning objectives.

In large interconnected power systems, storage systems contribute to balancing supply and demand across multiple time scales, from seconds to hours. At this level, optimization problems typically involve coordinating storage operation with conventional generation units, renewable resources, and transmission constraints. Storage can reduce the need for peaking generation, smooth net load variability, and improve system response to disturbances. Optimization frameworks at this scale often integrate storage into unit commitment or system dispatch models, where storage decisions affect generation scheduling and reserve allocation (Liao et al., 2025; Hamilton and Neises, 2023). Due to the size and complexity of these problems, simplified storage models are commonly adopted to maintain computational tractability.

Microgrids represent a more localized but increasingly important system-level application of energy storage. In both grid-connected and islanded modes, storage plays a central role in ensuring power balance, maintaining voltage and frequency stability, and enabling high penetration of local renewable generation (Asri et al., 2023). Optimization in microgrid applications typically involves joint scheduling of storage, distributed generation, and controllable loads. Objectives may include minimizing operating cost, reducing fuel consumption, or maximizing reliability during islanded operation. Compared to large power systems, microgrids allow for finer-grained modeling of storage behavior, but often face tighter operational constraints and higher sensitivity to uncertainty.

As energy storage deployment scales up at the regional or national level, optimization problems shift from individual asset performance to portfolio-level planning. Large-scale deployment studies aim to determine optimal storage capacity, technology mix, and spatial distribution across the system. At this level, optimization supports strategic decisions such as:

- where storage should be located,
- how much capacity is required to meet flexibility needs,
- how storage interacts with long-term generation and transmission expansion plans.

These studies often rely on long-term planning models with reduced temporal resolution and aggregated storage representations. Despite this abstraction, optimization remains essential to quantify system-level benefits and to avoid over- or underinvestment in storage infrastructure (Ahmadi et al., 2025; Rikkas and Lahdelma, 2021).

Utility-scale and system-level applications highlight the role of storage as a cornerstone of modern power systems. Whether integrated into bulk power systems, microgrids, or large-scale planning frameworks, optimized storage design contributes to improved efficiency, flexibility, and resilience. The diversity of application contexts underscores the importance of adaptable optimization models that can be tailored to different spatial and temporal scales. This breadth of applications sets the stage for the case study section, where optimization techniques are applied to a concrete scenario to illustrate practical implementation and quantitative benefits.

8.7 Case study: Optimization of a grid-connected PV–BESS system

8.7.1 Case study description

This case study aims to illustrate how the optimization frameworks and modeling approaches presented in the previous sections can be applied to a realistic yet tractable energy system configuration. Rather than focusing on a specific geographical location or market, the objective is to demonstrate general mechanisms and trade-offs that arise when optimizing the design and operation of an ESS in the presence of renewable generation and grid interaction.

The system under consideration is a grid-connected PV installation coupled with a BESS. This configuration has been selected because it represents one of the most common and practically relevant storage applications, spanning residential, commercial, and utility-scale contexts (Xu et al., 2023). At the same time, it remains sufficiently simple to allow clear interpretation of optimization results without requiring detailed network or market modeling. The PV–BESS system operates behind a single grid connection point and supplies a local electrical load. The system can exchange power with the grid, subject to standard import and export constraints. The battery is used to store excess PV generation, shift energy in time, and reduce reliance on grid imports during high-demand or high-price periods. No ancillary service provision is considered at this stage, allowing the analysis to focus on energy-based optimization. The case study addresses the following optimization objectives:

- minimization of total operating and degradation-related costs,
- improved utilization of locally generated PV energy,
- reduction of grid dependency through optimized energy shifting.

The optimization horizon corresponds to a representative operating period discretized into fixed time steps. This choice reflects a common practice in storage optimization studies, where representative days or weeks are used to capture typical system behavior while maintaining computational efficiency (Fernández-Muñoz and Pérez-Díaz, 2023). It is important to emphasize that the purpose of this case study is not to provide site-specific design recommendations, but rather to demonstrate how optimization alters storage behavior compared to heuristic or nonoptimized operation. All models, constraints, and cost formulations used in the case study are consistent with those introduced in Sections 8.3 and 8.4, ensuring methodological continuity throughout the chapter (Eyimaya et al., 2025). The following subsections describe the system modeling assumptions, optimization formulation, and resulting performance metrics in detail.

8.7.2 System modeling and assumptions

This subsection describes the system components, operating assumptions, and modeling simplifications adopted in the case study. The goal is to define a coherent and reproducible framework that captures the essential dynamics of a grid-connected PV–BESS system, while remaining consistent with the modeling approaches introduced earlier in the chapter. The considered system consists of three main components:

- a PV generation unit,
- a BESS,
- a local electrical load connected to the public grid.

All components are connected at a single point of common coupling. Power exchange with the grid is allowed in both directions, enabling energy import and export depending on local generation, demand, and storage operation. PV generation is modeled as a time-varying power source determined by a predefined irradiance profile. Rather than relying on site-specific meteorological data, a synthetic but realistic PV production profile is used to represent a typical clear-sky day. The PV output power at time step k is expressed as:

$$P_k^{PV} = \eta_{PV} P_{PV}^{\text{rated}} G_k \tag{8.38}$$

where P_k^{rated} is the rated PV capacity, η_{PV} is an aggregate efficiency factor and G_k is a normalized irradiance profile. This simplified representation is sufficient to capture the diurnal nature of PV generation, which is the dominant factor influencing storage operation in this application. The local electrical demand is represented by a fixed load profile with hourly resolution. The profile reflects typical daily consumption patterns, with higher demand during morning and evening periods. The load is assumed to be inelastic, meaning that demand is not influenced by price signals or storage operation. This assumption allows the analysis to focus on storage-driven flexibility rather than demand-side management effects. To illustrate the baseline operating conditions of the system, Fig. 8.1 shows a representative daily profile of PV generation and load demand. This profile highlights the inherent temporal mismatch between renewable energy production and electricity consumption, which constitutes the primary motivation for the integration of battery energy storage.

As shown in Fig. 8.1, PV generation is concentrated around midday, with zero output during night hours, while the load exhibits typical morning and evening peaks.

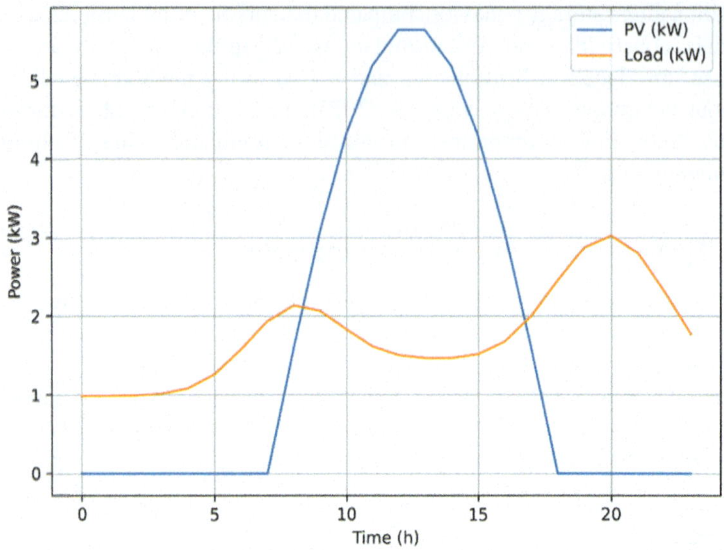

Figure 8.1 Representative daily profiles of photovoltaic generation and load demand used in the case study.

During several hours of the day, PV production exceeds local demand, leading to potential surplus energy, whereas during evening periods the load must be supplied by the grid. This temporal mismatch creates opportunities for energy shifting through battery storage and provides a clear rationale for the optimization problem formulated in the following section.

The BESS is characterized by its nominal energy capacity E_{nom} and maximum charging and discharging power P_{max}. Battery operation follows the SOC dynamics and efficiency models introduced in Section 8.3. The SOC evolution is given by:

$$SOC_{k+1} = SOC_k + \frac{\Delta t}{E_{nom}} \left(\eta_c P_k^{ch} - \frac{1}{\eta_d} P_k^{dis} \right) \tag{8.39}$$

Battery operation is constrained by:

$$SOC_{min} \leq SOC_k \leq SOC_{max} \tag{8.40}$$

$$0 \leq P_k^{ch}, P_k^{dis} \leq P_{max} \tag{8.41}$$

A simplified degradation model is included through an energy-throughput-based cost term, allowing the optimization to penalize excessive cycling without explicitly modeling electrochemical aging mechanisms. The grid is modeled as an infinite bus with time-varying electricity prices. A $time - of - use$ (TOU) tariff structure is assumed, with higher prices during peak demand hours and lower prices during off-peak periods.

Exported energy is compensated at a fixed feed-in tariff, assumed to be lower than the retail import price. This asymmetry reflects common regulatory frameworks and incentivizes local consumption and storage-based energy shifting. No grid capacity limits are imposed in this case study, allowing the analysis to focus on energy management rather than network congestion effects. The electricity pricing structure adopted in the case study is illustrated in Fig. 8.2. A time-of-use tariff is considered, with higher electricity prices during evening peak hours and lower prices during off-peak periods. A fixed feed-in tariff is assumed for energy exported to the grid.

The pricing structure shown in Fig. 8.2 introduces a clear economic incentive for temporal energy shifting. Electricity import prices are highest during evening hours, which typically coincide with increased load demand and reduced PV generation. In contrast, the export tariff remains significantly lower, favoring local consumption and storage of renewable energy rather than direct export. This price asymmetry plays a central role in shaping the optimal battery operating strategy. Several simplifying assumptions are deliberately adopted:

- constant efficiencies,
- absence of ancillary service participation,
- no uncertainty in PV generation or prices.

These choices are intentional and serve to isolate the core impact of optimization on storage behavior. More advanced aspects such as uncertainty handling, grid constraints, or multi-service operation can be incorporated using the frameworks discussed in Sections 8.4 and 8.5, but are beyond the scope of this illustrative case study.

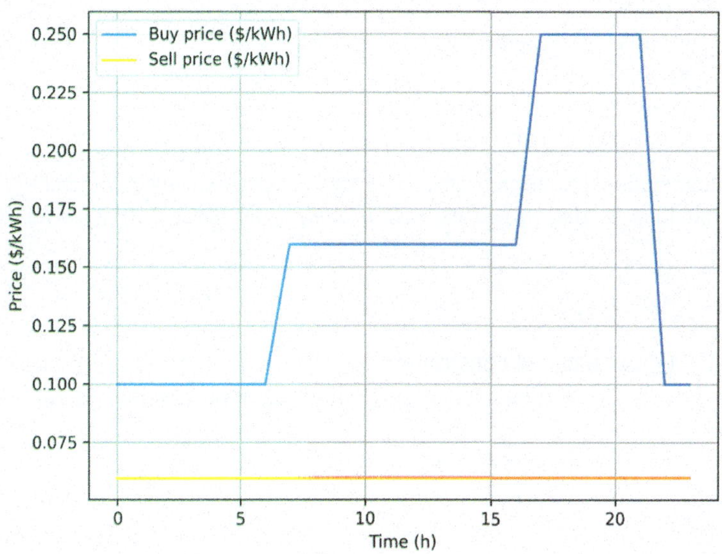

Figure 8.2 Time-of-use electricity prices for grid import and fixed feed-in tariff for grid export.

8.7.3 Optimization formulation

This subsection presents the optimization problem used to determine the optimal annual operating strategy of the grid-connected PV–BESS system described in Sections 8.7.1 and 8.7.2. The formulation aims to minimize the total electricity cost over the annual horizon while respecting battery operational constraints and accounting for battery usage through a degradation-aware penalty. The optimization problem is formulated over a discrete time horizon $k = 1, \ldots, N$, where $N = 8760$ corresponds to one full year with hourly resolution.

For each time step k, the following decision variables are defined:

- P_k^{ch}: battery charging power (kW).
- P_k^{dis}: battery discharging power (kW).
- SOC_k: battery SOC $(p.u.)$.
- P_k^{imp}: power imported from the grid (kW).
- P_k^{exp}: power exported to the grid (kW).

The battery energy capacity and power rating are assumed to be fixed parameters in this study. The optimization therefore focuses on operational scheduling rather than joint sizing and operation.

The objective is to minimize the total annual operating cost, defined as the sum of grid import costs, export revenues, and a degradation-aware battery usage penalty:

$$\min_x J = \sum_{k=1}^{N} \Delta t \left(\lambda_k^{buy} P_k^{imp} - \lambda_k^{sell} P_k^{exp} + \gamma \left(P_k^{ch} + P_k^{dis} \right) \right) \tag{8.42}$$

where:

- λ_k^{buy} is the electricity import price at time step k,
- λ_k^{sell} is the feed-in tariff,
- γ is a throughput-based degradation penalty coefficient,
- Δt is the time step duration (h).

The degradation term represents an aggregated cost associated with battery cycling and is intended to discourage excessive charge–discharge activity over the long horizon.

8.7.3.1 Constraints

8.7.3.1.1 Power balance constraint

At each time step, power balance at the point of common coupling is enforced:

$$P_k^{PV} + P_k^{dis} + P_k^{imp} = P_k^{\text{load}} + P_k^{ch} + P_k^{exp} \quad \forall k \tag{8.43}$$

The PV generation P_k^{PV} and load demand P_k^{load} are known time series generated as described in Section 8.7.2.

8.7.3.1.2 Battery SOC dynamics

Battery energy evolution follows the SOC balance equation introduced earlier in the chapter:

$$SOC_{k+1} = SOC_k + \frac{\Delta t}{E_{nom}} \left(\eta_c P_k^{ch} - \frac{1}{\eta_d} P_k^{dis} \right) \tag{8.44}$$

where E_{nom} is the nominal battery energy capacity, and η_c, η_d are charging and discharging efficiencies.

8.7.3.1.3 Operating limits

Battery operation is constrained by power and SOC limits:

$$SOC_{min} \leq SOC_k \leq SOC_{max} \quad \forall k \tag{8.45}$$

$$0 \leq P_k^{ch} \leq P_{max}, 0 \leq P_k^{dis} \leq P_{max} \quad \forall k \tag{8.46}$$

$$0 \leq P_k^{imp}, 0 \leq P_k^{exp} \quad \forall k \tag{8.47}$$

As in many long-horizon linear formulations, simultaneous charging and discharging is not explicitly prohibited. Instead, it is implicitly discouraged through the degradation-aware penalty term in the objective function.

8.7.3.1.4 Initial and terminal SOC conditions

To ensure realistic long-term operation and avoid boundary effects, the battery SOC is constrained at the beginning and end of the annual horizon:

$$SOC_1 = SOC_{\text{init}}, SOC_{N+1} = SOC_{\text{init}} \tag{8.48}$$

This condition ensures that the optimization does not exploit the end of the horizon by fully depleting or overcharging the battery. The resulting optimization problem is a large-scale LP problem, characterized by linear objective terms and linear constraints. Despite the long-time horizon, the problem remains tractable due to the simplicity of the battery model and the use of deterministic input profiles. Such formulations can be efficiently solved using modern LP solvers and are well suited for annual planning studies, sensitivity analyses, and comparative assessments of storage operation strategies.

The annual formulation highlights the trade-off between temporal resolution and computational burden. While hourly resolution allows seasonal and cumulative effects to be captured, further increases in resolution or model detail would require advanced decomposition techniques or rolling-horizon approaches, as discussed in Section 8.8.

8.7.4 Optimization results and discussion

This subsection presents and discusses the results obtained from the optimization of the PV–battery system over a representative daily horizon. The objective is to analyze how the optimized battery dispatch modifies the interaction between PV generation, load demand, and grid exchanges, and to assess the operational benefits of storage when compared to a baseline scenario without optimized control.

Fig. 8.3 illustrates the optimized daily power profiles of the PV generation, load demand, and battery charging and discharging. The results clearly show that the battery charging phase is mainly concentrated around midday, coinciding with periods of high PV generation and relatively low load demand. During these hours, a significant portion of the renewable surplus is absorbed by the battery rather than being exported to the grid. Conversely, battery discharging occurs primarily during the morning and evening periods, when PV production is insufficient to meet the load demand. These time intervals typically correspond to higher grid electricity prices, making battery discharge economically advantageous. The separation between charging and discharging periods indicates that the optimization respects the operational constraints of the storage system and avoids simultaneous charge–discharge behavior. Overall, the optimized battery dispatch reflects a clear energy-shifting strategy, where locally generated renewable energy is stored during surplus periods and released during demand peaks.

The corresponding SOC trajectory of the battery is shown in Fig. 8.4. At the beginning of the day, the SOC gradually decreases as the battery contributes to meeting early load demand. As PV generation increases, the SOC rises steadily due to sustained charging during the midday period. The maximum SOC is reached in the

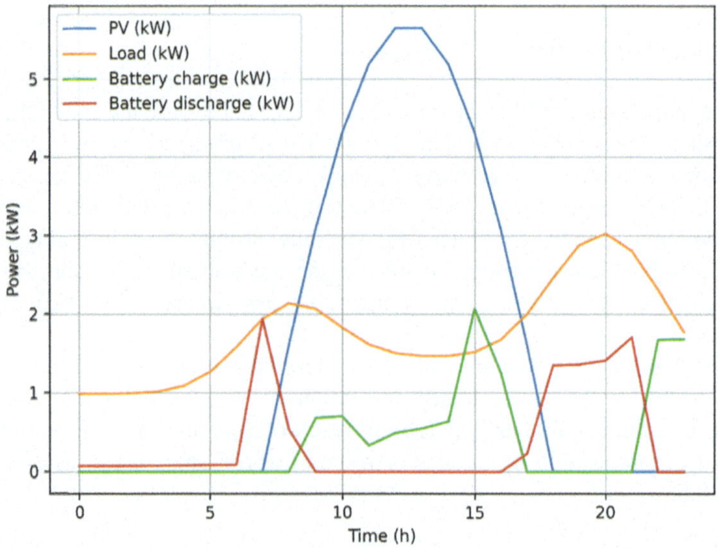

Figure 8.3 Optimized daily power profiles of PV generation, load demand, and battery charging and discharging.

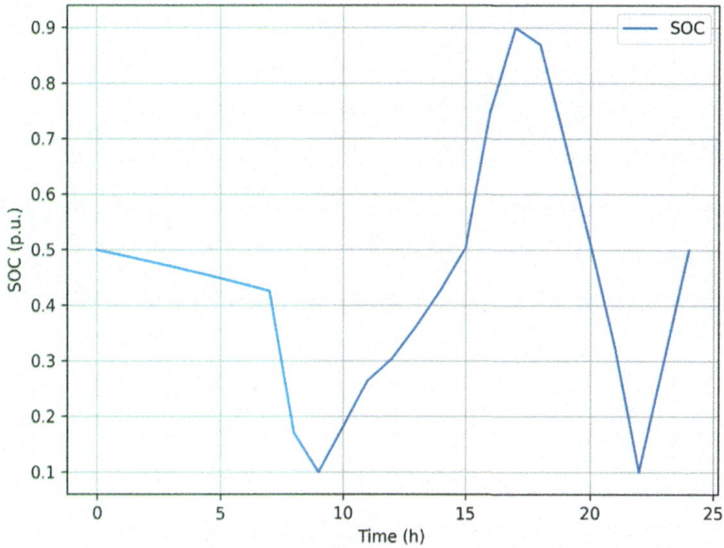

Figure 8.4 Optimized daily state-of-charge (SOC) trajectory of the battery.

late afternoon, after which the battery discharges to support the evening demand peak. Throughout the daily cycle, the SOC remains within the predefined minimum and maximum bounds, confirming that the optimization enforces the physical and operational limits of the storage system. The smooth evolution of the SOC also suggests that the optimization avoids overly aggressive cycling, which is consistent with the inclusion of efficiency considerations and degradation-aware constraints in the formulation. The SOC trajectory therefore provides a clear illustration of how the battery operates as an energy buffer, shifting renewable energy in time while maintaining feasible operating conditions.

Fig. 8.5 compares the grid import and export power for the baseline case (without optimized battery control) and the optimized scenario. In the baseline configuration, excess PV generation during midday leads to substantial power exports to the grid, while significant grid imports are observed during morning and evening demand peaks. Under optimized operation, these patterns are noticeably altered. Grid exports during periods of PV surplus are reduced, as a portion of the excess energy is stored in the battery. Similarly, grid imports during peak demand hours are significantly lowered due to battery discharging. This demonstrates the ability of the storage system to smooth grid interactions and reduce dependence on external energy supply during high-price periods. The reduction of both peak imports and low-value exports highlights the dual benefit of optimized storage operation: improved self-consumption of renewable energy and enhanced economic performance.

The results obtained over the representative daily horizon confirm the effectiveness of the proposed optimization framework in coordinating PV generation, battery storage, and grid interaction. Even at the daily scale, the optimization captures the essential

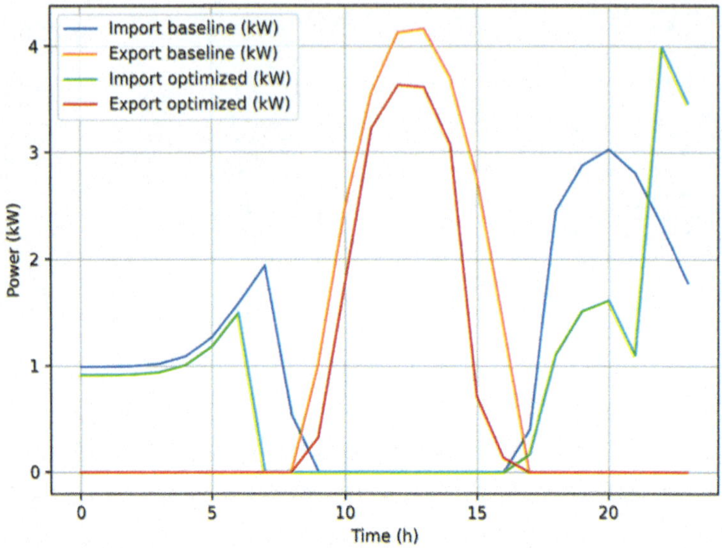

Figure 8.5 Comparison of grid import and export power under baseline and optimized operation.

mechanisms through which storage systems create value, namely energy shifting, peak shaving, and improved utilization of locally generated renewable energy. While the cost reductions achieved over a single day may appear limited in absolute terms, their repetition over extended operating periods can result in substantial cumulative benefits. This observation highlights the importance of considering storage optimization over longer time horizons, as discussed in subsequent sections. Overall, the daily case study provides a clear and intuitive demonstration of the operational principles of optimized storage control, while serving as a foundation for more advanced analyses incorporating longer horizons, uncertainty, and additional system constraints.

8.8 Challenges and future research directions

Despite the demonstrated benefits of optimized ESS design, several challenges remain that limit the direct transfer of optimization-based approaches to real-world deployment. These challenges span modeling fidelity, computational aspects, uncertainty management, and regulatory constraints. Addressing them represents an important direction for future research.

A fundamental challenge in storage system optimization lies in balancing model accuracy with computational tractability. Simplified linear models, such as those adopted in this chapter, enable efficient optimization and provide clear insights into system behavior. However, they inevitably rely on approximations, particularly regarding battery efficiency, degradation, and power electronics dynamics. More detailed electrochemical or thermal models can capture battery behavior with higher fidelity, but their integration into large-scale or long-horizon optimization frameworks remains challenging due to increased nonlinearity and computational burden (Tejada-Arango et al., 2025). Future research should focus on developing reduced-order or surrogate models that preserve key physical characteristics while remaining compatible with optimization-based control and planning.

As optimization horizons extend from daily operation to annual or multi-year planning, computational scalability becomes a critical concern. Even with linear or mixed-integer formulations, problem size grows rapidly with temporal resolution, system complexity, and the inclusion of multiple storage assets. This challenge is particularly relevant for applications such as distribution network planning, multi-microgrid coordination, or system-level studies involving high penetrations of renewable energy. Decomposition techniques, hierarchical optimization, and rolling-horizon strategies represent promising avenues to address scalability while maintaining solution quality (Cheng et al., 2025).

Battery degradation remains one of the most uncertain aspects of storage system modeling. While simplified degradation-aware formulations based on energy throughput or cycle depth are increasingly used in optimization studies, they do not fully capture the stochastic and history-dependent nature of battery aging (Dadash and Björsell, 2024). In practice, degradation is influenced by temperature, SOC trajectories, charge/discharge rates, and calendar effects, many of which are difficult to model accurately over long horizons. Future research should aim to integrate probabilistic or data-driven degradation models into optimization frameworks, allowing storage control strategies to explicitly account for uncertainty and risk.

The economic value of storage systems is strongly shaped by regulatory frameworks and market structures. Tariff design, access to ancillary service markets, capacity remuneration mechanisms, and grid connection rules vary significantly across regions and continue to evolve (Jafarizadeh et al., 2024). Optimization models must therefore be adaptable to changing regulatory conditions and capable of capturing multiple value streams. Future work should emphasize regulatory-aware optimization frameworks that can assess the impact of policy changes on storage profitability and system-level benefits.

Recent advances in AI and digital twin technologies offer new opportunities for storage system optimization. Machine learning techniques can be used to approximate complex system dynamics, forecast renewable generation and prices, or learn optimal control policies directly from data. Digital twins, combining physics-based models with real-time data, enable continuous system monitoring and adaptive optimization. The integration of AI-based methods with traditional optimization approaches represents a promising research direction, particularly for real-time control, predictive maintenance, and long-term asset management (Olabi et al., 2023).

In general, while optimization-based design and operation of ESSs have reached a high level of maturity, their practical deployment continues to face technical, computational, and regulatory challenges. Addressing these issues will require interdisciplinary approaches combining power system modeling, optimization theory, data-driven methods, and policy analysis.

8.9 Conclusion

This chapter has addressed the design and operation of ESSs from an optimization-oriented perspective. Rather than treating storage as a passive asset, the analysis has shown that its technical and economic value emerges primarily through coordinated and informed decision-making. Framing storage system design as an optimization problem provides a systematic way to evaluate trade-offs between investment cost, operational performance, and long-term system behavior. A central insight of the chapter is the importance of modeling choices and underlying assumptions. Energy balance formulations, efficiency representations, degradation models, and economic parameters all directly influence the outcome of optimization-based analyses. While simplified models enable tractable and interpretable solutions, their limitations must be clearly understood when extrapolating results to real-world applications. The case study has illustrated how even relatively simple models, when consistently applied, can capture the essential mechanisms through which storage systems create value.

The review of optimization techniques highlights the growing role of advanced methods in storage system design. Classical optimization approaches remain well suited for planning and scheduling problems with clear structure and constraints, while metaheuristic and hybrid techniques offer flexibility for handling nonlinearity, uncertainty, and multi-objective formulations. The integration of AI with optimization further expands the range of feasible applications, particularly for real-time operation and adaptive control. From a practical standpoint, the results emphasize that storage optimization must be considered within its broader system context. Market rules, grid constraints, and regulatory frameworks play a decisive role in shaping optimal storage strategies and achievable benefits. As energy systems continue to evolve toward higher shares of variable renewable generation, optimization-based storage design will remain a key tool for supporting flexibility, reliability, and economic efficiency. Overall, this chapter demonstrates that effective energy storage deployment is not solely a matter of technology selection or capacity sizing, but fundamentally a question of optimal integration. By combining appropriate models with suitable optimization techniques, storage systems can be designed and operated in a way that maximizes their contribution to modern power systems.

References

Abbasi, A., Gharehghani, A., Andwari, A., 2025. Real-time discharge curve and state of charge estimation of lithium-ion batteries via a physics-informed full homogenized macroscale model. J. Energy Storage 135, 118307.

Ahmadi, S.E., Kabir, E., Fattahi, M., Marzband, M., Li, D., 2025. Techno-economic planning of spatially-resolved battery storage systems in renewable-dominant grids under weather variability. Appl. Energy 401, 126706.

Ali, J.S., Qiblawey, Y., Alassi, A., Massoud, A.M., Muyeen, S.M., Abu-Rub, H., 2025. Power system stability with high penetration of renewable energy sources: challenges, assessment, and mitigation strategies. IEEE Access 13, 39912–39934.

AlJassmi, A., Mahmoud, M., Alami, A.H., 2026. Introduction to flywheel energy storage systems. In: Renewable Energy-Volume 4: Energy Storage Systems-Mechanical, Hydro, and Thermal. Academic Press, pp. 3–15.

Amini, M., Nazari, M.H., Hosseinian, S.H., 2023. Optimal energy management of battery with high wind energy penetration: a comprehensive linear battery degradation cost model. Sustain. Cities Soc. 93, 104492.

Ansari, M.S., Jalil, M.F., Bansal, R.C., 2023. A review of optimization techniques for hybrid renewable energy systems. Int. J. Model. Simulat. 43 (5), 722–735.

Asri, R., Aki, H., Kodaira, D., 2023. Optimal operation of shared energy storage on islanded microgrid for remote communities. Sustain. Energy Grids Netw. 35, 101104.

Assogna, R., Ciabattoni, L., Comodi, G., 2025. PSO-based supervisory adaptive controller for BESS-VSG frequency regulation under high PV penetration. Energies 18 (20), 5401.

Attia, E.M., Abdelsalam, H.A., Rashad, E.E.M., 2024. Energy management and SoC balancing of distributed batteries in AC microgrids using consensus tracking control. Sustain. Energy Grids Netw. 38, 101345.

Bamisile, O., Cai, D., Adun, H., Dagbasi, M., Ukwuoma, C.C., Huang, Q., Bamisile, O., 2024. Towards renewables development: review of optimization techniques for energy storage and hybrid renewable energy systems. Heliyon 10 (19).

Barbar, M., Mallapragada, D.S., Stoner, R., 2022. Decision making under uncertainty for deploying battery storage as a non-wire alternative in distribution networks. Energy Strategy Rev. 41, 100862.

Biggins, F.A.V., Homan, S., Ejeh, J.O., Brown, S., 2022. To trade or not to trade: simultaneously optimising battery storage for arbitrage and ancillary services. J. Energy Storage 50, 104234.

Cassano, S., Sossan, F., 2025. Scheduling power-intensive operations of Battery Energy Storage Systems and application to hybrid hydropower plants. Appl. Energy 386, 125559.

Castro, L.M., Guillen, D., 2026. Harnessing battery energy storage systems with CO_2 time-shift strategies for power grid emissions control. J. Energy Storage 145, 119923.

Chatzigeorgiou, N.G., Theocharides, S., Makrides, G., Georghiou, G.E., 2024. A review on battery energy storage systems: applications, developments, and research trends of hybrid installations in the end-user sector. J. Energy Storage 86, 111192.

Cheng, H., Chen, J., Jung, S., Kim, Y.B., 2025. Hierarchical rolling optimization strategy for hybrid electric-hydrogen system based on deep reinforcement learning. Energy 138952.

Cruz-De-Jesús, E., Martínez-Ramos, J.L., Marano-Marcolini, A., Gómez-Expósito, A., 2025. Economic assessment of battery energy storage systems for frequency regulation reserve provision: a case study of the Dominican Republic. Energy Sustain. Dev. 88, 101749.

Da Silva, D.J., Belati, E.A., López-Lezama, J.M., 2023. A mathematical programming approach for the optimal operation of storage systems, photovoltaic and wind power generation. Energies 16 (3), 1269.

Dadash, A.H., Björsell, N., 2024. A framework for designing a degradation-aware controller based on empirical estimation of the state–action cost and model predictive control. J. Manuf. Syst. 76, 599–613.

Das, P., Mathuria, P., Bhakar, R., Mathur, J., Kanudia, A., Singh, A., 2020. Flexibility requirement for large-scale renewable energy integration in Indian power system: technology, policy and modeling options. Energy Strategy Rev. 29, 100482.

De Lima, T.D., Faria, P., Vale, Z., 2025. Optimizing home energy management systems: a mixed integer linear programming model considering battery cycle degradation. Energy Build. 329, 115251.

Dong, Z., Tao, Y., Lai, S., Wang, T., Zhang, Z., 2025. Powering future advancements and applications of battery energy storage systems across different scales. Energy Storage Applic. 2 (1), 1.

Eingartner, A., Naumann, S., Schmitz, P., Worthmann, K., 2024. Adjustable robust energy operation planning under uncertain renewable energy production. Energies 17 (8), 1917.

Elalfy, D.A., Gouda, E., Kotb, M.F., Bureš, V., Sedhom, B.E., 2024. Comprehensive review of energy storage systems technologies, objectives, challenges, and future trends. Energy Strategy Rev. 54, 101482.

Eyimaya, S.E., Altin, N., Nasiri, A., 2025. Optimization of photovoltaic and battery storage sizing in a DC microgrid using LSTM networks based on load forecasting. Energies 18 (14), 3676.

Fan, J., Yan, R., He, Y., Zhang, J., Zhao, W., Liu, M., Ma, Q., 2025. Stochastic optimization of combined energy and computation task scheduling strategies of hybrid system with multi-energy storage system and data center. Renew. Energy 242, 122466.

Fernández-Muñoz, D., Pérez-Díaz, J.I., 2023. Optimisation models for the day-ahead energy and reserve self-scheduling of a hybrid wind–battery virtual power plant. J. Energy Storage 57, 106296.

Ferrucci, T., Fioriti, D., Poli, D., Barberis, S., Roncallo, F., Gambino, V., 2025. Battery energy storage systems for ancillary services in renewable energy communities. Appl. Therm. Eng. 260, 124988.

Foslie, S.S., Knudsen, B.R., Korpås, M., 2023. Integrated design and operational optimization of energy systems in dairies. Energy 281, 128242.

Gao, J., Sun, Y., Su, X., 2025. A risk preference-based optimization model for user-side energy storage system configuration from the investor's perspective. Electricity 6 (1), 3.

García-Miguel, P.L.C., Alonso-Martinez, J., Gómez, S.A., Rodríguez-Amenedo, J.L., 2024. Impact of risk measures and degradation cost on the optimal arbitrage schedule for battery energy storage systems. Int. J. Electr. Power Energy Syst. 157, 109883.

Gervaz, S., Favre, F., 2024. Identifying key parameters in building energy models: sensitivity analysis applied to residential typologies. Buildings 14 (9), 2804.

Gholami, M., Muyeen, S.M., Mousavi, S.A., 2024. Optimal sizing of battery energy storage systems and reliability analysis under diverse regulatory frameworks in microgrids. Energy Strategy Rev. 51, 101255.

Gomes, E., Pereira, L., Esteves, A., Morais, H., 2024. Metaheuristic optimization methods in energy community scheduling: a benchmark study. Energies 17 (12), 2968.

Gómez-Quiles, C., Paramo-Balsa, P., Riquelme-Santos, J., Gómez-Expósito, A., 2025. Minimum-time MILP-based EV charging scheduling for prosumers. Int. J. Electr. Power Energy Syst. 172, 111293.

Gonçalves, A., Cavalcanti, G.O., Feitosa, M.A., Dias Filho, R.F., Pereira, A.C., Jatobá, E.B., Gómez-Malagón, L.A., 2023. Optimal sizing of a photovoltaic/battery energy storage system to supply electric substation auxiliary systems under contingency. Energies 16 (13), 5165.

Guo, J., Ping, P., Ren, J., Ren, X., Gao, W., Kong, D., Feng, Z., 2025. High-temperature calendar aging at low state-of-charge: electrochemical degradation, thermal safety implications, and optimal SOC ranges for lithium-ion battery storage and transport. J. Energy Storage 125, 116988.

Gusain, C., Tripathi, M.M., Nangia, U., 2023. Study of meta-heuristic optimization methodologies for design of hybrid renewable energy systems. Therm. Sci. Eng. Prog. 39, 101711.

Hachez, J., Latiers, A., Berger, B., Bram, S., 2025. Multi-energy systems fast optimization: a new formulation in linear programming for temperatures and magnitudes of thermal power flows in heating systems. Energy Build. 336, 115618.

Hamilton, W.T., Neises, T.W., 2023. Dispatch optimization of electric thermal energy storage within system advisor model. J. Energy Storage 61, 106786.

Hassan, M.U., Saha, S., Haque, M.E., Islam, S., Mahmud, A., Mendis, N., 2022. A comprehensive review of battery state of charge estimation techniques. Sustain. Energy Technol. Assessments 54, 102801.

Hassan, A., Hollweg, G.V., Su, W., Zhou, X., Wang, M., 2025. Degradation-aware Bi-level optimization of second-life battery energy storage system considering demand charge reduction. Energies 18 (15), 3894.

Heroual, S., Belabbas, B., Allaoui, T., Denai, M., 2024. Performance enhancement of a hybrid energy storage systems using meta-heuristic optimization algorithms: genetic algorithms, ant colony optimization, and grey wolf optimization. J. Energy Storage 103, 114451.

Hong, H., Jiang, M., Yen, G.G., 2024. Boosting scalability for large-scale multiobjective optimization via transfer weights. Inf. Sci. 670, 120607.

Hu, C., Geng, M., Yang, H., Fan, M., Sun, Z., Yu, R., Wei, B., 2024. A review of capacity fade mechanism and promotion strategies for lithium iron phosphate batteries. Coatings 14 (7), 832.

Huo, D., Gu, C., Greenwood, D., Wang, Z., Zhao, P., Li, J., 2021. Chance-constrained optimization for integrated local energy systems operation considering correlated wind generation. Int. J. Electr. Power Energy Syst. 132, 107153.

Indramaya, E., Suyanto, S., 2021. Comparative study of recent swarm algorithms for continuous optimization. Procedia Comput. Sci. 179, 685–695.

Jafarizadeh, H., Yamini, E., Zolfaghari, S.M., Esmaeilion, F., Assad, M.E.H., Soltani, M., 2024. Navigating challenges in large-scale renewable energy storage: barriers, solutions, and innovations. Energy Rep. 12, 2179–2192.

Jamroen, C., 2022. The effect of SoC management on economic performance for battery energy storage system in providing voltage regulation in distribution networks. Elec. Power Syst. Res. 211, 108340.

Jaradat, T.E., Khatib, T., 2025. A review of battery energy storage system for renewable energy penetration in electrical power system: environmental impact, sizing methods, market features, and policy frameworks. Future Batteries 100106.

Kampker, A., Späth, B., Song, X., Wang, D., 2025. Modelling of battery energy storage systems under real-world applications and conditions. Batteries 11 (11), 392.

Kelly, J.J., Leahy, P.G., 2020. Optimal investment timing and sizing for battery energy storage systems. J. Energy Storage 28, 101272.

Khalili, S., Lopez, G., Breyer, C., 2025. Role and trends of flexibility options in 100% renewable energy system analyses towards the Power-to-X Economy. Renew. Sustain. Energy Rev. 212, 115383.

Killingtveit, Å., 2019. Hydropower. In: Managing Global Warming. Academic Press, pp. 265–315.

Kim, S.H., Shin, Y.J., 2023. Optimize the operating range for improving the cycle life of battery energy storage systems under uncertainty by managing the depth of discharge. J. Energy Storage 73, 109144.

Koubar, M., Lindberg, O., Lingfors, D., Huang, P., Berg, M., Munkhammar, J., 2025. Techno-economical assessment of battery storage combined with large-scale Photovoltaic power plants operating on energy and Ancillary Service Markets. Appl. Energy 382, 125200.

Kurz, T., Gradl, P., Kriechbaum, L., Solic, G., Pfleger-Schopf, K., Kienberger, T., 2024. Approaches to simplify industrial energy models for operational optimisation. J. Clean. Prod. 452, 141848.

Kuzhiyil, J.A., Damoulas, T., Planella, F.B., Widanage, W.D., 2025. Lithium-ion battery degradation modelling using universal differential equations: development of a cost-effective parameterisation methodology. Appl. Energy 382, 125221.

Lekhel, C.E.S., Mbayed, R., Hokmabad, H.N., Husev, O., Velihorskyi, O., Monmasson, E., 2026. Joint sizing and energy management optimization with dynamic battery lifetime modeling in residential PV–battery systems. Renew. Energy 260, 125188.

Leon, J.I., Dominguez, E., Wu, L., Alcaide, A.M., Reyes, M., Liu, J., 2020. Hybrid energy storage systems: concepts, advantages, and applications. IEEE Ind. Electron. Mag. 15 (1), 74–88.

Li, Z., Deusen, D., 2025. Role of energy storage technologies in enhancing grid stability and reducing fossil fuel dependency. Int. J. Hydrogen Energy 102, 1055–1074.

Liao, C., He, L., Tan, Z., Li, Y., Chen, C., Cao, Y., 2025. Optimal planning method for scalable energy storage station in power system considering multiple application scenarios. Int. J. Electr. Power Energy Syst. 172, 111140.

Lin, T., Zhang, L., Wan, J., Chen, C.M., Li, J., 2025. Energy price uncertainty and renewable energy technological innovation: evidence from listed Chinese firms. Renew. Sustain. Energy Rev. 213, 115447.

Liu, Y., Chen, Y., Xin, H., Tu, J., Zhang, L., Song, M., Zhu, J., 2025. System strength constrained grid-forming energy storage planning in renewable power systems. IEEE Trans. Sustain. Energy 16 (2), 981–994.

Lucaferri, V., Valentini, M., De Lia, F., Laudani, A., Presti, R.L., Schioppo, R., Fulginei, F.R., 2023. Modeling and optimization method for Battery Energy Storage Systems operating at variable C-rate: a comparative study of Lithium technologies. J. Energy Storage 73, 109232.

Malik, F.H., Hussain, G.A., Alsmadi, Y.M., Haider, Z.M., Mansoor, W., Lehtonen, M., 2025. Integrating energy storage technologies with renewable energy sources: a pathway toward sustainable power grids. Sustainability 17 (9), 4097.

Manassaldi, J.I., Incer-Valverde, J., Mussati, S.F., Morosuk, T., Mussati, M.C., 2024. Optimization of liquid air energy storage systems using a deterministic mathematical model. J. Energy Storage 102, 113940.

Martínez, M., Mateo, C., Gómez, T., Alonso, B., Frías, P., 2024. Distributed battery energy storage systems for deferring distribution network reinforcements under sustained load growth scenarios. J. Energy Storage 100, 113404.

McPherson, M., McBennett, B., Sigler, D., Denholm, P., 2020. Impacts of storage dispatch on revenue in electricity markets. J. Energy Storage 31, 101573.

Minh, N.Q., Van Dai, T., Hoang, P.M., Hai, P.H., Chau, L.T.M., 2025. A mixed-integer linear programming model for BESS sizing optimization considering aging effects and emission costs. Smart Energy 100225.

Mirzaei, M.A., Yazdankhah, A.S., Mohammadi-Ivatloo, B., Marzband, M., Shafie-khah, M., Catalão, J.P., 2019. Stochastic network-constrained co-optimization of energy and reserve products in renewable energy integrated power and gas networks with energy storage system. J. Clean. Prod. 223, 747–758.

Mulky, L., Srivastava, S., Lakshmi, T., Sandadi, E.R., Gour, S., Thomas, N.A., Sudhakar, K., 2024. An overview of hydrogen storage technologies–Key challenges and opportunities. Mater. Chem. Phys. 325, 129710.

Naseri, N., Ghiassi-Farrokhfal, Y., Ketter, W., Collins, J., 2023. Understanding and managing the participation of batteries in reserve electricity markets. Decis. Support Syst. 165, 113895.

Nicoli, M., Faria, V.A.D., de Queiroz, A.R., Savoldi, L., 2024. Modeling energy storage in long-term capacity expansion energy planning: an analysis of the Italian system. J. Energy Storage 101, 113814.

Ochoa-Barragán, R., Ponce-Ortega, J.M., Tovar-Facio, J., 2023. Long-term energy transition planning: integrating battery system degradation and replacement for sustainable power systems. Sustain. Prod. Consum. 42, 335–350.

Odoi-Yorke, F., Abbey, A.A., Frimpong, T.A., Asante, E., Amewornu, E.M., Davis, J.E., Atepor, L., 2024. A bird's eye view of pumped hydro energy storage: a bibliometric analysis of global research trends and future directions. J. Energy Storage 103, 114339.

Olabi, A.G., Abdelghafar, A.A., Maghrabie, H.M., Sayed, E.T., Rezk, H., Al Radi, M., Abdelkareem, M.A., 2023. Application of artificial intelligence for prediction, optimization, and control of thermal energy storage systems. Therm. Sci. Eng. Prog. 39, 101730.

Ong, T.C., Sarvghad, M., Bell, S., Will, G., Steinberg, T.A., Yin, Y., Lewis, D., 2024. Review on the challenges of salt phase change materials for energy storage in concentrated solar power facilities. Appl. Therm. Eng. 238, 122034.

Oosterom, J.P., Hall, C.A., 2022. Enhancing the evaluation of Energy Investments by supplementing traditional discounted cash flow with Energy Return on Investment analysis. Energy Policy 168, 112953.

Pan, T., Tan, Z., Wang, Z., Liu, B., Jin, X., Wang, L., 2025. Optimal planning method for energy storage system based on power allocation of integrated energy system. Int. J. Green Energy 1–17.

Parvathareddy, S., Yahya, A., Amuhaya, L., Samikannu, R., Suglo, R.S., 2025. A hybrid machine learning and optimization framework for energy forecasting and management. Results Eng. 26, 105425.

Piat, C., Sari, A., Viton, C., 2025. Emerging battery technologies: the main aging mechanisms and challenges. Batteries 10.

Prasanthi, A., Shareef, H., Errouissi, R., Alahmad, A.K., 2026. Stochastic optimization of energy storage systems in hybrid electric vehicle under uncertainty. J. Energy Storage 142, 119516.

Rajah, I., Sowe, J., Schimpe, M., Barreras, J.V., 2024. Degradation-aware derating of lithium-ion battery energy storage systems in the UK power market. Electronics 13 (19), 3817.

Razzhivin, I.A., Suvorov, A.A., Ufa, R.A., Andreev, M.V., Askarov, A.B., 2023. The energy storage mathematical models for simulation and comprehensive analysis of power system dynamics: a review. Part I. Int. J. Hydrogen Energy 48 (58), 22141–22160.

Reibsch, R., Blechinger, P., Kowal, J., 2023. The importance of battery storage systems in reducing grid issues in sector-coupled and renewable low-voltage grids. J. Energy Storage 72, 108726.

Ren, L., Yuan, M., Jiao, X., 2023. Electric vehicle charging and discharging scheduling strategy based on dynamic electricity price. Eng. Appl. Artif. Intell. 123, 106320.

Ricardo, C.H., Adriana, L.H., Nelson, D.A., 2023. Energy management supported on genetic algorithms for the equalization of battery energy storage systems in microgrid systems. J. Energy Storage 72, 108510.

Rikkas, R., Lahdelma, R., 2021. Energy supply and storage optimization for mixed-type buildings. Energy 231, 120839.

Rojas-Innocenti, S., Baeyens, E., Frechoso, F., Martín-Crespo, A., Saludes-Rodil, S., 2025. A robust optimization framework for flexible industrial energy scheduling: application to a cement plant with market participation. Energy Rep. 14, 4570–4588.

Safavi, V., Vaniar, A.M., Bazmohammadi, N., Vasquez, J.C., Keysan, O., Guerrero, J.M., 2025. A battery degradation-aware energy management system for agricultural microgrids. J. Energy Storage 108, 115059.

Sahoo, S., Timmann, P., 2023. Energy storage technologies for modern power systems: a detailed analysis of functionalities, potentials, and impacts. IEEE Access 11, 49689–49729.

Saxena, V., Manna, S., Rajput, S.K., Sharma, B., Kumar, P., Diwania, S., Kim, M.K., 2025. Evolutionary computing in energy systems integration: techniques and applications. Energy Rep. 13, 6450–6478.

Seward, W., Qadrdan, M., Jenkins, N., 2022. Revenue stacking for behind the meter battery storage in energy and ancillary services markets. Elec. Power Syst. Res. 211, 108292.

Shabani, M., Yan, J., 2025. Optimization and evaluation of operational and economic performance of grid-connected battery storage at different stages of battery health. Energy Convers. Manag. X 101113.

Shahbazbegian, V., Dehghani, F., Shafiyi, M.A., Shafie-khah, M., Laaksonen, H., Ameli, H., 2023. Techno-economic assessment of energy storage systems in multi-energy microgrids utilizing decomposition methodology. Energy 283, 128430.

Shuja, A., Khan, H.R., Murtaza, I., Ashraf, S., Abid, Y., Farid, F., Sajid, F., 2024. Supercapacitors for energy storage applications: materials, devices and future directions: a comprehensive review. J. Alloys Compd. 1009, 176924.

Su, X., Sun, B., Wang, J., Ruan, H., Zhang, W., Bao, Y., 2023. Experimental study on charging energy efficiency of lithium-ion battery under different charging stress. J. Energy Storage 68, 107793.

Tarashandeh, N., Karimi, A., 2021. Utilization of energy storage systems in congestion management of transmission networks with incentive-based approach for investors. J. Energy Storage 33, 102034.

Tasneem, O., Tasneem, H., Xian, X., 2025. Lithium-ion battery technologies for grid-scale renewable energy storage. Research 100297.

Tejada-Arango, D.A., Kiviluoma, J., Morales-España, G., 2025. Debunking the speed-fidelity trade-off: speeding-up large-scale energy models while keeping fidelity. Int. J. Electr. Power Energy Syst. 168, 110674.

Ullah, A., Nobanee, H., Ullah, S., Iftikhar, H., 2024. Renewable energy transition and regional integration: energizing the pathway to sustainable development. Energy Policy 193, 114270.

Vallese, L., Javadi, H., Badenes, B., Urchueguia, J.F., Lombardo, G., Menegazzo, D., Fedele, L., 2026. A comprehensive review of thermal energy storage technologies and their applications: creation of a database. Renew. Sustain. Energy Rev. 225, 116133.

Van Someren, C., Visser, M., Slootweg, H., 2025. How battery energy storage impacts grid congestion–An evaluation of the trade-offs between different battery control strategies. J. Energy Storage 134, 118100.

Wang, C., Wang, R., Li, J., Li, Z., Yu, Q., 2025. Cycle-Efficient modeling for degradation staging and early life prediction of lithium batteries. Green Energy Intell. Transp. 100338.

Wang, Y., Wang, H., Cao, Y., 2024. Comprehensive review of storage optimization techniques in blockchain systems. Appl. Sci. 15 (1), 243.

Wang, X., Ying, L., Wen, K., Lu, S., 2022. Bi-level non-convex joint optimization model of energy storage in energy and primary frequency regulation markets. Int. J. Electr. Power Energy Syst. 134, 107408.

Wang, X., Yu, X., 2024. Differential evolution algorithm with three mutation operators for global optimization. Mathematics 12 (15), 2311.

Wang, Q., Li, X., 2025. Evaluation of battery energy storage system to provide virtual transmission service. Elec. Power Syst. Res. 244, 111570.

Wu, F.B., Yang, B., Ye, J.L., 2019. Grid-scale Energy Storage Systems and Applications. Academic Press.

Xie, W., Zhang, L., Zhao, X., Zhang, D., 2025. Analysis of thermal characteristics and energy efficiency of lithium-ion batteries under different charging conditions. Therm. Sci. Eng. Prog. 61, 103577.

Xu, Y., Gao, W., Li, Y., Xiao, F., 2023. Operational optimization for the grid-connected residential photovoltaic-battery system using model-based reinforcement learning. J. Build. Eng. 73, 106774.

Yagci, M.C., Richter, O., Behmann, R., Bessler, W.G., 2025. Degradation modes of large-format stationary-storage LFP-based lithium-ion cells during calendaric and cyclic aging. J. Energy Storage 124, 116774.

Yang, H., Wang, J., Ma, Y., Zhang, D., Guerrero, J.M., 2024. Battery energy storage system planning for promoting renewable energy consumption using a constraint boundarization technique. IET Renew. Power Gener. 18 (15), 2867–2879.

Yang, C., Xia, X., Sun, C., Peng, B., Wu, J., Wang, Z., 2025. Single-objective and multi-objective optimization of the thermally integrated-pumped thermal energy storage system based on the novel optimization method coupling the orthogonal design and intelligent algorithm. Appl. Therm. Eng. 272, 126404.

Zhai, C., Nguyen, H.D., Xiao, G., 2020. A robust optimization approach for protecting power systems against cascading blackouts. Elec. Power Syst. Res. 189, 106794.

Zhang, T., Strbac, G., 2025. Artificial intelligence applications for energy storage: a comprehensive review. Energies 18 (17), 4718.

Zhang, Y., Yang, Z., 2021. Comparative study on optimized round-trip efficiency of pumped thermal and pumped cryogenic electricity storages. Energy Convers. Manag. 238, 114182.

Zhang, Y., Tian, J., Guo, Z., Fu, Q., Jing, S., 2025. Uncertainty-aware economic dispatch of integrated energy systems with demand-response and carbon-emission costs. Processes 13 (6), 1906.

Zhao, C., Andersen, P.B., Træholt, C., Hashemi, S., 2023. Grid-connected battery energy storage system: a review on application and integration. Renew. Sustain. Energy Rev. 182, 113400.

Zhou, J., Wang, Y., Chiu, L.N., Ghabraie, K., 2025. On the suitability of simplified sensitivity estimation for partial elements in topology optimization. Eng. Optim. 1–23.

Microgrid optimization model: Component cost and constraints

9

Mohammed Amine Hoummadi[1], *Badre Bossoufi*[1], *Mohammed Karim*[1] *and Mohammed Alenezi*[2]
[1]LIMAS Laboratory, Faculty of Sciences Dhar El Mahraz, Sidi Mohammed Ben Abdellah University, Fez, Morocco; [2]Wolfson Centre for Magnetics, School of Engineering, Cardiff University, Cardiff, United Kingdom

9.1 Introduction

Microgrids have emerged as one of the most significant developments in today's energy landscape. They're reshaping the way electricity is produced, delivered, and managed. As cities expand, global energy demand rises, and environmental concerns grow more serious, the old model of relying on large, centralized power plants is becoming less practical. Frequent grid failures, transmission losses, and continued dependence on fossil fuels have pushed the shift toward more decentralized and renewable-based systems. In this context, microgrids offer a flexible and dependable solution: they can operate alongside the main grid or independently from it, while drawing on a mix of different energy sources (Hoummadi et al., 2023).

The rise of renewable energy sources such as solar, wind, and biomass has created new opportunities to generate power in a sustainable way. At the same time, it has brought technical and economic challenges, since these sources aren't always consistent and their output depends on weather and location. Microgrids help address these issues by combining distributed energy resources with advanced control systems, energy storage, and demand-side management. This allows them to balance production and consumption in real time, making the system more reliable and efficient while also reducing the overall carbon footprint (Hoummadi et al., 2024a).

Despite their advantages, the deployment and operation of microgrids face significant barriers, primarily related to cost and complexity. The optimization of microgrid configuration requires balancing multiple objectives—minimizing energy costs, maximizing renewable utilization, ensuring supply reliability, and maintaining technical stability. The initial investment cost, including the capital expenditures for renewable units such as photovoltaic (PV) panels, wind turbines, converters, batteries, and auxiliary systems like biomass plants, often represents a large share of the total project budget. Moreover, operational and maintenance (O&M) costs, replacement cycles, and local economic conditions further influence the system's financial viability. As such, a holistic approach is needed to evaluate the interplay between technical and economic parameters to achieve an optimal configuration (Hoummadi et al., 2025).

Energy Storage Technologies. https://doi.org/10.1016/B978-0-443-45858-3.00010-3

Computational and mathematical optimization models have emerged as essential resources for microgrid system design and evaluation in recent years. For example, researchers and engineers can use techniques like artificial intelligence (AI)-based heuristics, mixed-integer linear programming (MILP), and multi-objective optimization to simulate and evaluate different scenarios that take into account market dynamics, demand profiles, energy production, and storage. These models make it possible to find the most economical setups that meet technical requirements and advance policy and environmental objectives. To compare various system architectures and assess the economic viability of renewable microgrids in comparison to traditional energy systems, levelized cost of energy (LCOE) analysis in particular offers a standardized metric (Hoummadi et al., 2025).

This study focuses on creating a mathematical and economic optimization model for a hybrid microgrid, which brings together wind turbines, solar panels, biomass plants, battery storage, and power converters. The goal is simple but important: to figure out the setup that minimizes the total cost of energy (TCOE) while respecting each component's limitations and performance targets. But it's not just about numbers—financial and policy factors, such as local resource availability, electricity tariffs, and government incentives, also play a big role in whether a microgrid is truly viable. By combining these technical and economic perspectives, this work sheds light on the cost dynamics and optimization strategies needed to build microgrids that are not only efficient and sustainable but also practical in the real world (Hoummadi et al., 2024b).

Ultimately, this study emphasizes how crucial it is to combine technical optimization with economic modeling when developing the next generation of resilient energy systems. The goal is to support local energy independence and drive technological innovation, while also helping policymakers, engineers, and other decision-makers craft strategies that align with broader sustainability goals. By relying on careful analysis and data-driven modeling, the study provides practical insights that can guide real-world planning and implementation (Hajizadeh and Aliakbar Golkar, 2007).

9.2 Fundamental economic and technical factors

The initial investment cost, one of the most important factors influencing the viability and overall performance of the system, is a crucial component of microgrid optimization. It covers capital expenditures for the acquisition and installation of renewable energy sources, including wind turbines, PV panels, battery storage systems, power converters, and, in certain cases, backup sources like diesel generators or biomass. Since these expenses account for a sizable portion of the overall investment, they play a critical role in determining the LCOE, one of the most significant metrics for comparing the economic efficiency of various energy systems. Proper CAPEX estimation is essential to ensure long-term cost-effectiveness and investment stability because renewable technologies typically have high initial costs and very low operating costs (Hangaragi, 2015).

Beyond the capital phase, operation and maintenance costs have a strong impact on the economics of microgrids. These are the ongoing costs needed to maintain system reliability and performance, including periodic inspections, the replacement of components, software updates, and technical support. O&M costs also become relevant in hybrid systems where multiple technologies interact, since the complexity of the system integration might increase maintenance needs. Battery degradation over time and inverter efficiency losses must be considered to ensure real financial projections, for instance. A well-structured maintenance strategy extends not only the lifetime of the critical components but also minimizes downtime for stable energy delivery, thereby improving ROI (An Luu, 2014; Vinayagam et al., 2018).

Fuel costs, which are frequently influenced by regional availability and market volatility, are a consequence of integrating fuel-based generating sources, like diesel or biomass, into a microgrid. Biomass systems, for instance, may be dependent on organic waste or agricultural residues; their supply chains and prices may vary seasonally or in response to local demand pressures. Despite their dependability, diesel-based systems are susceptible to changes in the world's oil markets, transportation expenses, and environmental regulations. For accurate cost estimates and risk reduction, it is crucial to incorporate these stochastic variables into an optimization model. In order to attain the best possible economic balance, planners can use this modeling to look into alternate operational strategies, such as giving priority to renewable generation during times of high fuel prices.

From a financial view, government policies and incentives have an equally significant influence on the overall cost structure and viability of microgrids. Mechanisms like subsidies, feed-in tariffs, tax credits, and renewable energy grants can dramatically reduce the effective cost per kilowatt-hour (kWh), thereby encouraging investment in sustainable energy infrastructure. In developing regions, where upfront financing can be a major barrier, access to favorable policy frameworks or international development funds often determines whether a microgrid project becomes feasible. Consequently, the optimization process must integrate these policy variables to accurately reflect the true cost-benefit dynamics of renewable energy deployment (Vinayagam et al., 2018; Xu et al., 2015).

Tariff structures that specify how producers and consumers engage with the energy market create additional complicating factors. For example, TOU tariffs impose variable rates according to the time of consumption, imposing higher prices during periods of high demand and offering reduced rates during off-peak hours. By encouraging users to adjust loads to better match the profiles of renewable generation, such mechanisms contribute to increased grid stability and overall system efficiency. Demand-based tariffs encourage investment in energy storage and load management technologies by penalizing consumers for excessive instantaneous power consumption. These pricing strategies influence consumer behavior and play a key role in maintaining supply and demand equilibrium in the microgrid (Olivares et al., 2014).

Ultimately, geographic location is a key factor in determining technical performance and financial returns for microgrid systems. Natural resources, such as solar irradiation, wind potential, and biomass feedstock, differ significantly based on location, which directly impacts generation capacity and the cost per unit of energy.

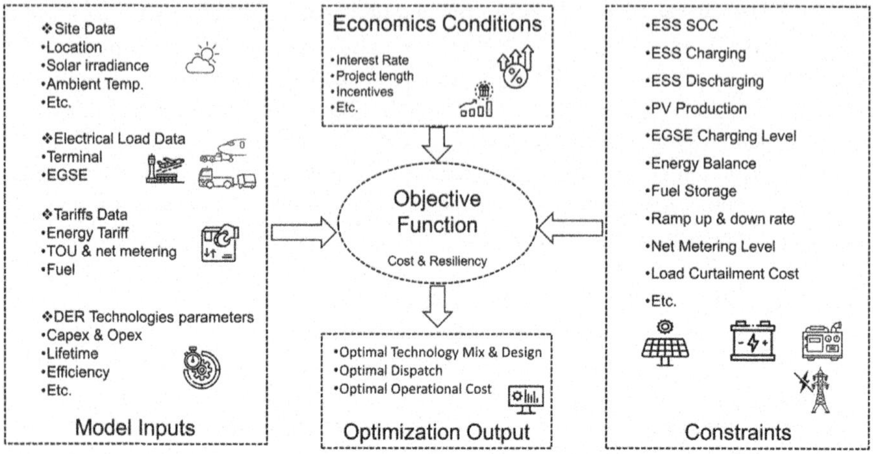

Figure 9.1 The main parameters that influence energy optimization in a microgrid.

Similarly, local infrastructure conditions, labor costs, import duties on renewable equipment, and climate patterns all contribute to variations in the installation and maintenance expenses. Hence, site-specific modeling is necessary to ensure that the optimization framework reflects the true economic and environmental realities of each project, as illustrated in Fig. 9.1.

In short, a multifaceted analysis that incorporates capital investments, operating costs, fuel variability, policy incentives, tariff mechanisms, and geographic considerations is necessary for the economic optimization of microgrids. Energy planners and decision-makers can only create technically sound, economically viable, and resilient systems in the face of environmental and market uncertainties by taking such a thorough and dynamic approach (Lombardi et al., 2009).

9.3 Optimization model

The optimization model aims to minimize the TCOE expressed as:

$$TCOE = Min[(\Sigma Costs)/Pmoy]$$

With Pmoy = 8.2 GWh as the total average power production. The objective function can be expressed as:

$$(Nwind*Cwind + Npanel*Cpanel + Nbatt*Cbatt + Cconvert + Cbiomass)/Pmoy \quad (9.1)$$

Where the following component costs and limits apply:

- <Pmoy> = 47 kW × 365.24 × 20 = 8.2 GWh
- <Pmoy> = Nwind × <Pwind> + Npanel × <Ppanel>

- Cpanel = 1200 + 1200 + 4×20 = 2480 USD/kW
- Cwind = 2300 + 1500 + 20×4 = 3880 USD/kW
- Cbatt = 234×4 + 1.67 + 20 = 970 USD/unit
- Cconvert = 150×20 + 127×150 = 22,000 USD
- Cbiomass = 3880×50 = 194,000 USD
- Constraints: Nwind ≤19; Npanel ≤125; Nbatt ≤14

9.3.1 Objective: Minimize the total cost of energy

The core objective of the optimization is to minimize the TCOE, expressed as the sum of component investments divided by the total produced energy over the evaluation period.

In our simplified formulation, the numerator contains the capital costs of the major components (wind, PV, batteries, converter, biomass). The denominator is the total average energy production over the planning horizon (Lombardi et al., 2009; Nema et al., 2009).

All costs are expressed in USD and ⟨Pmoy⟩ in kWh (or converted to GWh consistently).

9.3.2 Definition of the denominator (average production)

This represents the total energy demand over the 20-year horizon at a constant 47 kW average load. In practice, the model can replace this scalar with a time series of hourly demand D(t) for higher fidelity, but the scalar is useful for conceptual cost comparisons (NaitMalek et al., 2017).

9.3.3 Assumptions and unit consistency

1. **Units:** Typical practice: convert the numerator to the present value of lifetime costs (USD) and denominator to lifetime energy produced (kWh) to compute USD/kWh (LCOE).
2. **Capacity factor:** ⟨Ppanel⟩ = capacity (kW) × hours/year and same for other components.
3. **Lifetimes:** If component lifetimes differ (e.g., batteries 10 years, PV 25 years), either choose a common horizon (20 years here) and include replacement costs or discount future replacements to present value.
4. **O&M and fuel:** Current Eq. (9.3) uses only CAPEX. For a realistic LCOE, add O&M and fuel costs.

9.3.4 Constraints (technical, operational, policy)

In addition to capacity maxima, the optimization must enforce technical and policy constraints, for example:

- **Energy balance (per time step):** The dynamic demand and produced energy.
- **State of charge (SOC) dynamics for battery:** Use the battery only if the dynamic SOC allows it.

- **Generator limits:** The renewable maximum capacity of production in peak.
- **Reserve and reliability constraints:** Minimum spinning reserve, outage probabilities, or a reliability index (e.g., loss of load probability $\leq$ target).
- **Policy constraints:** Local content rules, maximum allowed fuel consumption, or required renewable share.

9.3.5 Solution approach and recommended algorithms

- **Deterministic MILP:** If using linear approximations (piecewise linear battery efficiencies, linearized generation constraints), model as MILP. Good solvers: Gurobi, CPLEX, or open-source CBC.
- **Nonlinear programming/MINLP:** If you keep nonlinearities (battery degradation models, nonlinear cost curves), use MINLP solvers or nonlinear solvers (IPOPT, BONMIN).
- **Metaheuristics:** For very large or highly nonlinear models, genetic algorithms, particle swarm optimization or simulated annealing may be used to find near-optimal solutions.
- **Stochastic optimization:** To handle fuel price and renewable resource uncertainty, use scenario-based stochastic programming or robust optimization (chance constraints).

9.3.6 Sensitivity and scenario analysis

Run sensitivity analyses on key parameters:

- **Discount rate r:** 3%–10% range (capturing public versus private finance).
- **Capacity factors:** ±20% to model site variability.
- **Fuel price scenarios:** nominal, high, low.
- **Subsidy/tariff scenarios:** presence/absence of incentives, different feed-in rates, TOU tariffs.
- **Component cost declines:** PV/wind capex fall over time (technology learning).

Report results as tornado charts or contour plots of LCOE versus two key variables.

9.3.7 Extensions and remarks

- **Include revenue streams:** P2P energy sales, capacity payments, and crypto-mining income can be modeled as negative costs (revenues) in the numerator or as a separate revenue term to calculate net present value.
- **Include degradation and replacement:** battery cycle life, inverter replacement costs at year TrepT_{\text{rep}}Trep.
- **Multi-objective optimization:** minimize LCOE and maximize renewable share or minimize CO_2 emissions simultaneously (Pareto front analysis) (Kharbouch et al., 2018).

9.3.8 Practical checklist to implement the model

1. **Data preparation:** obtain capacity factors (hourly or monthly), cost breakdowns, O&M schedules, fuel price projections.

2. **Choose time resolution:** hourly for dispatch modeling; annual average for CAPEX-only sizing.
3. **Choose solver and language:** Python + Pyomo/GLPK/CBC/Gurobi, or MATLAB + YALMIP.
4. **Model verification:** sanity checks (extreme cases), unit tests, conservation of energy.
5. **Run base, high/low scenarios** and produce sensitivity charts.
6. **Document assumptions** clearly (lifetimes, discount rates, replacement policy).

9.4 Simulation on an example of 100 houses in Fes, Morocco

The city of Fez, located in northern Morocco, is one of the country's main urban and industrial hubs. Like most Moroccan regions, its electricity is mainly supplied through the national grid managed by the Office National de l'Électricité et de l'Eau Potable (ONEE). Currently, the average retail price of electricity for medium-load consumers in Fez is about 0.115 USD per kilowatt-hour (kWh). This price reflects Morocco's partly fossil fuel-dependent energy mix, where imported natural gas and coal still make up a large part of electricity generation.

This value serves as the baseline cost for evaluating the benefits of renewable microgrid integration. By comparing this baseline with simulation results obtained from HOMER Pro and AI-based optimization models, we can assess the potential economic and environmental improvements achievable through hybrid renewable systems, as shown in Fig. 9.2.

- Y-axis: Electricity price (USD/kWh)
- X-axis: Energy source or model → "Fez Grid (0.115)," "HOMER Raw (0.264)," "HOMER Adjusted (0.06)," "AI-Optimized (0.037–0.055)"
- Purpose: To visually demonstrate how optimization techniques can significantly reduce electricity costs compared to the grid.

9.4.1 Simulation results using HOMER

The simulation carried out in HOMER Pro (Hybrid Optimization of Multiple Energy Resources) provided crucial insights into the economic performance of hybrid microgrid configurations.

The model was designed to satisfy a peak load demand of 210 kW while maintaining system reliability and minimizing operational costs. Under these conditions, the software estimated a LCOE of 0.264 USD/kWh.

However, this result corresponds to the system's peak demand period. In practical scenarios, the average load is significantly lower, approximately 47 kW, which better represents daily operational patterns. To correct the cost value to reflect real-world energy use, the LCOE is proportionally adjusted according to the load ratio:

Cadjusted $= 0.264 \times 47{,}210 = 0.06$ USD/kWh

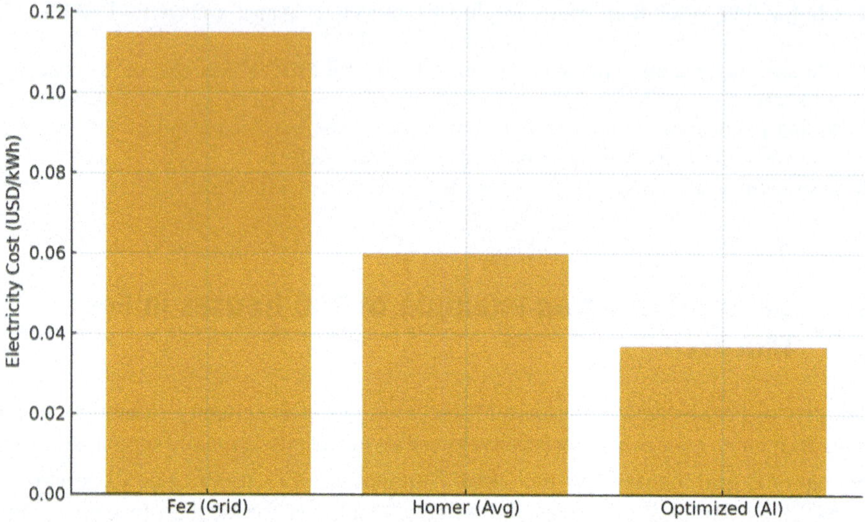

Figure 9.2 Average electricity price comparison.

This adjustment reveals that under typical operating conditions, the effective cost of energy produced by the optimized microgrid is only 0.06 USD/kWh—almost half the price of grid-supplied electricity.

This outcome illustrates HOMER's sensitivity to the load profile and the significance of accurate demand modeling in achieving realistic results. It also demonstrates that microgrids can be economically competitive even before AI optimization is applied (Fig. 9.3).

- X-axis: Load demand (kW)
- Y-axis: Energy cost (USD/kWh)
- Curves: "HOMER Raw (Peak-Based)" versus "Adjusted Load-Based"
- Interpretation: The figure should show a steep decline in cost as the load decreases, stabilizing near 0.06 USD/kWh around 47 kW average load.

9.4.2 Comparative analysis and discussion

9.4.2.1 Cost reduction potential

By integrating renewable energy sources—specifically solar PV, wind turbines, and battery storage—the total cost of electricity can be significantly reduced from 0.115 USD/kWh (conventional grid) to a range between 0.037 and 0.055 USD/kWh.

The minimum achievable cost of 0.037 USD/kWh corresponds to an optimized configuration that utilizes renewable resources with AI-assisted optimization of component sizing, scheduling, and energy flow control.

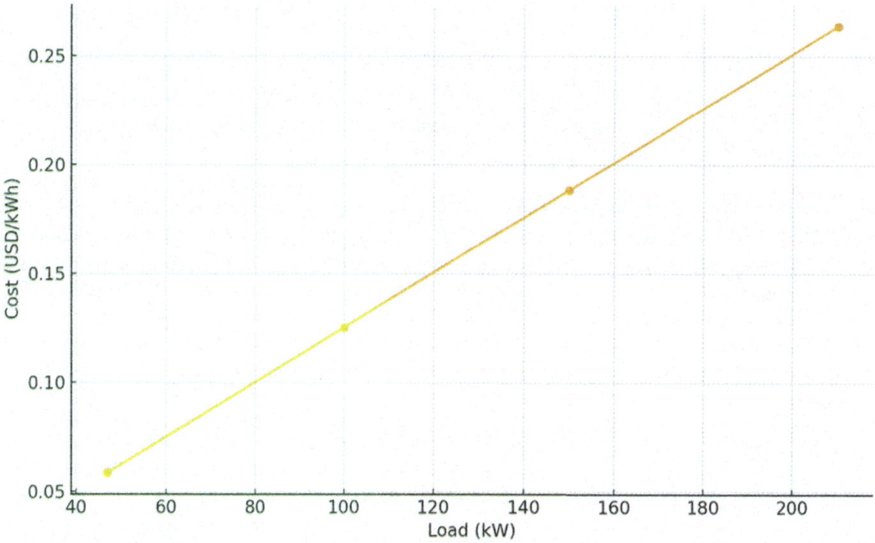

Figure 9.3 Relationship between load and cost per kWh.

This represents a 67% reduction in energy costs, demonstrating that renewable-based microgrids, when correctly optimized, can outperform traditional power supply systems both economically and environmentally (Fig. 9.4).

- Bars: "Conventional Grid," "Renewable Only," "HOMER Optimized," "AI + Renewable Optimized"
- Y-axis: Electricity cost (USD/kWh)
- Key takeaway: The combination of AI and renewables achieves the steepest cost decline.

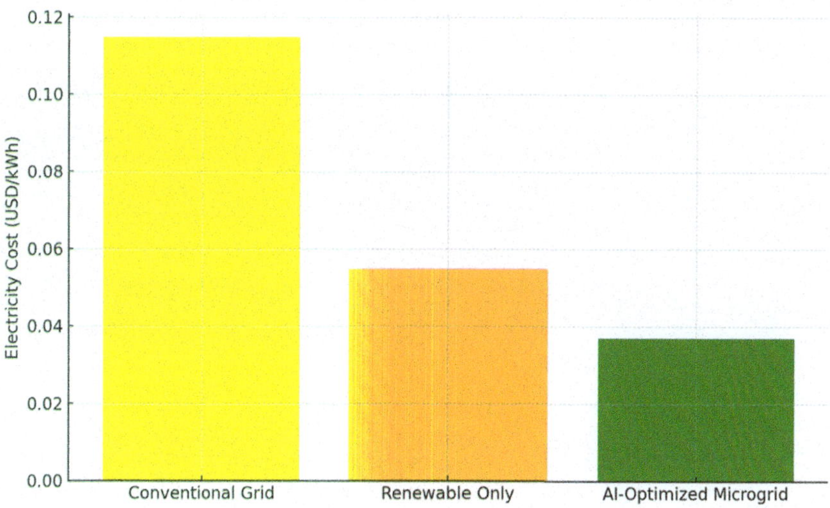

Figure 9.4 Cost reduction achieved by renewable and AI integration.

9.4.2.2 Role of artificial intelligence in microgrid optimization

Traditional optimization methods are limited in handling nonlinear, dynamic, and multi-objective systems such as microgrids. To overcome these challenges, our study introduces AI as a key enabler for next-generation energy management.

Advanced algorithms such as:

- Genetic Algorithm (GA)—for global search of optimal system configurations.
- Artificial Bee Colony (ABC)—for efficient resource allocation and local optimization.
- Ant Colony Optimization (ACO)—for adaptive power flow and routing under varying conditions.

These methods enable dynamic, real-time adaptation of the microgrid to fluctuating energy production (e.g., changing solar irradiation or wind speed) and variable demand profiles.

Unlike static models, AI-driven microgrids can autonomously learn, forecast, and adjust their operation—minimizing costs, maximizing renewable penetration, and reducing CO_2 emissions simultaneously (Fig. 9.5).

- Axes: Optimization efficiency parameters
- Algorithms compared: Classical, GA, ABC, ACO
- Insight: AI-based algorithms (GA, ABC, ACO) show superior adaptability and cost savings versus classical approaches.

9.4.2.3 Adaptability and reusability of microgrid components

One of the innovative contributions of this work is the adaptability of microgrid configurations. Even existing systems can be retrofitted or re-optimized by upgrading

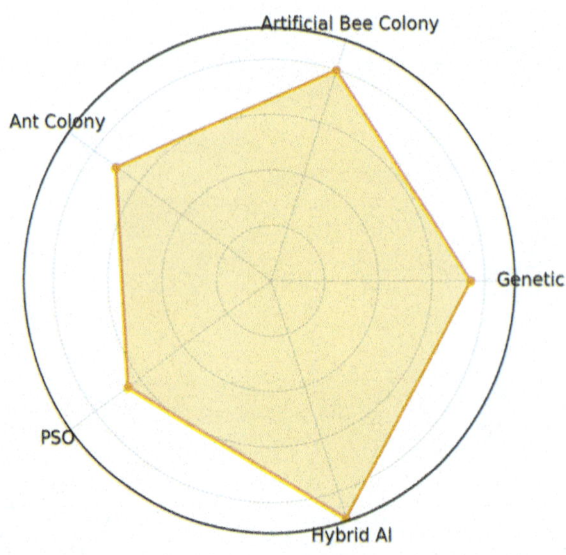

Figure 9.5 Radar chart comparing optimization algorithms.

certain components—such as inverters, batteries, or controllers—to match new optimization results.

Furthermore, these enhanced components can be transferred or reused in other microgrids, located in different regions with unique resource constraints. For example, a microgrid designed for Fez could be partially replicated in another Moroccan city with minor adjustments in solar and storage capacity, leveraging the same optimized architecture.

This approach not only improves cost-efficiency but also supports the scalability and sustainability of renewable infrastructure. Through AI-assisted planning, each microgrid can evolve continuously, reconfiguring itself to maintain peak efficiency in response to changing conditions.

In summary, the combination of HOMER-based modeling and AI-enhanced optimization establishes a robust framework for designing cost-efficient and sustainable microgrids.

For Fez, Morocco, the transition from grid electricity (0.115 USD/kWh) to an optimized renewable microgrid (0.037–0.055 USD/kWh) represents not only a substantial cost reduction but also a step forward toward national energy independence and carbon neutrality.

By embedding intelligence into energy systems, this study showcases how AI-driven microgrid management can revolutionize the local and global energy landscape—making energy cleaner, cheaper, and smarter.

9.5 Discussion and interpretation of funding

So, the equation in (9.1) is basically a quick way to look at a microgrid design by only checking the CAPEX and how much energy we want it to deliver. It's fine when you just want a rough idea of the investment, but honestly, it's too limited if we want something realistic. A microgrid doesn't stop costing money after the installation, so only looking at the upfront cost doesn't tell the whole story. In real life, you still need to count the O&M stuff, fuel for generators if you have any, and even the fact that money today isn't worth the same as money in a few years (so discounting matters). On top of that, renewables don't behave the same every day, and the load isn't constant either, so there's a lot of randomness that affects the final cost. Once you include all these things, the "TCOE" becomes more like a standard LCOE calculation. And honestly, that gives a way better idea of how the microgrid will perform economically and helps you pick a design that actually works in the real world, not just on paper (Elkhoukhi et al., 2018).

The results of the optimization highlight the importance of balancing investment and operational costs across multiple energy sources to achieve both efficiency and sustainability. Wind and solar generation typically dominate the renewable energy contribution due to declining costs and favorable generation patterns, while battery storage enhances system reliability, flexibility, and resilience by mitigating the intermittency of renewables. Power electronic converters ensure seamless compatibility between AC and DC subsystems, allowing diverse generation and storage assets to operate harmoniously within the same microgrid. Biomass or other dispatchable generation units provide a reliable backup, enabling continuous operation even under periods of low renewable generation or high demand. This multidimensional optimization demonstrates that efficient microgrid design is not simply a matter of maximizing renewable penetration but requires careful consideration of how all components interact technically and economically.

A significant insight from the model is that the optimal combination of energy sources and storage assets is not confined to new microgrid designs. Existing microgrids can also benefit from this framework through retrofitting and component replacement. By selectively upgrading storage systems, converters, or generation units, an existing microgrid can be aligned more closely with the optimal configuration identified by the model, improving efficiency, reducing operational costs, and extending system lifespan without necessitating a full redesign. Components replaced from one installation can be repurposed in other microgrids with different operational requirements or geographic constraints, maximizing resource efficiency and providing flexibility in the deployment of multiple systems across varied contexts.

Beyond technical performance, economic and policy factors remain critical determinants of microgrid viability. Financial incentives, local electricity tariffs, and volatile fuel prices can profoundly influence the cost-effectiveness of different configurations. Without integrating these parameters into the optimization framework, even technically optimal microgrids may fail to achieve long-term financial sustainability. Thus, a truly effective approach requires a multidisciplinary perspective, combining technical design, economic analysis, and policy considerations to deliver solutions that are both practically and financially viable (Bakhouya et al., 2017; Guerrero et al., 2011).

The use of advanced AI tools and modern optimization methods can actually push this whole process much further. Instead of relying on the usual "test and adjust" way of designing a microgrid, things like machine learning models, genetic algorithms, or even simple heuristic methods can scan through thousands of possible configurations very quickly. They help find setups that are close to optimal—or at least good enough—under different technical limits and uncertainties. And because AI can simulate real-time situations (like sudden drops in solar output, weird load spikes, or even equipment failure), planners can make decisions earlier and adapt the system before problems appear. In simple words, AI makes it a lot easier to test many scenarios fast, compare components, and see what design gives the best performance without blowing the budget (Xu and Chen, 2011).

Another important point is that this AI-based approach makes it easier to share "what works" between different microgrids. For example, if one site finds a combination of components that works really well, the same logic can guide the design of another microgrid—even if the region has different sunlight levels, different loads, or different local rules. Also, when a component is reused from an old microgrid, AI models already know how it behaves, so it can be placed in a new system more confidently. This creates a sort of dynamic cycle where parts are reused, adapted, and reoptimized instead of thrown away. It cuts costs, avoids unnecessary waste, and speeds up the rollout of strong and sustainable energy systems (Elmouatamid et al., 2017).

To put everything together, the mix of economic modeling, technical optimization, and AI-based decision tools forms a pretty solid blueprint for designing future microgrids—both brand new ones and old ones being upgraded. This approach lets planners look at the full picture: investment costs, operating expenses, system reliability, and even policy or financial incentives. The goal is to end up with designs that make sense technically, stay affordable in the long run, and can be adapted to different regions or conditions. On top of that, the possibility to retrofit old systems and reuse components in new sites means we get more value out of every resource. Overall, this combined method helps engineers and investors build microgrids that meet energy needs, reduce costs, and remain flexible in a fast-changing energy environment.

9.6 Conclusion

This study basically builds a full microgrid optimization model that looks at both cost and performance at the same time. The idea is simple: keep the system reliable and running well, but also push the overall energy cost down as much as possible. The model mixes different sources like solar, wind, biomass, storage units, and converters, so it can find setups that make sense financially and also work well for the environment. One thing that really shows up in the analysis is how important storage and converters are. Without them, the renewable part becomes too random. So the best design is really about balancing the money you invest, how much it costs to operate, and how flexible the system stays.

The model isn't only for new microgrids. You can use it on old ones, too, and upgrade them piece by piece instead of replacing everything. Some parts can even be moved or reused in other microgrids that have different needs or locations. This makes the whole thing more practical and cheaper for large-scale deployment. It's not just theory—it helps both technical decisions and long-term investment planning.

For the future, the model can be improved a lot. Things like demand response, real-time electricity prices, or even carbon pricing can give a more realistic picture. Using ML or AI could help run fast scenario tests and handle uncertainty. Adding features like grid-interactive operation, predictive maintenance, or even coordination between several microgrids would make the system smarter and more resilient.

So, in short, the framework is a solid tool. It gives planners and investors clear insights by mixing technical, economic, and environmental points together. It helps design microgrids that are efficient, sustainable, and flexible enough to follow the changes in energy demand. The work here builds a strong base for future research and real applications, especially for the shift toward cleaner and more resilient energy systems.

References

An Luu, N., 2014. Control and Management Strategies for a Microgrid. Grenoble Institute of Technology, Grenoble.

Bakhouya, M., NaitMalek, Y., Elmouatamid, A., et al., 2017. Towards a context-driven platform using IoT and big data technologies for energy efficient buildings. In: Proceedings of the 2017 3rd International Conference of Cloud Computing Technologies and Applications (CloudTech), Rabat, Morocco, 24–26 October 2017. IEEE, New York, pp. 1–5.

Elkhoukhi, H., NaitMalek, Y., Berouine, A., et al., 2018. Towards a real-time occupancy detection approach for smart buildings. Procedia Comput. Sci. 134, 114–120.

Elmouatamid, A., Ouladsine, R., Bakhouya, M., et al., 2017. Modeling and performance evaluation of photovoltaic systems. In: Proceedings of the 2017 International Renewable and Sustainable Energy Conference (IRSEC), Tangier, Morocco, 4–7 December. IEEE, New York, pp. 1–7.

Guerrero, J.M., Vasquez, J.C., Matas, J., et al., 2011. Hierarchical control of droop-controlled AC and DC microgrids—a general approach toward standardization. IEEE Trans. Ind. Electron. 58 (1), 158–172.

Hajizadeh, A., Aliakbar Golkar, M., 2007. Intelligent power management strategy of hybrid distributed generation system. Int. J. Elec. Power 29, 783–795.

Hangaragi, G.B., 2015. Recent Integration of a PV-wind energy system with enhanced efficiency. Indian J. Sci. Res. 11 (1), 72–78.

Hoummadi, M.A., Bouderbala, M., Alami Aroussi, H., Bossoufi, B., El Ouanjli, N., Karim, M., 2023. Survey of sustainable energy sources for microgrid energy management: a review. Energies 16, 3077. https://doi.org/10.3390/en16073077.

Hoummadi, M.A., Bossoufi, B., Karim, M., Hatatah, M., H. Alghamdi, T.A., Alenezi, M., 2024a. AI-enhanced energy management systems for efficient flow control in microgrids. IET Renew. Power Gener. 19 (1), e70039. https://doi.org/10.1049/rpg2.70039.

Hoummadi, M.A., Aroussi, H.A., Bossoufi, B., Karim, M., 2024b. Constraints and adjustable parameters in microgrids for cost and CO2 emission reduction. In: 2024 16th International Conference on Electronics, Computers and Artificial Intelligence (ECAI). Iasi, Romania, pp. 1–5. https://doi.org/10.1109/ECAI61503.2024.10607579.

Hoummadi, M.A., Bossoufi, B., Karim, M., Althobaiti, A., Alghamdi, T.A., Alenezi, M., 2025. Advanced AI approaches for the modeling and optimization of microgrid energy systems. Sci. Rep. 15 (1), 1–23. https://doi.org/10.1038/s41598-025-96145-w.

Kharbouch, A., El Khoukhi, H., NaitMalek, Y., et al., 2018. Towards an IoT and Big data analytics platform for the definition of diabetes telecare services. In: Proceedings of the Smart Application and Data Analysis for Smart Cities (SADASC'18), Casablanca, Morocco. SSRN, pp. 27–28. https://doi.org/10.2139/ssrn.3186346.

Lombardi, P., Powalko, M., Rudion, K., 2009. Optimal operation of a virtual power plant. In: Proceedings of the 2009 IEEE Power and Energy Society General Meeting (PES'09), Calgary, AB, Canada, 26–30 July. IEEE, New York.

NaitMalek, Y., Kharbouch, A., El Khoukhi, H., et al., 2017. On the use of IoT and Big data technologies for real-time monitoring and data processing. Procedia Comput. Sci. 113, 429–434.

Nema, P., Nema, R.K., Rangnekar, S., 2009. A current and future state of art development of hybrid energy system using wind and PV-solar: a review. Renew. Sustain. Energy Rev. 13 (8), 2096–2103.

Olivares, D.E., Mehrizi-Sani, A., Etemadi, A.H., et al., 2014. Trends in microgrid control. IEEE Trans. Smart Grid 5, 1905–1919.
Vinayagam, A., Alqumsan, A.A., Swarna, K.S.V., et al., 2018. Intelligent control strategy in the islanded network of a solar PV microgrid. Elec. Power Syst. Res. 155, 93–103.
Xu, L., Chen, D., 2011. Control and operation of a DC microgrid with variable generation and energy storage. IEEE Trans. Power Deliv. 26 (4), 2513–2522.
Xu, Y., Zhang, W., Hug, G., et al., 2015. Cooperative control of distributed energy storage systems in a microgrid. IEEE Trans. Smart Grid 6, 238–248.

Global energy storage policies, markets, and investment trends: Driving the transition to a sustainable energy future

10

G.M. Lionus Leo[1]*, S. Murugapoopathi*[2]*, Deepanraj Balakrishnan*[3] *and I.M. Rizwanul Fattah*[4]

[1]St. Joseph's College of Engineering, Chennai, Tamil Nadu, India; [2]PSNA College of Engineering and Technology, Dindigul, Tamil Nadu, India; [3]Department of Mechanical Engineering, College of Engineering, Prince Mohammad Bin Fahd University, Al-Khobar, Saudi Arabia; [4]University of Technology Sydney, Ultimo, NSW, Australia

10.1 Introduction

The global energy industry is accelerating the adoption of the sustainable energy practices. The major factor driving this is the powerful forces of nature, such as the advent of renewable energy sources—solar PV and wind power—interconnection, and the urgent need for climate action (or global warming) in the case of the latter (Mohammed et al., 2026). By the end of this year, the newly installed capacity of solar and wind power plants worldwide will total an astonishing 3372 GW. Nevertheless, the variability and uncertainty of wind and solar power have become major issues, making it very hard to keep the grid and electricity supply secure and reliable. Thus, energy storage systems (ESSs) are very important in the energy transition as they are supposed to support the energy produced during the high generation periods and give it during the peak demand periods or when the renewable energy output is low (Mahadevan et al., 2025; Bragagni et al., 2025; Mitali et al., 2022). They stabilize the grid, cut the waste that comes with renewable energy sources, and facilitate the integration of variable renewable energy sources (VRESs) into the global electricity grids.

Energy storage technologies consist of various solutions and range from electrochemical storage (like lithium-ion and flow batteries) through mechanical storage (pumped hydro and flywheels), thermal storage (like molten salt and phase change materials), and chemical storage (like hydrogen) (Elalfy et al., 2024; Vaghela et al., 2023; Schischke et al., 2024; Kandhasamy et al., 2024; Eswanto, Hamidi et al., 2026; Schwaegerl, 2025). Each technology comes with its special advantages and disadvantages in energy density, scalability, and response time, thus serving applications such as short-term frequency regulation and long-duration seasonal storage (Bhaduri and Singh, 2025). The rapid reductions in pricing in some cases especially for battery

Energy Storage Technologies. https://doi.org/10.1016/B978-0-443-45858-3.00012-7

energy storage systems (BESSs), together with the advancement of new materials and control methods have made these systems suitable for grid-scale, microgrid, and behind-the-meter applications (Prakash et al., 2022; Hatziargyriou and Joos, 2025).

The energy storage business is largely dependent on the policy frameworks and market forces that dictate the short- and long-term conditions of the market. In particular, as part of Sustainable Development Goals (SDGs), governments worldwide are incentivizing energy storage technology adoption by providing financial assistance, reforming regulations, and supporting research and development targeting SDG 7 (Affordable and Clean Energy) and SDG 13 (Climate Action) (Chen, 2025). The U.S. Investment Tax Credit (ITC) and China's 14th Five-Year Plan have positively impacted the market to a great extent through policy measures. Still, there are difficulties like expensive capital and fragmented regulations that the industry has to deal with which are the main reasons why these regions are lagging behind in the renewable energy and electrification investments (Li et al., 2025). Besides this, such issues are the reasons for the prolonged period in the establishment of new energy storage markets in Asia–Pacific, Latin America, and Africa (Mohd Razif et al., 2024; Jean et al., 2025).

This chapter examines the central role of energy storage and its related technologies in achieving a sustainable energy future. The technological landscape is introduced, and special emphasis is put on battery energy storage, pumped hydro, and thermal storage innovations. In addition, it interrogates global policies and market dynamics in the process of exposing the main trends, challenges, and opportunities. The objective of this chapter mainly is to provide a better understanding of the significant contribution that energy storage technologies, which are firmly supported by strong policies and market mechanisms, can and will play in energizing the transition of the world to a decarbonized and resilient energy system through minimizing the time required for the transition to take place.

10.1.1 Scope and objectives

The climate and the future of the planet are inextricably tied to the global energy sector, which is fully reliant on nature and the sun. Such flexibility needs very robust ESSs to ensure that the grid is trustworthy, increase the utilization of renewables, and set up an energy paradigm that can easily absorb fluctuations to be helpful to the scenario mentioned above. To this end, the key objective of this chapter is to provide a comprehensive overview of the technical advancements, policy frameworks, and market dynamics that are influencing the energy storage of the world.

This chapter is structured around three key aspects:

1. **Energy storage technologies**: This section presents an in-depth analysis on the energy storage technologies such as the BESS, PHS (PHS), and thermal energy storage (TES). They are given with more discussion about their technical properties, their application, and their cost-effectiveness and, with reference, they mention notable innovations including enhancement of lithium-ion batteries, low-head Pumped-Storage Hydropower (PHES), and novel phase change materials made for TES.

2. **Global policy frameworks**: This chapter broadens perspective on policy actions in different parts of the world (North America, Europe, Asia–Pacific, Latin America, the Middle East, and Africa) and supports the necessary ESS with financial inducements, regulatory changes, and renewable energy targets. It also evaluates adherence to the SDGs 7 (Affordable and Clean Energy) and 13 (Climate Action).
3. **Market dynamics and investment trends**: This chapter discusses the economic and market drivers of ESSs, which include but are not limited to involvement in different marketplaces, such as competing in various markets, bidding strategies, and the economic side of hybrid systems. It identifies the role of market setups and policies in determining profitability and also highlights the most current topics like peer-to-peer trading and shared-storages.

This chapter investigates these aspects of the above in terms of the peer-reviewed papers, industry reports and public policies until June 2025, and give an international view as well. It also addresses high upfront costs, inconsistent policies, and political uncertainty and also highlights the field that is innovative and for growth.

Major objectives of this chapter are mainly aimed at the following:

1. To consider the technical advancements in BESS, pumped hydro energy storage (PHES), and TES it is necessary to look into their performance metrics, applications, and limitations in making renewable energy integration and grid stability possible.
2. To analyze global policy surrounding the energy storage, to reveal the key drivers, the differences among the regions, and the existing gaps that influence the deployment and scalability.
3. To conduct an inquiry into the market dynamics and investment trends through the lens of economic feasibility, the innovative ways of market participation, and their contribution to carbon neutrality.
4. To articulate the recommendations that will aid in overcoming the technical, economic, and policy barriers, while at the same time recognizing the crucial role of sustainable materials, digital technologies, and aligned regulations in the transition to a green energy future.

10.2 Energy storage technologies landscape

The future of the world moving toward renewable energy sources depends heavily on handling the intermittent nature and variability of power generation from renewable energy sources, which are mainly solar photovoltaic (PV) systems and wind power. Technological advances in power storage not only help stabilize the grid but also allow renewables to be added and so fulfil the decarbonization objectives. Technological options include electrochemical storage (e.g., lithium-ion batteries, flow batteries, and sodium-based batteries), mechanical storage (e.g., flywheels, pumped hydro, and compressed air), thermal storage (e.g., molten salt and phase change materials), chemical storage (e.g., hydrogen) (Njema et al., 2024; Donnelly, 2023; Schwaegerl, 2025; Kandhasamy et al., 2024). While the classification of technologies is one thing, on the other hand, each technology is endowed with its own set of strengths regarding energy density, power output, response time, and scalability, which makes them ideal for different applications like short-term frequency regulation, long-term energy arbitrage, and seasonal storage (Guo et al., 2024; Gür, 2018; De Carne et al., 2026).

10.2.1 BESS

BESS have become the leading technology, thanks to their flexibility, fast reaction time, and continuously dropping prices. BESS are highly employed in grid-scale projects, microgrids, and behind-the-meter setups to control peak demands, reduce renewable energy wastage, and improve the reliability of the grid (AhmadiAhangar et al., 2024; Zhao et al., 2023a). Among other innovations, hydrogen-based storage and Carnot batteries are already on the way to becoming widely acknowledged technologies in the field of energy storage for long-duration and hybrid energy systems applications (Hayati et al., 2024; Shiraishi et al., 2024). The specific requirements of the application, the economic aspect of it, and the local energy policies are the factors to determine the storage technology selection. This section highlights BESS as a critical technology in the energy transition due to its significant advantages. Fig. 10.1 shows the schematic diagram of a BESS integrated with the electric grid.

BESSs are made up of electrochemical systems, which can chemically store and release electrical energy when needed. Lithium-ion (Li-ion) batteries are the most important technology among BESS because they have the highest energy density, efficiency, and lifetime. However, the possibility of using sodium-based batteries and flow batteries is still under consideration, and they might take part in more applications (Conzen et al., 2023; Killer et al., 2020). BESS provides the necessary support for the integration of renewables by offering such services as frequency regulation, voltage support, energy arbitrage, and peak shaving. Technological advances in designs, control techniques, and hybrid setups have contributed to improving the performance, safety, and economic viability of BESS, thus making them an indispensable part of contemporary energy systems (Saxena et al., 2024; Prakash et al., 2022).

10.2.1.1 Technological advancements

Recent research studies underline the direction in which BESS materials, thermal management, and various optimizations are headed. Round-trip efficiencies of lithium-ion batteries range between 85% and 95%, thus keeping the lead in market share, thanks to the innovations in electrode materials and electrolytes that on the one hand, enhance energy density and cycle life while on the other hand, cut costs (Huang et al., 2025; Li et al., 2020). Sodium-based batteries prove to be a

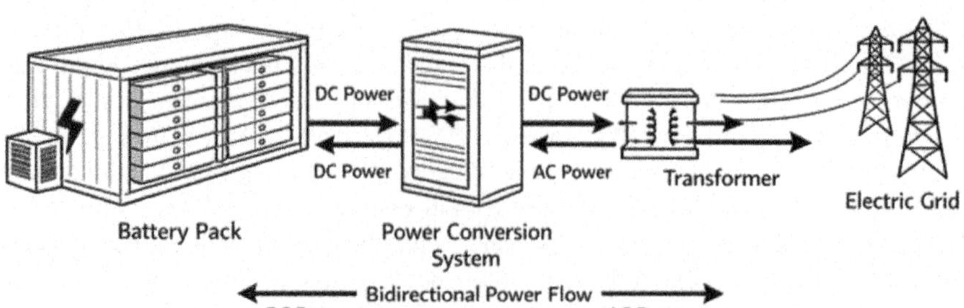

Figure 10.1 Diagram of battery energy storage system (BESS) grid integration.

cost-efficient alternative for large-scale power storage, besides granting safer operation (Xu and Wang, 2022; Wu et al., 2024; Zhao et al., 2023b). In terms of long-term storage, flow batteries are the most viable option due to their energy and power capacities being separate; however, the initial expenses remain a big obstacle.

Proper thermal management is paramount to BESS performance and safety. Researchers are proposing a hybrid method that combines liquid cooling with fin-enhanced phase change materials (Mubashir et al., 2025; Cui et al., 2023; Lu et al., 2024; Lin et al., 2025; Xiao et al., 2023). This new technology lowers the temperature variations by 0.218–0.273 depending on the discharge rates. Such advancements are said to weaken the occurrence of thermal runaway (Rasool et al., 2024; Han et al., 2025). In addition, researchers have harnessed the power of both electrochemical and thermal modeling, allowing for the safe use of air cooling and a maximum temperature of 308.60 K (Dhungana et al., 2025).

The loss of the ability of batteries to maintain charge is the reason for shortening the life span and economic impacts. Newer studies show that calendar and cycle aging can be offset by improved control and modeling of the state-of-health (SOH) (Zhang et al., 2024a; Torregrosa et al., 2024). From an electric vehicle point of view, the incorporation of ultracapacitors in hybrid storage systems can lower Li-ion battery degradation by approximately 23%, resulting in a prolonged battery lifetime increased by nearly 47%; this approach also appears to work on grid-connected ESSs (Pang et al., 2024; Adeyinka et al., 2024). These innovations strongly suggest the adoption of a holistic and performance, safety, and durability-oriented approach to design.

10.2.1.2 Applications and integration

BESS systems are applied in large-scale power generation, microgrid, and backside-meter projects. The table of significant applications and performance metrics from recent studies is presented in Table 10.1, which also highlights the versatility of BESS.

Recent work evaluating battery ESSs for frequency regulation in the Dominican Republic reports a 10% reduction in operational costs and a 12% increase in renewable energy penetration, with payback periods as short as 1.2 years (Cruz-De-Jesús et al., 2025). The research introduces a stochastic model predictive control (SMPC)-based energy management system for residential facilities, which not only optimally runs the BESS to save money but also makes the grid more stable during unpredictable conditions (Saleem et al., 2024).

HESS has combined BESS with other energy sources like hydrogen, flywheels, or supercapacitors to improve the overall performance. The integration of a hybrid ESS that consists of both hydrogen and batteries is going to help reduce the fluctuations in PV power to the extent that the grid requirements are fully met and the renewable energy utilization is consequently improved (Adeyinka et al., 2024). On their part, have mixed iron-air redox batteries with solar power and have thus achieved a round-trip efficiency of 48.77% with an additional power supply capacity of 72.55%. These hybrid systems make use of the different strengths of their technologies to overcome BESS's limitations with respect to long-duration storage.

Table 10.1 Key BESS applications and performance metrics.

Application	Study	Key metrics	Context
Frequency regulation	Khaleel and Yusupov (2026); Cruz-De-Jesús et al. (2025)	10% reduction in operational costs, 12% increase in renewable penetration, 1.2-year payback period	Dominican Republic island grid
Residential energy management	Saleem et al. (2024)	Cost reduction, improved voltage stability under uncertainty	Stochastic MPC for PV-BESS-EV charging
PV power fluctuation suppression	Zhang et al. (2023a)	Meets grid standards, improves PV utilization	Hydrogen-battery HESS with adaptive wavelet decomposition
Microgrid energy storage	Qi et al. (2025)	Minimized costs, managed SoC uncertainties	Adaptive robust optimization for hydrogen-battery HESS
Hybrid thermal-electrical storage	Tetteh et al.(2021); Al-Ghussain et al. (2022); Wang et al. (2025)	48.77% round-trip efficiency, 72.55% increase in power supply capacity	Iron-air battery with CSP-CaL system

In microgrids and off-grid applications, BESS supports energy independence. Recent studies report that improved optimization of university campus PV-BESS systems can increase on-site solar power self-consumption by up to 9.09%, with an estimated payback period of about 8.46 years for residential applications. Recent work proposes the use of adaptive robust optimization for hydrogen–battery hybrid ESSs in microgrids to reduce operating costs while effectively managing uncertainties in the state of charge. Collectively, these findings underscore the broad applicability of BESSs across diverse energy scenarios (Craparo et al., 2017; Linjuan et al., 2023).

10.2.1.3 Economic and policy considerations

The aspects of BESS that are economically viable are capital costs, operational efficiency, and the structure of the market. To support this, recent studies claim that the optimization of BESS capacity in buildings results in a 9.2% reduction in electricity costs with a 6.8-year period for recovering the investment (Su et al., 2025). Recent studies indicate that the use of second-life batteries in hydrogen production reduces levelized costs by taking advantage of the fluctuating power purchase agreements (Förster et al., 2025). However, recent analyses point out that regulations regarding self-consumption might lead to very low welfare gains and thus suggest different policies. The economic effects of the studies are presented together in Table 10.2.

Table 10.2 Economic impacts of BESS deployment.

Economic impact	Context	Study
9.2% reduction in electricity bills, 6.8-year payback period	Building BESS capacity optimization	Su et al. (2025)
Reduced levelized cost of hydrogen	Second-life batteries in hydrogen production	Förster et al. (2025)
LCOE correlation coefficient for PV-BESS	ToU tariff analysis	Komilov et al. (2025)
Low welfare benefits from self-consumption regulation	Household BESS policy evaluation	Semmelmann et al. (2024)

The deployment of BESS is impacted by policy frameworks. The arbitrage is made possible by the Time-of-use (ToU) tariffs, supports the development of a coefficient for assessing the BESS impact on PV LCOE. Studies suggest a bi-level optimization strategy to harmonize the motivations of regulators and businesses regarding hybrid storage while also pointing out the necessity of collaboration in incentives for safety and cost optimization. This interaction between technology, economics, and policy is highlighted in BESS advancement.

10.2.1.4 Challenges and future directions

Battery energy storage systems continue to face key challenges related to safety, scalability, and sustainability. Among these, fire hazards remain a critical concern, underscoring the need for more stable, reliable, and mature technological solutions. Durability of long-duration storage that can be applied in large-scale projects is limited; thus, hydrogen and Carnot batteries are being considered as alternatives (Lykas et al., 2026). Material shortage and recycling are among the sustainability-related issues, a good solution being advanced recycling for material recovery.

The main research in the future should be to control algorithms improving battery life, and this would lead to the enhancement of grid stability and longevity (Sayed et al., 2025). AI can help pinpoint and streamline the state estimation and control processes. The suggestion of the BESS union with cement-based storage paves the way for smart infrastructures to get support. Moreover, the development of BESS will be encouraged by the regulatory support for hybrid systems and second-life batteries.

10.2.2 Pumped hydro storage

Pumped hydro energy storage is a conventional and recognized technology for energy storage that accounts for more than 90% of the global capacity for long-duration energy storage (Ali et al., 2023; Rehman et al., 2015). The process is done by converting excess electric power into gravitational potential energy and pumping water from the lower reservoir to the higher one. Therefore, during the times of peak power demand, this stored energy is set free through the turbines to produce electricity. PHES has

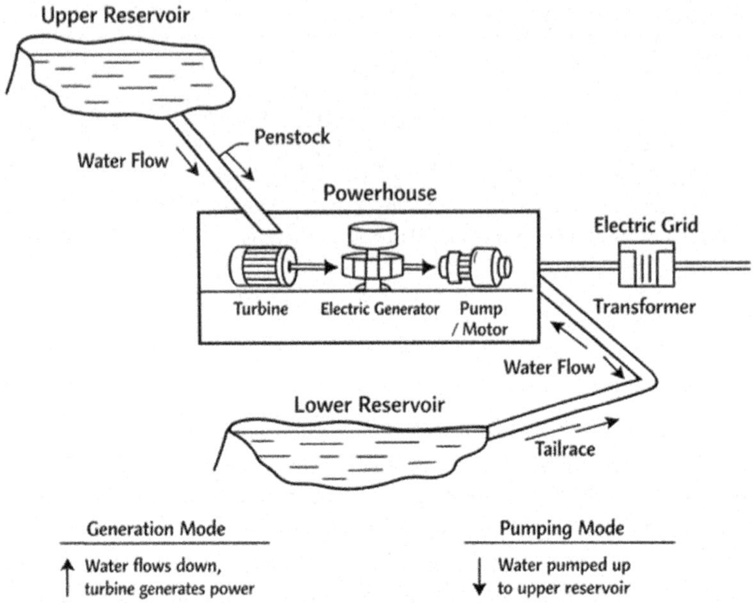

Figure 10.2 Schematic line diagram of a typical pumped hydro energy storage (PHES) system.

been a crucial element in the electricity grid's reliability, the use of clean energy, and carbon-neutrality initiatives because its energy delivered is flexible and can be controlled, even over very long periods. Fig. 10.2 shows the schematic line diagram of a typical pumped hydro ESS operating in pumping and generation modes.

10.2.2.1 Operational principles and configurations

PHES systems are commonly made up of two reservoirs at different heights, connected through a penstock and a reversible pump-turbine or distinct units for pumping and turbine. In times of low demand, the surplus electricity generated by renewable sources like wind or solar is used to pump water to the top reservoir. The same water can then be released to turn the turbines and generate electricity during peak hours or when the wind is calm (Hassan et al., 2023). The latest technologies include variable-speed PHES (VS-PHS) and ternary PHES (T-PSHPG), which, although their versatility and green power generation still remain to be discussed, are the systems of the future. The doubly fed variable-speed pump-turbines of the VS-PHS systems, for example, allow for better frequency regulation and institutionalized efficiency across a broader range of operational settings (Zhang et al., 2023b). Ternary systems that include hydraulic short-circuit modes facilitate concurrent pumping and generation, thus improving small-signal stability (Satapathy et al., 2024).

In addition to the aforementioned, low-head and micro-PHES, etc., are also some of the innovative configurations that circumvent geographical restrictions. To give the best possible performance across most types of terrain, the LH-PHS generates very high efficiencies by using contra-rotating pump-turbines (CRPTs). ELH-HS can achieve around 88.6 and 86.1% efficiency in pump and turbine mode, respectively, producing power outputs in the range of 13.0–15.4 kW for more than 80% efficiencies (Ansorena Ruiz et al., 2025). Micro-PHES, which is based on using the agricultural reservoirs, presents the most economical solution for the rural regions, with costs projected at $0.215/kWh, on par with lithium-ion batteries for smaller demand (Cui et al., 2024). Offshore PHES like the Ocean Battery use the hydrostatic pressure of the seabed to reach the round-trip efficiencies of conventional PHES while lessening the ecological footprint (Nienhuis et al., 2023).

10.2.2.2 Environmental and economic performance

Energy storage systems based on pumped hydro (PHES) show environmental superiority over alternatives like compressed air energy storage (CAES) and lithium-ion batteries. The life cycle assessment (LCA) of the PHES project at Liaoning, China, indicated that most of the environmental impact categories were mainly affected by grid electricity, mostly sourced from coal, which accounted for more than 99% of the total environmental effect. The use of electricity generated from renewable energies, such as wind or solar, results in a significant drop in environmental impacts; on the contrary, it may contribute to metal depletion issue, as resource-intensive requirements for renewable infrastructures lead to such depletion (Osman et al., 2023). It is very important to keep the plant running at high efficiency, as higher efficiencies are able to counteract the environmental impacts much better than prolonging the plant's lifecycle.

Pumped hydro energy storage incurs large up-front capital costs and unfulfilled markets that are especially pronounced in deregulated electricity networks like Australia's National Electricity Market (NEM). Policy measures such as semi-regulated models can effectively bring about almost 40% lower financing costs, which in turn sizes down the market prices for end-users (Šeho and Hodžić, 2024). In the case of mixed systems, if PHES is used together with PV and wind, it gains a lot in terms of economic feasibility. For instance, a solar-PHES project in Northern China had a power supply guarantee rate of over 90% with under 15% curtailment rate by using the optimal PV overbuilding ratios of 1.33 resulting in 5.79% cost reduction as compared to simpler models. In the case of Portugal by 2030, PHES is going to play a role in reaching the target of 85% renewable energy integration and the resulting CO_2 emissions would be only about 1 Million Tons of CO_2eq.

10.2.2.3 Role in renewable integration and grid stability

PHES is the key technology that can resolve the issue of renewable curtailment and stabilize the grid. In the Northwest China scenario, a 10% increase in PHES capacity will contribute to the renewable curtailment reduction. On the contrary, if the planning is not synchronized, certain regions may suffer a 1.1% increase in curtailment due to the mismatch of supply and demand. Large reservoir PHES in Brazil can utilize this method to generate power while flood control by storing 13 km^3 of water at an estimated cost of $2.7 billion. Wind-PV-PHES kind of hybrid systems ensure stability of frequency with a recovery rate of 98% at the level of 62% of the variable renewable energy (VRE) penetration which is helping to reduce the insufficiency of flexibility by 15% for fixed-speed PHES and by 4% for variable-speed PHES.

10.2.2.4 Challenges and innovations

Despite the high capital costs and the need for specific sites, the barriers to entering the market are still quite high. The implementations of Abandoned Mine Pumped Hydro Energy Storage (AMPHS), among other things, have been able to overpass land limits and at the same time, to remove the fluctuations coming from PV generation. The advanced design of isobaric liquid carbon dioxide PHES systems will give a rise to 52.95% efficiencies in round-trip and 3.85 kWh/m^3 in energy storage densities which will in turn provide economic performance of $0.131/kWh with a cost of energy storage being flat. Also, the low-head PHES systems are described to have virtual clocking effects induced by gravity which leads to the need for more efficient designs to eliminate flow separation and hydraulic instability (Wang et al., 2022).

10.2.2.5 Policy and market implications

The implementation of PHES is greatly influenced by policy frameworks. Semi-regulated policies in Australia cater to the market's needs, thus they conveniently and effectively bring down financing costs (Abdul Qadir et al., 2024). On the other hand, in China, the capacity tariff mechanism for PV energy systems (PHES) as part of the integrated renewable energy systems (IRESs) distribution has been a very efficient way to balance the benefits which, in turn, has attracted more investors. All these grants support the power of coordinating to fully exploit the benefits of PHES.

10.2.3 Thermal energy storage

Thus, the TES systems not only store but also manage the supplied energy and the energy needed in the situation of the use of renewable sources. The TES systems are mostly used as a means to increase the efficiency and reliability of solar power, district heating, and many industrial processes. However, different TES technology approaches exist, and they are categorized into three main groups: sensible heat storage, latent heat storage, and thermochemical storage, each of which has its own advantages that suit specific application needs. This chapter discusses the recent advancements in TES technology, exploring new materials, system designs, and performance improvements, backed up by major research findings.

10.2.3.1 Sensible heat storage

Sensible heat storage is a process where thermal energy is stored by raising the temperature of a storage medium, which can be water, molten salts, or even solid materials like concrete and rocks. To train on this subject, investigations have mainly pointed out to materials that possess good thermal conductivity as well as being economical and environmentally friendly. To this end, recent research has examined the thermophysical characteristics of 15 representative rock varieties in Hong Kong (Liu et al., 2024). The researchers found that basalt would be the best candidate rock for high-temperature sensible heat storage because it was thermally stable at 850°C and had good thermal conductivity. The introduction of ceramic fillers within the molten salt storage system not only enhances thermal stratification but also remarkably reduces operating costs because of the 92% utilization rates in cyclic operations made possible by thermal stratification, which is a direct consequence of the latter (Weiss et al., 2024).

10.2.3.2 Latent heat storage

Latent heat storage relies on phase change materials (PCMs), which not only store energy during transitions of state but also have an energy density that is considerably higher than that of sensible heat storage. The advancements in PCMs have predominantly focused on enhancing their thermal characteristics, shape stability, and durability in cycling applications. Recent work reports the development of a latent TES module with complex fin structures, achieving a high volumetric thermal storage density of 145.3 kWh/m^3 and a round-trip energy efficiency of 92.1% (Zhang et al., 2025). In a separate investigation, Tao et al. (2025) analyzed high-temperature latent heat TES with the use of a steel slag/chloride composite PCM and showed an 86% increase in energy storage density and a 2.76- to 9.46-fold increase in heat storage/release capacity compared to traditional steel slag (Tao et al., 2025). Besides, Yadav et al. (2023) engineered a peg (polyethylene glycol)-polymethyl methacrylate (PMMA) composite incorporated with multiwalled carbon nanotubes (MWCNTs), securing 92.3% augmentation in thermal conductivity and solid thermal reliability above 500 cycles (Yadav et al., 2023).

10.2.3.3 Thermochemical storage

Thermochemical storage systems use reversible chemical reactions for energy storage, thus offering high energy density and very low heat loss in the course of the storage process. Hu et al. (2024) merged thermochemical storage with pumped thermal energy storage (PTES), the process of methanol decomposition was used to convert low-grade thermal energy into high-grade chemical energy, and through this, a round-trip efficiency of 63.7% and a storage density of 8.10 kWh/m^3 were achieved (Hu et al., 2024).

10.2.3.4 Aquifer thermal energy storage

The underground aquifers that are part of ATES systems are used for storing thermal energy, which allows groundwater to be used for seasonal energy retention. According to a recent study, with the coupling of the thermo-hydro-mechanical processes in ATES, it was indicated that though thermal stress boosts the efficiency of storage by 0.5%, it has the potential to also result in a surface deformation of 1.5 mm/year.

The application of grey correlation analysis by Shi et al. (2023) in the study of ATES has shown that the thickness of the aquifers and spacing of the wells have the greatest influence on the rates of thermal loss and recovery. Their scheme led to a thermal loss reduction of 12.69% (Shi et al., 2023).

10.2.3.5 Innovative applications and system integration

The big trend in the use of TES systems is their application together with renewable energy sources and industrial processes. As an example, recent study reported that copper slag application for stratified TES in cement works could retain waste heat to the extent of 1.04% reduction in heat demand, with less than half a year for payback. In addition, a study presented the case of hybrid thermal integration in PTES, where solar energy and waste heat were combined to give the round-trip efficiency of 96.1% and cut the levelized cost of storage by 12.69% (Kurşun et al., 2024).

10.2.3.6 Performance metrics and optimization

In Table 10.3, the main performance metrics for the chosen TES systems are summarized and compared, giving special attention to energy density, and efficiency, identifying their application domains. It is the optimization strategies, which include finned structures, composite PCMs, and advanced control algorithms, that have been playing a significant role in improving the performance of TES systems.

Table 10.3 Performance metrics of selected TES systems.

TES type	Material/System	Energy density (kWh/m^3)	Efficiency (%)	Application
Sensible heat	Molten salt with ceramic fillers	21.43–53.77	70.4–97.3	CSP, industrial heat
Latent heat	Steel slag/ chlorides C-PCMs	~86% increase	Not specified	High-temperature TES
Latent heat	PEG-PMMA/ MWCNT composite	Not specified	70.89 (electro-thermal)	Thermal management, electronics
Thermochemical	Methanol decomposition in PTES	8.10	63.7	Renewable energy storage
ATES	Aquifer with optimized well spacing	Not specified	12.69% loss reduction	Seasonal energy storage
Sensible heat	Hong Kong basalt	Not specified	Not specified	High-temperature TES

10.2.3.7 Challenges and future directions

Even though advancements have made, TES was still faced with problems such as corrosion of materials, low thermal conductivity of PCMs, and the site limitations for ATES. Future research should focus on the creation of sustainable, bio-based PCMs, smart energy grids system integration would be another area for advancement. It will be very important to carry out life cycle analyses and do cost optimization since that will be one of the key facilitators for the progress of TES technologies.

10.3 Global policy framework for energy storage

The global transition to a sustainable energy future depends mainly on energy storage technologies' successful implementation. These technologies serve as the backbone of the renewable energy sector by allowing electric power generation from renewables, thus making electric power available for others consuming nonrenewable energy sources, and setting new carbon neutral targets. Worldwide energy storage policy, with the main initiatives, the differences between regions and their connection to the Sustainable Development Goals (SDGs), especially SDG 7 (Affordable and Clean Energy) and SDG 13 (Climate Action), will be the focus of this section. Recent studies will help us to see how policies are pushing the innovation of energy storage, its deployment, and market growth, using data and case studies as support.

10.3.1 Overview of global energy storage policies

Energy storage technologies, comprising lithium-ion batteries, pumped hydro, thermal storage, and cutting-edge hydrogen systems, are crucial in the renewable energy systems to overcome the problem of intermittency and supply power of good quality and reliability. Storage usually policies promote the entrance of various energy types and the gradual establishment and expansion of energy storage solution through proper market structure and practices; they also consist of the formulation of regulatory policies, the allocation of financial incentives, and the backing of research and development (R&D). The latter shifts toward the larger goals of reduced carbon emissions and increased energy resilience. Populated on all corners of the world, the climate goals, along with associated requirements for their compliance, will be the most impactful tool to pull energy storage policies, as they will be completely governed by the scenarios stated in the Paris Agreement, the COP28 guidelines, as well as other similar frameworks emphatically calling for prompt renewable energy use and large-scale emissions reduction. For instance, stringent environmental regulations have been correlated with the faster development of clean energy technologies, such as hydrogen storage. A case in point is that a 1% rise in the strictness of environmental regulations equals a long-term increase of 0.212% in the production of hydrogen-related technologies (Cui and Wang, 2022).

10.3.2 Regional policy analysis

10.3.2.1 North America

The Federal Energy Regulatory Commission's (FERC) Order 841 in the United States is the principal frame for energy storage and mandates that regional grid operators allow storage systems to participate in the wholesale electricity markets. The Inflation Reduction Act (2022) has been a significant factor in the growth of the storage market by prolonging the Investment Tax Credits (ITC) for independent storage projects, contributing thus to a great market expansion. Encouraging clean energy and provincial actions, such as Ontario's Independent Electricity System Operator (IESO) storage procurement, have resulted in the rapid adoption of battery storage in Canada.

10.3.2.2 Europe

The Renewable Energy Directive of the European Union (RED III) underlines the importance of energy storage for the integration of renewable energies, setting up ambitious storage capacity targets for the year 2030. Countries such as Germany and the Netherlands have implemented various methods such as feed-in tariffs and capacity payments to promote the development of ESSs. In addition, the EU's Emissions Trading System (ETS) is indirectly benefiting energy storage by increasing the price of carbon-intensive electricity and simultaneously rendering low-carbon solutions like battery and hydrogen storage more competitive (Zhang et al., 2024b).

10.3.2.3 Asia–Pacific

Among the targets laid out in China's 14th Five-Year Plan (2021–25), the energy storage sector is promised the installation of 30 GW of new storage capacity by the year 2025 accompanied by suitable regulations for pumped hydro and battery systems. Also, India's ESS Roadmap converts into reality a remarkable 50 GW of storage capacity by 2030 with the help of PLI for local battery manufacturing plants. Conversely, Australia's Australian Renewable Energy Agency (ARENA) is literally funding the most creative storage projects which include large-scale battery development and hydrogen storage.

10.3.2.4 Latin America

Latin American countries, which include Brazil and Chile, are incorporating storage technologies along with solar and wind projects into the renewable energy auctions. One of Brazil's energy auctions in 2021 did offer hybrid projects that combined solar energy with battery storage, regulatory schemes aimed at grid stability being the main support. On the other hand, Mexico's Energy Transition Law is encouraging the building of storage facilities to improve the use of renewable energy, while the country's dependence on fossil fuels still hampers the development of the sector.

10.3.2.5 Middle East and Africa

In the Middle East, the United Arab Emirates (UAE) and Saudi Arabia are allocating significant funds to hydrogen storage to transform their economies through their long-term strategy. The UAE H_2 Strategy predicts that the country will hold 25% of the global low-carbon hydrogen market by 2030 (Meighan et al., 2025). On the other hand, in Africa, South Africa's Renewable Energy Independent Power Producer Procurement Program (REIPPPP) is co-creating off-grid storage systems that are conducive to rural electrification and that help to lessen the problems of energy access (Anekwe et al., 2025; Leigland and Eberhard, 2018).

10.3.3 Policy impacts on energy storage markets

The energy storage market is subject to regulatory incentives and prohibitions and largely governed by them. For instance, it was determined that by employing storage technologies the CO_2 emissions would be the least harmful to the environment, thus making their eco-friendliness very pronounced on a large scale. The coupling of strict environmental policies and better material efficiencies has allowed for the use of renewables, one of which is storage solutions, in the OECD regions. Nonetheless, fiddling with geopolitics can still take place through trade bans, for example, and it would be an expensive and uncertain deployment of storage technologies.

10.3.4 Challenges and policy gaps

While there have been considerable improvements, global energy storage policy still faces several issues:

1. Policy fragmentation: One of the reasons that the global adoption of good operational practices is greatly hindered is the policy frameworks of different countries are very different from each other. Such diversity in energy supply characteristics across Latin America makes it difficult to create common storage policies.
2. Economic barriers: The high initial investment and lack of financial resources for energy storage projects in developing countries remain the primary barriers and the hardest obstacles to overcome.
3. Regulatory misalignment: Contradictions in nonuniform sets of rules on grid access and restrictions to integrating storage in certain regions also exist.
4. Geopolitical risks: The authors also predicted that the pervasive economic and political instability in recent times could seriously affect and even slow down the deployment of storage technologies in all areas, especially lithium and cobalt which are considered the "key" minerals.

10.3.5 Future directions and recommendations

- **Harmonize regulations:** The adoption of universal norms for storage system integration will not only make operating in multiple countries easier for companies and organizations to collaborate but also help them to grow and to offer more services to clients.

- **Enhance financial mechanisms:** Increasing the number of subsidies, tax credits, and green bonds can eliminate financial obstacles to storage projects in developing countries.
- **Leverage digital technologies:** Integration of artificial intelligence with the Internet of Things is also a key initiative for the improvement of storage technology as exemplified by the hydrogen storage system.
- **Align with SDGs:** Guarantee that all policies align with SDGs 7 and 13 by supporting circular supply chains of battery storage and at the same time decreasing the use of fossil fuels.

10.4 Energy storage market dynamics

The market for global energy storage is subject to a quick evolution, which is mainly caused by the larger use of VRES, mainly wind and solar, to an extent that their integration into the grid is a decisive factor. Mankind's need for a reliable grid, the elimination of carbon emissions, and energy supply security also hasten this transformation. The importance of ESSs, particularly BESS, in managing the instability of VRES, equalizing power supply and demand, and resulting in access to different electricity markets, namely energy, ancillary services, and carbon markets, is highlighted by this section. Using recent studies and data, the section discusses the key forces that are influencing the energy storage market by looking at market structures, bidding strategies, economic feasibility, and policy factors.

10.4.1 Market structures and bidding strategies

Energy storage systems are slowly but surely becoming a part of the electricity markets, and they are involved in buying and selling of energy, offering different services like frequency regulation and flexible ramping which are called ancillary services and are also participating in peer-to-peer (P2P) markets. The currently dominant market structure has a big impact on the dissemination and earnings of the ESS. For instance, previous work applied a system dynamics framework to mimic ESS bidding strategies in imperfectly competitive markets and flexible ramping services to shed light on the utility of storage in promoting market competition and curtailing the market power of dominant participants. These simulations suggested that ESSs can be regarded as stabilizers of the market-clearing process. At the same time, penalties imposed by independent system operators for capacity withholding or price manipulation are essential to ensuring fair competition (Liang et al., 2025).

10.4.2 Economic viability and multimarket participation

Energy storage systems financial capacity and utilization largely depend on different markets for their products and services (energy, capacity, and ancillary services). Therefore, applying the Markov Decision Process to manage the BESS operation enables the participation of market factors and technical limitations.

10.4.3 Impact on grid stability and renewable integration

Energy storage systems make an important contribution to stabilizing power grids; they help to control the ups and downs of the use of VRESs and also minimize the price volatility. A simulation study has demonstrated recently that storage systems with green power not only make supply more reliable but also help realize better prices through less market price fluctuations (Hussain and Ghafoor, 2022). The impact, however, diminishes with the higher capacity of storage. Likewise, Baltputnis et al. (2024) proposed an elaborate management policy for BESS focusing on the European balancing markets aimed at guaranteeing compliance to frequency containment reserve (FCR) and automatic frequency restoration reserve (aFRR) standards. Their successful validation conducted in Germany and Finland has shown that ESS can safeguard the grid despite the most difficult conditions (Baltputnis et al., 2024).

With respect to carbon markets, Song et al. (2025) developed a mixed-integer nonlinear model for a hybrid wind-solar-thermal system along with a combined ESS, achieving a power generation cost reduction of 9.95% and a 97.43% capacity for wind and solar energy (Song et al., 2025). The establishment of linkage between the Energy Saving Scheme and carbon trading systems will serve as an effective approach toward greater acceptance of renewable energy and, thus, reduce the emissions.

10.4.4 Policy and regulatory influences

Political, regulatory, and policy discourses are essential for ESS. The research done by Umair and Guliyeva (2025) highlights the power of a bi-level optimization model in estimating the government-support ESS investments in Western Europe, where the situation of perfect competition comes out as yielding €7.06 million in welfare gain while Cournot oligopoly conditions only capture €0.68 million (Umair and Guliyeva, 2025). Hence, the requirement of regulatory policies to mitigate the effect of market power distortions becomes apparent.

The paper by Papadaki et al. (2025) points out that for islands in Europe, the elimination of licensing hurdles is a driver of hybrid renewable energy systems (HRESs) adoption, awarding it the score of 4.6 out of 5. Their SWOT, coming from the analysis, gave the estimation that strong regulatory frameworks and the participation of stakeholders in an active role, assist in the installation of ESS which in turn can cut down the islands' carbon emissions by as much as 40% (Papadaki et al., 2025). According to Han et al. (2024), the transition in the electricity market and the staggered pricing in spot markets are the factors that together help the installation of low-carbon technologies in China, although rapid thermal power growth could make it an uphill task (Han et al., 2024).

The recent studies mentioned in Table 10.4 provide the main policy recommendations essential to the development of the ESS market.

Table 10.4 Policy recommendations for energy storage market development.

Region	Recommendation	Expected impact	Study
Western Europe	Targeted subsidies and competition frameworks	Enhanced social welfare (€7.06M under perfect competition)	Umair and Guliyeva (2025)
European Islands	Simplified licensing and stakeholder engagement	Up to 40% emission reduction	Papadaki et al. (2025)
China	Differentiated pricing and capacity control	Reduced subsidies for low-carbon technologies	Han et al. (2024)

10.4.5 Challenges and future directions

The new ESS market has good potential for growth; however, it has its share of difficulties such as getting rid of the initial investment costs, going through the regulatory processes, and figuring out the market design aspects. To put it differently, the same study made by Sirin et al. (2025) pointed out that the accidental failure of VRESs leads to price fluctuations over a short period of time, and thus, it is necessary to strengthen the grid flexibility along with demand response tactics (Sirin et al., 2025). On the other hand, Frischmuth et al. (2024) argued that hydrogen storage infrastructure for 2050 must be planned very quickly at the same time he pointed out that the majority (75%) of the storage capacity envisaged would be new buildings (Frischmuth et al., 2024).

The future market might be influenced greatly by the development of new multi-market participation models as Tostado-Véliz et al. (2025) suggest such cooperative equilibrium-based frameworks for energy communities and grid-scale BESSs among multi-market participation models (Tostado-Véliz et al., 2025).

10.5 Conclusion

The successful adoption of ESSs is one of the main sources of the transition to a sustainable energy future, as they will solve the unreliable nature of renewable energy sources, and at the same time make the grid more reliable and assist in decarbonization, among other things. This chapter has explored the diverse range of energy storage technologies, BESSs, PHS, and TES, for instance, each having on the plus side the specific case of application, going from short-term frequency regulation to long-duration energy arbitrage. Modern battery systems have not only established the use of lithium-ion with the highest efficiency and the craze for sodium-based and flow battery technologies but have also made it clear that they are the backbone of today's energy systems, being unlocked by grid-scale, microgrid, and behind-the-meter use cases. PHS is still the king in the long-term storage department and is even getting a shot in the arm from cutting-edge trends like low-head and offshore configurations that broaden its market. Meanwhile, the contribution of renewables to industrial and heating applications has been boosted by TES technologies, which now include sensible, latent, and thermochemical storage.

The global energy policy frameworks have been the major driving force behind energy storage adoption, with different regions having their own initiatives that impose ambitious requirements for storage and renewable energy integration. The combination of policies, financial incentives, and regulatory reforms resulted in a massive market expansion. Although, the supplied critical minerals' loss risk from unstable regions, particularly under developing areas, and others like high entry costs and policy subdivisions, are some of the continuing problems.

The economic feasibility of energy storage is the main concern of the power markets. They are supported by multimarket participation in energy, ancillary services, and peer-to-peer trading which is leading to the increase in profitability. In addition, hybrid systems combining batteries with hydrogen or thermal storage are gaining popularity due to their efficiency and cost-effectiveness. Nevertheless, the grid stability and the renewable energy penetration aspects of energy storage are not fully appreciated due to the complexity of market design and the regulatory constraints.

A global synchronized action is a must for the energy storage industry to overcome the barriers of technology, economy, and policy in the future. It is the right time to introduce recyclable batteries and bio-based phase change materials as these are the only resources besides the already existing ones that the industry can use in the fight against waste and environmental concerns. Also, the synchronization of the international regulations, fortification of the financing methods, and infusion of the digital technologies like AI and the Internet of Things (IoT) into energy storage will continue to enhance the performance and the scalability of the storage system. Developing synergies with the Sustainable Development Goals will ensure that energy storage remains in the forefront heading to a cleaner, more robust, and fairer energy system. The supporting technological advancements, favorable policies, and strong market structures will pave the way for ESSs to be key influencers in the realization of a sustainable energy future by 2050.

Nomenclature

ATES	Aquifer thermal energy storage
BESS	Battery energy storage system
CSP	Concentrated solar power
DRL	Deep reinforcement learning
FCR	Frequency Containment Reserve
FERC	Federal Energy Regulatory Commission
HESS	Hybrid energy storage system
HRES	Hybrid renewable energy system
IoT	Internet of Things
IRES	Integrated Renewable Energy System

ITC	Investment Tax Credit
Li-ion	Lithium-ion
PCM	Phase change material
PHES	Pumped hydro energy storage
PTES	Pumped thermal energy storage
PV	Photovoltaic
SDG	Sustainable Development Goal
TES	Thermal energy storage
VRES	Variable renewable energy sources

References

Abdul Qadir, S., Ahmad, F., Al-Wahedi, A.M.A.B., Iqbal, A., Ali, A., 2024. Navigating the complex realities of electric vehicle adoption: a comprehensive study of government strategies, policies, and incentives. Energy Strategy Rev. 53. https://doi.org/10.1016/j.esr.2024.101379.

Adeyinka, A.M., Esan, O.C., Ijaola, A.O., Farayibi, P.K., 2024. Advancements in hybrid energy storage systems for enhancing renewable energy-to-grid integration. Sustain. Energy Res. 11 (1). https://doi.org/10.1186/s40807-024-00120-4.

AhmadiAhangar, R., Plaum, F., Haring, T., Drovtar, I., Korotko, T., Rosin, A., 2024. Impacts of grid-scale battery systems on power system operation, case of Baltic region. IET Smart Grid 7 (2), 101–119. https://doi.org/10.1049/stg2.12142.

Al-Ghussain, L., Darwish Ahmad, A., Abubaker, A.M., Hassan, M.A., 2022. Techno-economic feasibility of thermal storage systems for the transition to 100% renewable grids. Renew. Energy 189, 800–812. https://doi.org/10.1016/j.renene.2022.03.054.

Ali, S., Stewart, R.A., Sahin, O., Vieira, A.S., 2023. Integrated GIS-AHP-based approach for off-river pumped hydro energy storage site selection. Appl. Energy 337. https://doi.org/10.1016/j.apenergy.2023.120914.

Anekwe, I.M.S., Ashiegbu, D., Isa, Y.M., Potgieter, H., 2025. Circular economy and energy transition: a review of South Africa's 5-year just energy transition investment plan (JET-IP). Mitig. Adapt. Strat. Glob. Change 30 (6). https://doi.org/10.1007/s11027-025-10230-z.

Ansorena Ruiz, R., Schürenkamp, D., Truijen, D.P.K., Hoffstaedt, J.P., Jarquin Laguna, A., Fahlbeck, J., Nilsson, H., Joseph, M., Zangeneh, M., De Kooning, J.D.M., Oldeweme, J., Schwarz, H., Goseberg, N., 2025. Performance evaluation of a variable-speed contra-rotating pump-turbine for low-head pumped hydro energy storage: an experimental study. Energy Convers. Manag. X 26. https://doi.org/10.1016/j.ecmx.2025.101009.

Baltputnis, K., Broka, Z., Cingels, G., Sīlis, A., Junghāns, G., 2024. Robust market-based battery energy storage management strategy for operation in European balancing markets. J. Energy Storage 102. https://doi.org/10.1016/j.est.2024.114082.

Bhaduri, A., Singh, S., 2025. Evolution of energy storage system and its life-saving applications. Eng. Mater. 1023, 281–306. https://doi.org/10.1007/978-981-96-7952-2_12.

Bragagni, M., Xhaferraj, L., Bragagni, G., Checchi, M., 2025. Recharging the transition to low carbon economy: the role of battery energy storage systems. Sci. J. Energy Eng. 13 (1), 1–23. https://doi.org/10.11648/j.sjee.20251301.11.

Chen, J., 2025. The Role of Energy Storage: Enabling the Renewable Energy Transition. Springer Science and Business Media LLC, pp. 317–370. https://doi.org/10.1007/978-981-96-7975-1_10.

Conzen, J., Lakshmipathy, S., Kapahi, A., Kraft, S., DiDomizio, M., 2023. Lithium ion battery energy storage systems (BESS) hazards. J. Loss Prev. Process. Ind. 81. https://doi.org/10.1016/j.jlp.2022.104932.

Craparo, E., Karatas, M., Singham, D.I., 2017. A robust optimization approach to hybrid microgrid operation using ensemble weather forecasts. Appl. Energy 201, 135–147. https://doi.org/10.1016/j.apenergy.2017.05.068.

Cruz-De-Jesús, E., Martínez-Ramos, J.L., Marano-Marcolini, A., Gómez-Expósito, A., 2025. Economic assessment of battery energy storage systems for frequency regulation reserve provision: a case study of the Dominican Republic. Energy Sustain. Dev. 88. https://doi.org/10.1016/j.esd.2025.101749.

Cui, M., Wang, N., 2022. The effects of industry growth and government efficiency on environmental quality: the global perspective. Emerg. Mark. Finance Trade 58 (12), 3516–3525. https://doi.org/10.1080/1540496X.2022.2054324.

Cui, X., Ruan, Q., Zhuo, X., Xia, X., Hu, J., Fu, R., Li, Y., Wang, J., Xu, H., 2023. Photothermal nanomaterials: a powerful light-to-heat converter. Chem. Rev. 123 (11), 6891–6952. https://doi.org/10.1021/acs.chemrev.3c00159.

Cui, J.Z., Xie, C., Wu, W., Widijatmoko, S.D., Hong, Y., Li, Y., Leeke, G.A., 2024. Techno-economic analysis of deploying a short or mixed energy storage strategy in a 100 % green power grid. J. Energy Storage 99. https://doi.org/10.1016/j.est.2024.113333.

De Carne, G., Maroufi, S.M., Morandi, A., Simonazzi, M., 2026. Power-intensive energy storage systems. Lecture Notes Energy 47, 89–119. https://doi.org/10.1007/978-3-031-97755-8_5.

Dhungana, A., Hudson Doyle, E., McDonald, G., Prasanna, R., 2025. Navigating scientific modelling and uncertainty: insights from hazard, risk, and impact scientists in disaster risk management (DRM). Int. J. Disaster Risk Reduct. 118. https://doi.org/10.1016/j.ijdrr.2025.105260.

Donnelly, K.B., 2023. Storing the future of energy: navigating energy storage policy to promote clean energy generation. Environ. Prog. Sustain. Energy 42 (2). https://doi.org/10.1002/ep.14062.

Elalfy, D.A., Gouda, E., Kotb, M.F., Bureš, V., Sedhom, B.E., 2024. Comprehensive review of energy storage systems technologies, objectives, challenges, and future trends. Energy Strategy Rev. 54. https://doi.org/10.1016/j.esr.2024.101482.

Eswanto, Hamidi, M.A., Daud, S., Yusop, A.F., Allehaibi, M., 2026. Hydrogen fuel addition in diesel engines: current and future research trends through bibliometric analysis. J. Comput. Mech. Manag. 5 (3), 311–333. https://doi.org/10.57159/jcmm.5.3.26566.

Förster, R., Eiser, N., Kaiser, M., Buhl, H.U., 2025. Leveraging synergies for energy-flexible operated electrolysis: a techno-economic analysis of power purchase agreement procurement with battery energy storage systems for renewable hydrogen production. Appl. Energy 393. https://doi.org/10.1016/j.apenergy.2025.126029.

Frischmuth, F., Berghoff, M., Braun, M., Härtel, P., 2024. Quantifying Seasonal Hydrogen Storage Demands under Cost and Market Uptake Uncertainties in Energy System Transformation Pathways. SSRN, Germany. https://doi.org/10.2139/ssrn.4803453.

Guo, L., Peng, Y., Li, W., Yu, H., Gu, T., Wang, K., 2024. Economic analysis of the energy storage systems for frequency regulation. Lecture Notes Electr. Eng. 1118, 147–153. https://doi.org/10.1007/978-981-99-8878-5_16.

Gür, T.M., 2018. Review of electrical energy storage technologies, materials and systems: challenges and prospects for large-scale grid storage. Energy Environ. Sci. 11 (10), 2696–2767. https://doi.org/10.1039/c8ee01419a.

Han, J., Tan, Q., Ding, Y., Liu, Y., 2024. Exploring the diffusion of low-carbon power generation and energy storage technologies under electricity market reform in China: an agent-based modeling framework for power sector. Energy 308. https://doi.org/10.1016/j.energy.2024.133060.

Han, D., Wang, J., Yin, C., Zhao, Y., 2025. Advances in early warning of thermal runaway in lithium-ion battery energy storage systems. Adv. Sens. Res. 4 (5). https://doi.org/10.1002/adsr.202400165.

Hassan, Q., Algburi, S., Sameen, A.Z., Salman, H.M., Jaszczur, M., 2023. A review of hybrid renewable energy systems: solar and wind-powered solutions: challenges, opportunities, and policy implications. Results Eng. 20. https://doi.org/10.1016/j.rineng.2023.101621.

Hatziargyriou, N., Joos, G., 2025. Aggregation of Battery Energy Storage and Distributed Energy Resources (DER), Including Solar PV. Springer Science and Business Media LLC, pp. 1–44. https://doi.org/10.1007/978-3-030-87693-7_6-1.

Hayati, M.M., Safari, A., Nazari-Heris, M., Oshnoei, A., 2024. Hydrogen-incorporated sector-coupled smart grids: a systematic review and future concepts. Green Energy Technol. 2414, 25–58. https://doi.org/10.1007/978-3-031-52429-5_2.

Hu, Y., Yao, E., Zhong, L., Wu, S., Zou, H., Xi, G., 2024. Techno-economic analysis of thermochemical-integrated pumped thermal energy storage system. J. Energy Storage 104. https://doi.org/10.1016/j.est.2024.114394.

Huang, J., Li, C., Jiang, D., Gao, J., Cheng, L., Li, G., Luo, H., Xu, Z.L., Shin, D.M., Wang, Y., Lu, Y., Kim, Y., 2025. Solid-state electrolytes for lithium metal batteries: state-of-the-art and perspectives. Adv. Funct. Mater. 35 (1). https://doi.org/10.1002/adfm.202411171.

Hussain, A.S.T., Ghafoor, D.Z., 2022. Efficient power generation using the hybrid design of wind & solar systems. Int. J. Ambient Energy 43 (1), 6829–6838. https://doi.org/10.1080/01430750.2022.2048266.

Jean, R.J., Kim, D., Cavusgil, E., Tsuteaki, H., Ojala, A., 2025. How do virtual and relational ties affect international opportunity realization for emerging market born-global firms? A contingency model. Int. Bus. Rev. 34 (4). https://doi.org/10.1016/j.ibusrev.2025.102399.

Kandhasamy, M., Duvaragan, B.K., Kamaraj, S., Shanmugam, G., 2024. An overview on classification of energy storage systems. ACS (Am. Chem. Soc.) Symp. Ser. 1484, 1–25. https://doi.org/10.1021/bk-2024-1484.ch001.

Khaleel, M., Yusupov, Z., 2026. Advancing sustainable energy transitions: insights on finance, policy, infrastructure, and demand-side integration. Unconvent. Resour. 9. https://doi.org/10.1016/j.uncres.2025.100274.

Killer, M., Farrokhseresht, M., Paterakis, N.G., 2020. Implementation of large-scale Li-ion battery energy storage systems within the EMEA region. Appl. Energy 260. https://doi.org/10.1016/j.apenergy.2019.114166.

Komilov, A., Rakhimov, E., Eshchanov, B., Nasrullayev, Y., Davronov, S., Li, K., 2025. A holistic approach to understanding the impact of battery energy storage systems on the levelized cost of energy of photovoltaic systems. J. Energy Storage 125. https://doi.org/10.1016/j.est.2025.116996.

Kurşun, B., Ökten, K., Özak, B., 2024. Reducing the cost of electricity storage with a novel solar thermal support scenario in pumped thermal energy storage (PTES). Appl. Therm. Eng. 241. https://doi.org/10.1016/j.applthermaleng.2023.122317.

Leigland, J., Eberhard, A., 2018. Localisation barriers to trade: the case of South Africa's renewable energy independent power program. Dev. South. Afr. 35 (4), 569–588. https://doi.org/10.1080/0376835X.2018.1487829.

Li, W., Garg, A., Xiao, M., Peng, X., Le Phung, M.L., Tran, V.M., Gao, L., 2020. Intelligent optimization methodology of battery pack for electric vehicles: a multidisciplinary perspective. Int. J. Energy Res. 44 (12), 9686–9706. https://doi.org/10.1002/er.5600.

Li, W., Li, J.P., Wang, Y.F., Stan, S.E., 2025. Is artificial intelligence an impediment or an impetus to renewable energy investment? Evidence from China. Energy Econ. 147. https://doi.org/10.1016/j.eneco.2025.108550.

Liang, Y., Zhao, J., Srinivasan, D., 2025. Handling and application of big data in modern power systems for situational awareness and operation. In: Smart Cyber-Physical Power Systems: Solutions from Emerging Technologies: Volume 2. wiley, China, pp. 209–253. https://doi.org/10.1002/9781394334599.ch10.

Lin, X.W., Zhou, Z.F., Liu, M., Chen, B., Lu, Y.J., Jing, D.W., Guo, L., 2025. Thermal management performance and optimization of a hybrid system integrating liquid cooling and fin-enhanced phase change material for large-capacity energy storage battery pack. Energy Convers. Manag. 336. https://doi.org/10.1016/j.enconman.2025.119889.

Linjuan, Z., Han, F., Zhiheng, Z., Shangbing, W., Jinbin, Z., 2023. Site selection and capacity determination of charging stations considering the uncertainty of users' dynamic charging demands. Front. Energy Res. 11. https://doi.org/10.3389/FENRG.2023.1295043/FULL.

Liu, Z., Wong, L.N.Y., Chang, S.C., 2024. Experimental investigation of major rocks in Hong Kong as potential sensible thermal energy storage medium. Eng. Geol. 343. https://doi.org/10.1016/j.enggeo.2024.107763.

Lu, D., Cui, N., Zhou, J., Li, C., 2024. Hybrid cooling system with phase change material and liquid microchannels to prevent thermal runaway propagation within lithium-ion battery packs. Appl. Therm. Eng. 247. https://doi.org/10.1016/j.applthermaleng.2024.123118.

Lykas, P., Bellos, E., Tzivanidis, C., 2026. Comprehensive review and performance assessment of Carnot battery storage systems with multiple energy outputs. Energy Convers. Manag. 348. https://doi.org/10.1016/j.enconman.2025.120702.

Mahadevan, V., Raja, S.S., Rusho, M.A., Yishak, S., 2025. Critical review of energy storage systems: a comparative assessment of mechanisms, advantages, challenges, and integration with renewable energy. Results Eng. 26. https://doi.org/10.1016/j.rineng.2025.105589.

Meighan, E., Joyce, S., Godfrey Nnabuife, S., Oko, E., 2025. Low carbon hydrogen readiness in the United Arab Emirates: status, SWOT analysis and strategy matrix. Sustainability 6, 100212. https://doi.org/10.1016/j.nxsust.2025.100212.

Mitali, J., Dhinakaran, S., Mohamad, A.A., 2022. Energy storage systems: a review. Energy Storage Sav. 1 (3), 166–216. https://doi.org/10.1016/j.enss.2022.07.002.

Mohammed, A., Singh, R.S., Aslam, S., 2026. Humidity-aware hybrid transformer-LSTM framework for IoT-enabled photovoltaic power prediction. J. Comput. Mech. Manag. 5 (2), 25–38. https://doi.org/10.57159/jcmm.5.2.25394.

Mohd Razif, A.S., Ab Aziz, N.F., Ab Kadir, M.Z.A., Kamil, K., 2024. Accelerating energy transition through battery energy storage systems deployment: a review on current status, potential and challenges in Malaysia. Energy Strategy Rev. 52. https://doi.org/10.1016/j.esr.2024.101346.

Mubashir, M., Xu, J., Guo, Z., Wang, X., Shi, C., Mei, X., 2025. Hybrid metallic phase change materials based battery thermal management system embedded with circular hollow fins for higher discharge rates. J. Energy Storage 137. https://doi.org/10.1016/j.est.2025.118703.

Nienhuis, R.M., van Rooij, M., Prins, W.A., Jayawardhana, B., Vakis, A.I., 2023. Investigating the efficiency of a novel offshore pumped hydro energy storage system: experimental study on a scale prototype. J. Energy Storage 74. https://doi.org/10.1016/j.est.2023.109374.

Njema, G.G., Ouma, R.B.O., Kibet, J.K., Ponnusami, V., 2024. A review on the recent advances in battery development and energy storage technologies. J. Renew. Energy 2024, 1–35. https://doi.org/10.1155/2024/2329261.

Osman, A.I., Chen, L., Yang, M., Msigwa, G., Farghali, M., Fawzy, S., Rooney, D.W., Yap, P. S., 2023. Cost, environmental impact, and resilience of renewable energy under a changing climate: a review. Environ. Chem. Lett. 21 (2), 741–764. https://doi.org/10.1007/s10311-022-01532-8.

Pang, B., Zhu, H., Tong, Y., Dong, Z., 2024. Optimal design and control of battery-ultracapacitor hybrid energy storage system for BEV operating at extreme temperatures. J. Energy Storage 101. https://doi.org/10.1016/j.est.2024.113963.

Papadaki, A., Savvakis, N., Sifakis, N., Arampatzis, G., 2025. Analysis of hybrid renewable energy systems for European islands: market dynamics, opportunities and challenges. Sustain. Futures 9. https://doi.org/10.1016/j.sftr.2025.100601.

Prakash, K., Ali, M., Siddique, M.N.I., Chand, A.A., Kumar, N.M., Dong, D., Pota, H.R., 2022. A review of battery energy storage systems for ancillary services in distribution grids: current status, challenges and future directions. Front. Energy Res. 10. https://doi.org/10.3389/fenrg.2022.971704.

Qi, N., Huang, K., Fan, Z., Xu, B., 2025. Long-term energy management for microgrid with hybrid hydrogen-battery energy storage: a prediction-free coordinated optimization framework. Appl. Energy 377, 124485. https://doi.org/10.1016/j.apenergy.2024.124485.

Rasool, G., Xinhua, W., Sun, T., Hayat, T., Sheremet, M., Uddin, A., Shahzad, H., Abbas, K., Razzaq, I., Yuexin, W., 2024. Recent advancements in battery thermal management system (BTMS): a review of performance enhancement techniques with an emphasis on nano-enhanced phase change materials. Heliyon 10 (17). https://doi.org/10.1016/j.heliyon.2024.e36950.

Rehman, S., Al-Hadhrami, L.M., Alam, M.M., 2015. Pumped hydro energy storage system: a technological review. Renew. Sustain. Energy Rev. 44, 586–598. https://doi.org/10.1016/j.rser.2014.12.040.

Saleem, M.I., Saha, S., Izhar, U., Ang, L., 2024. A stochastic MPC-based energy management system for integrating solar PV, battery storage, and EV charging in residential complexes. Energy Build. 325, 114993. https://doi.org/10.1016/j.enbuild.2024.114993.

Satapathy, S., Nahak, N., Patra, S., 2024. A new integrated modelling and control of ternary pumped storage hydro power generation for small signal stability studies. J. Energy Storage 103. https://doi.org/10.1016/j.est.2024.114179.

Saxena, V., Kumar, N., Nangia, U., 2024. Computation and optimization of BESS in the modeling of renewable energy based framework. Arch. Comput. Methods Eng. 31 (4), 2385–2416. https://doi.org/10.1007/s11831-023-10046-7.

Sayed, K., Aref, M., Almalki, M.M., Mossa, M.A., 2025. Optimizing fast charging protocols for lithium-ion batteries using reinforcement learning: balancing speed, efficiency, and longevity. Results Eng. 25. https://doi.org/10.1016/j.rineng.2025.104302.

Schischke, E., Grevé, A., Ehrenstein, U., Doetsch, C., 2024. Overview of Energy Storage Technologies besides Batteries. Springer Science and Business Media LLC, pp. 53–68. https://doi.org/10.1007/978-3-031-48359-2_4.

Schwaegerl, C., 2025. Basics of Energy Storage Technologies. Springer Science and Business Media LLC, pp. 1–39. https://doi.org/10.1007/978-3-030-87693-7_5-1.

Šeho, M., Hodžić, S., 2024. The role of the financial sector in energy justice. Sustain. Dev. Goals Ser. 3606, 109–121. https://doi.org/10.1007/978-981-97-6059-6_8.

Semmelmann, L., Konermann, M., Dietze, D., Staudt, P., 2024. Empirical field evaluation of self-consumption promoting regulation of household battery energy storage systems. Energy Policy 194. https://doi.org/10.1016/j.enpol.2024.114343.

Shi, Y., Cui, Q., Yang, Z., Song, X., Liu, Q., Lin, T., 2023. Optimizing the thermal energy storage performance of shallow aquifer based on gray correlation analysis and multi-objective optimization. Nat. Gas. Ind. 43 (6), 156–168. https://doi.org/10.3787/j.issn.1000-0976.2023.06.016.

Shiraishi, K., Park, W.Y., Kammen, D.M., 2024. The role of hydrogen as long-duration energy storage and as an international energy carrier for electricity sector decarbonization. Environ. Res. Lett. 19 (8). https://doi.org/10.1088/1748-9326/ad5856.

Sirin, S.M., Elshurafa, A.M., Cakmak, N., 2025. The impact of variable renewables on price dynamics within wholesale electricity markets. Util. Policy 95. https://doi.org/10.1016/j.jup.2025.101928.

Song, X., Wu, H., Zhang, J., Zhao, C., Peña-Mora, F., 2025. Optimal operation of shared energy storage-assisted wind–solar–thermal power generation systems under the electricity-carbon markets. Energy 330. https://doi.org/10.1016/j.energy.2025.136614.

Su, L., Gang, W., Liu, M., Ling, Z., Zhang, Y., Dong, S., 2025. A capacity optimization method for the battery energy storage system based on historical electricity data of existing buildings. Energy Rep. 13, 6132–6147. https://doi.org/10.1016/j.egyr.2025.05.060.

Tao, M., Cai, J., Zhang, T., Song, J., Jiang, F., Ling, X., 2025. Thermal performance and design optimization for high-temperature ($\geq$500 °C) latent heat thermal energy storage device: using modified steel slag/chlorides composite phase change materials. Appl. Therm. Eng. 277. https://doi.org/10.1016/j.applthermaleng.2025.127091.

Tetteh, S., Yazdani, M.R., Santasalo-Aarnio, A., 2021. Cost-effective electro-thermal energy storage to balance small scale renewable energy systems. J. Energy Storage 41. https://doi.org/10.1016/j.est.2021.102829.

Torregrosa, A.J., Broatch, A., Olmeda, P., Agizza, L., 2024. A semi-empirical model of the calendar ageing of lithium-ion batteries aimed at automotive and deep-space applications. J. Energy Storage 80. https://doi.org/10.1016/j.est.2023.110388.

Tostado-Véliz, M., Borghetti, A., Siano, P., Jurado, F., 2025. Privacy-preserving energy management in local energy communities with EVs – an enhanced benders-like solution strategy. Sustain. Cities Soc. 132. https://doi.org/10.1016/j.scs.2025.106823.

Umair, M., Guliyeva, S., 2025. Optimizing welfare and market power: energy storage strategies in renewable-integrated power markets. J. Energy Storage 118. https://doi.org/10.1016/j.est.2025.116315.

Vaghela, P., Pandey, V., Sircar, A., Yadav, K., Bist, N., Kumari, R., 2023. Energy storage techniques, applications, and recent trends: a sustainable solution for power storage. MRS Energy Sustain. 10 (2), 261–276. https://doi.org/10.1557/s43581-023-00069-9.

Wang, M., Li, Y., Yuan, J., Yuan, S., 2022. Effects of different vortex designs on optimization results of mixed-flow pump. Eng. Appl. Comput. Fluid Mech. 16 (1), 36–57. https://doi.org/10.1080/19942060.2021.2006091.

Wang, Z., Nie, B., He, N., Chen, Q., Zhang, Y., Yao, L., 2025. Realizing hybrid electrical and thermal energy storage by integrating an iron-air redox battery into the CSP-CaL system. Energy Convers. Manag. 340. https://doi.org/10.1016/j.enconman.2025.119997.

Weiss, J., Hübner, F., Fluri, T., 2024. Experimental characterisation of a molten salt thermal energy storage with filler. J. Energy Storage 103. https://doi.org/10.1016/j.est.2024.114420.

Wu, Y., Shuang, W., Wang, Y., Chen, F., Tang, S., Wu, X.L., Bai, Z., Yang, L., Zhang, J., 2024. Recent progress in sodium-ion batteries: advanced materials, reaction mechanisms and energy applications. Electrochem. Energy Rev. 7 (1). https://doi.org/10.1007/s41918-024-00215-y.

Xiao, J., Zhang, X., Bénard, P., Yang, T., Zeng, J., Long, X., 2023. Fin structure and liquid cooling to enhance heat transfer of composite phase change materials in battery thermal management system. Energy Storage 5 (6). https://doi.org/10.1002/est2.453.

Xu, Z., Wang, J., 2022. Toward emerging sodium-based energy storage technologies: from performance to sustainability. Adv. Energy Mater. 12 (29). https://doi.org/10.1002/aenm.202201692.

Yadav, A., Samykano, M., Pandey, A.K., Aljafari, B., Tyagi, V.V., 2023. Synthesis and characterization of polyethylene glycol- polymethyl methacrylate infused multiwalled carbon nanotube nanocomposite as an efficient thermal energy storage. Mater. Today Commun. 37. https://doi.org/10.1016/j.mtcomm.2023.107472.

Zhang, Y., Wu, Y., Li, L., Liu, Z., 2023a. A hybrid energy storage system strategy for smoothing photovoltaic power fluctuation based on improved HHO-VMD. Int. J. Photoenergy 2023. https://doi.org/10.1155/2023/9633843.

Zhang, Y., Zhao, X., Wang, X., Li, A., Wu, X., 2023b. Multi-objective optimization design of a grid-connected hybrid hydro-photovoltaic system considering power transmission capacity. Energy 284. https://doi.org/10.1016/j.energy.2023.128683.

Zhang, G., Shen, W., Wei, X., 2024a. Lithium-ion battery thermal safety evolution during high-temperature nonlinear aging. Fuel 362. https://doi.org/10.1016/j.fuel.2023.130845.

Zhang, R., Lu, Z., Shahidehpour, M., Li, Z., Yan, L., 2024b. Deciphering the electricity–carbon market nexus: challenges and prospects of electricity–carbon coupling. Energy Internet 1 (1), 34–51. https://doi.org/10.1049/ein2.12003.

Zhang, S., Liu, G., Zhen, H., Hu, W., Hong, Y., Pu, L., 2025. Experimental study on improving the rapid response performance of modular latent thermal energy storage by topological fins. Appl. Therm. Eng. 277. https://doi.org/10.1016/j.applthermaleng.2025.126976.

Zhao, C., Andersen, P.B., Træholt, C., Hashemi, S., 2023a. Grid-connected battery energy storage system: a review on application and integration. Renew. Sustain. Energy Rev. 182. https://doi.org/10.1016/j.rser.2023.113400.

Zhao, L., Zhang, T., Li, W., Li, T., Zhang, L., Zhang, X., Wang, Z., 2023b. Engineering of sodium-ion batteries: opportunities and challenges. Engineering 24, 172–183. https://doi.org/10.1016/j.eng.2021.08.032.

11 Materials innovation for advanced energy storage

Gaydaa AlZohbi [1], *Rohitha Weerasinghe* [2] *and Deepanraj Balakrishnan* [1]
[1]Department of Mechanical Engineering, College of Engineering, Prince Mohammad Bin Fahd University, Al-Khobar, Saudi Arabia; [2]School of Computing, Engineering and Creative Industries, University of Bedfordshire, Luton, United Kingdom

11.1 Introduction

The worldwide shift toward a cleaner and more stable energy system is increasing the demand for the advanced energy storage technologies that are often seen as the enabler of such a change. This demand is mainly driven by the need for decarbonization, integration of variable renewable energy sources, electrification of transport, and grid quality/stability. To mitigate the climate change, the world must act together in drastically cutting the greenhouse gases (GHGs) emissions. Energy storage is a technology that can make a big difference in this respect by capturing leftover renewable energy generation from the time of low demand and releasing it when demand is high. This in turn would accelerate the implementation of renewable energy mainly solar and wind and deal with the biggest issue of these energy sources, their intermittency and variability. Storage is no longer seen as the main problem to be solved, rather as a necessary buffer that enables a reliable incorporation of renewable energy in the grid and thus ensures a stable supply even with fluctuations in energy generation. Besides this, the global transition to a clean transport sector via electrification is one of the main factors driving the need for the development of energy storage technologies. Battery systems that are efficient and durable are what is required not only to provide power to BEVs but also to cope with peak demand and renewable energy usage related to the charging infrastructure. Besides that, on the energy grids' side, modern energy grids are facing several challenges in terms of their reliability as well as the balance between supply and demand. The energy storage system (ESS) can offer a set of ancillary services such as that of frequency regulation, backup power, and voltage control during the local loss of the power grid. Thus, it contributes to the improvement of the grid stability and at the same time enabling better load management.

The prevailing energy storage technology is the battery, in particular, lithium-ion batteries (LIBs). A battery has become a major instrument in different areas, for example, electronics and electric vehicles. LIBs are attractive in many areas because of their high energy density, well-established technology, and ability to charge and discharge quickly (Placke et al., 2017). The high energy density of LIBs is what makes longer usage times and larger ranges for EVs possible. The maturity of LIBs

Energy Storage Technologies. https://doi.org/10.1016/B978-0-443-45858-3.00014-0

has led to a minimized cost owing to the economies of scale. The enhancement of high-power delivery and the capability of rapid charging of LIBs made them more suitable for application that is demanding rapid energy transfer. In addition to the benefits mentioned above, LIBs must deal with some challenges such as lithium and cobalt resources being finite, thermal runaway risks, energy density, and costs (Wang et al., 2023b). The raw materials for manufacturing of LIBs are not plentiful, and their extraction has negative impacts on the environment and is associated with bad labor practices. The thermal runaway of LIBs may result in overheating and fire, thus, in large-scale applications where safety assurance is critical, it is a considerable risk. The rise in the size of energy storage which is needed for the larger capacities increases the cost and limits the practicality of the scaling up of the present technologies. New technologies in materials science will determine the next major breakthrough in energy storage technology. The principal fields of innovations in materials for energy storage are electrolytes, electrodes, and new chemistries.

First of all, improvements in solid-state electrolytes (SSEs) could raise the energy density, provide safety, and alleviate the risk of heat (Zhao et al., 2020). New materials are being explored as alternative to liquid electrolytes, enhancing safety and stability of battery. Advanced electrode materials could considerably enhance energy capacity and charging rate. Nanostructured materials are being researched to be used in electrodes with the aim of enhancing the energy absorption, and the efficiency of the cycle (Zhang and Liu, 2017). Additionally, exploring new chemistries to replace lithium ion, including sodium ion, multi-valent ions, and potassium-ion (Sada et al., 2023). These alternatives could mitigate the availability issue of some materials, reduce the cost, enhance safety, and minimize environmental impact.

The main goals of materials innovation include higher energy density, enhanced safety, extended lifetime, reduced cost, and contribution to a circular economy. This chapter aims at presenting the last advancement in materials sciences and innovation for an efficient, and a more sustainable ESS.

11.2 Development of solid electrolytes for safer batteries

11.2.1 The role of conventional Liquid Organic Electrolytes in safety failure

The safety and stability issues in conventional LIBs arise primarily from the intrinsic characteristics of their liquid organic electrolyte (LOE) (Li et al., 2023b). Li salts dissolved in highly volatile, flammable organic solvents, are used in conventional LIBs to streamline ion transport. However, safety concern resulted from cascading thermal runaway event is raised. This event is initiated by uncontrolled internal short circuits or external abuse, resulting in a quick increase in the temperature. The initial heat may arise from electrode decomposition, but the electrolyte primarily dictates the later stages of failure. The LOE is highly flammable, and when the cell vents, the combination of decomposition gases, air, and volatile solvents can ignite, leading to a catastrophic failure (Li et al., 2023b). According to analysis performed on this event, the major source of energy release is the combustion of the vented solvents, with the ability to yield energy several times higher than that released by the cell contained reaction enthalpies of the electrode materials themselves. This means that LOE is the major energy multiplier in a thermal event. The flammability dynamics are intricate, relying mainly on the localized ratio of fuel to air, pressure, and temperature. The conventional approaches used to address the thermal event include adding flame retardant additives that use chemical radical scavenging or physical char-forming processes, but they restricted applicability (Mishra et al., 2024). Without addressing the main risk, they try to forcibly put out the fire of a highly volatile and energetic material, which results in an external safety measure that is most of the time not enough under severe conditions.

11.2.2 Inherent safety and stability benefits of all-solid-state batteries

The deployment of all-solid-state batteries (ASSBs) constitutes a major shift from chemical safety controls to structural safety engineering. Replacing the flammable liquid solvent with a solid material—typically a ceramic or polymer—effectively eliminates the primary cause of battery fires, making the battery safer. This modification in the structure is described as having two main benefits:

- Nonflammability and high thermal stability: Solid electrolytes (SEs) are a nonflammable, and they also have higher thermal stability inherently than liquids. As a result of this stability, the solid can isolate the fire at the most basic level thus the occurrence of an uncontrolled chain reaction called thermal runaway is highly unlikely (Famprikis et al., 2019). In a case of extreme abuse where there may still be some risks, the total removal of the volatile liquid thus the fire will be much less severe and will occur with a much lower probability.
- Dendrite suppression: SEs, ceramics in particular, have a high Young's modulus and have a good mechanical strength. This stiffness can prevent the metallic lithium dendrites that come from the anode from physically forming and growing (Tang et al., 2021). In the case of LIBs, dendrite penetration is thus the main cause of those failures that lead to internal short circuits. On the one hand, by stopping the growth physically, ASSBs solve the problem of the trigger mechanism (short circuits) and on the other hand, they prevent the disastrous result (flammable fuel). In this way, the SE is not only ion transport but it is also a very important mechanical/thermal safety protector (Hoang and Kim, 2022).

11.2.3 Types of solid electrolytes

The focus of the research and innovation of SE is around three material categories, each defined by a distinct compromise between chemical stability, mechanical properties, and manufacturing simplicity. The primary catalyst of SEs is attaining the benchmark of sim 10^{-3} $S \cdot cm^{-1}$ conductivity under room temperature, required for high-rate applications (Lee et al., 2022).

11.2.3.1 Ceramic/inorganic SEs

Inorganic SEs are preferred mainly because of their inherent stability and very high ionic conductivities, which, in some cases, can be comparable to those of liquid electrolytes. However, they complicate the production process due to their inflexibility.

- Garnets

Garnet-type substances like $Li_7La_3Zr_2O_{12}$ (LLZO) have attracted a lot of attention due to their beneficial property combinations. They have outstanding lithium-ion conductivity, which is usually between 10^{-4} and 10^{-3} S·cm^{-1} for the doping condition used, and they have a broad electrochemical stability window (ESW) (Narla, 2023). In addition, LLZO is very stable chemically when it is in direct contact with metallic lithium, thus, it is different from a lot of oxide systems (Connell et al., 2020). Nonetheless, the inflexible ceramic character of LLZO brings up serious difficulties in terms of engineering. It is still very difficult to obtain low-resistance ionic contact at the interface of two solids because of bad wetting and solid-solid contact (Lv et al., 2021). Additionally, LLZO is a material that is reactive with the environment even though it is a long-lasting compound. In particular, it reacts with water and atmospheric CO_2, thus producing lithium insulating layers that are highly resistive and that cover its surface. It results in performance degradation and the need for a careful and specially designed synthesis and handling environment.

- Sulfide-based electrolytes

Sulfide-based superionic conductors such as Argyrodites (e.g., Li_6PS_5Cl) and Li-Germanium–Phosphorus-Sulfides (LGPS-types) are essentially the leading models of ion bulk transport in terms of performance (Zhang et al., 2019). Ion conductivity at room temperature in these materials is reported to be as high as 10^{-2} S·cm^{-1}, with some doped LGPS-type materials reaching $1.5*10^{-3}$ S·cm^{-1} and even higher (Xu et al., 2019). These values are on par with or go beyond the conductivity goals for high-power applications. In spite of their conductivity, sulfides suffer from two significant drawbacks that limit their use: chemical instability and electrochemical instability. Sulfide electrolytes are very vulnerable to moisture, as they hydrolyze because the sulfur atom in the PS_4^{3-} moiety is a Lewis-base active site (Li et al., 2025). Moisture exposure initiates a degradation route that yields poisonous hydrogen sulfide (H_2S) gas, thus inducing significant safety and operational problems during material preparation and cell manufacturing. This kind of sensitivity to the environment imposes strict and costly dry room manufacturing requirements. Furthermore, sulfide systems may generate positive charges at the high potentials of high-voltage cathodes, leading to subsequent oxidation to elemental sulfur and lithium polysulfides (Li et al., 2017; Xiang et al., 2022). This contamination of the interface between the electrode and electrolyte causes a decline in cell performance. Consequently, the sulfide systems become deteriorated.

- Oxide-based electrolytes

NASICON-type materials like $LiAlTiO(PO_4)_3$(LATP) are thermally and chemically stable and also have moderate ionic conductivities which are often close to 10^{-4} S·cm^{-1} at room temperature (Yin et al., 2023). The main limitation of NASICONs, however, is their incompatibility in terms of thermodynamics with metallic lithium (Zhang et al., 2023). When LATP is placed against a Li metal anode, the transition metal in the cathode gets chemically reduced, specifically Ti^{4+} is reduced to Ti^{3+} This reduction reaction thus drastically changes the electronic conductivity of the LATP material from a desirable $3*10^{-9}$ S·cm^{-1} to a highly detrimental $2.9*10^{-6}$ S·cm^{-1}(Ai et al., 2024). The result of this electronic shorting is the degradation of the material and the eventual failure of the cell, thus making NASICONs not suitable for high-energy density Li-metal ASSBs unless complicated protective buffer layers are used (Ai et al., 2024). The instability that can be seen at the different inorganic classes suggests that the problem is not the same for all: sulfides have kinetic instability, garnets have processing/mechanical instability, and NASICONs have thermodynamic instability for the anode material of choice.

11.2.3.2 Polymer SEs

- Polyethylene oxide (PEO)-based systems

Polymer SEs, in particular, PEO-based ones, raise no doubts about their advantages in terms of processability, flexibility, and the general ease of fabricating films (Zhang et al., 2021). On the other hand, PEO is an inherently low ionic conductivity device at room temperature that is between 10^{-8} and 10^{-6} S·cm^{-1} in most cases (Johari et al., 2021). The origin of this limitation is mainly because at ambient temperatures the PEO matrix is highly crystalline, and this restricts the mobility of lithium ions as these ions are transported through the dynamic movement (hopping or diffusion) in the amorphous polymer regions (Su et al., 2024). Thus, PEO-based systems are usually allowed to work at elevated temperatures (sim 60°C–80°C) so that they can have a conductivity of sim 10^{-4} S·cm^{-1}. Further dissection shows that the conductivity depends on the cation identification and its interaction with the polymer host. Li^+ ions form a stronger interaction with PEO chains than, for example, Na^+ ions. This stronger interaction not only raises the local viscosity of the PEO matrix but also decreases Li^+ mobility (Zardalidis et al., 2013). This finding indicates that the difference in performance is not only due to the limitation of crystallization but also to the increase in hydrodynamic radius and local viscosity caused by the specific Li^+-polyether coordination.

- Hybrid SEs

Hybrid solid electrolytes (HSEs) are a concept to combine the advantages of polymers' mechanical flexibility and ceramics' high ionic conductivity. The idea is to incorporate highly conductive inorganic ceramic fillers (e.g., LLZO nanoparticles) into a polymer matrix (usually PEO or a cross-linked derivative) (Pan et al., 2020). Such nanocomposites exhibit significantly increased ionic conductivities at room temperature, and, in some instances, they can go up to 10^{-4} S·cm^{-1} (Arya and Sharma, 2020). Besides that, the addition of the ceramic component can result in a simultaneous enhancement of the Li^+ transference number (the fraction of current carried by Li^+), with values reported to be raised up to 0.7 (Zhai et al., 2022). The reason for this synergistic effect is that the inorganic fillers, being Lewis's acid sites, bind preferentially to the mobile anions, thereby releasing a higher fraction of Li^+ cations to carry the current. The dual improvement in conductivity and transference number is therefore a great advantage of HSEs as a very promising route to flexible, high-rate ASSBs functioning at room temperature. Table 11.1 provides a brief overview of the performance features and main engineering trade-offs of the major SE classes.

11.2.4 Interfacial challenges and solutions

Resistance at the solid-solid interface (SSI) between the SE and the solid electrode is what drastically limits the performance of ASSBs, even with materials showing high bulk ionic conductivity (Banerjee et al., 2020). This problem is the major technological barrier to commercialization, which is why the ASSB bottleneck is more about how to keep low-impedance contact under changing mechanical and chemical stress than bulk ion kinetics. The high interfacial resistance stems primarily from two related issues:

- Mechanical mismatch: Ceramic SEs have a high Young's modulus which makes them stiff. The contact between the electrodes and the pressing surface is not close, thus large voids and poor wetting are formed (Dai et al., 2023). These mechanical gaps limit the ionic pathway very severely, so the area-specific resistance is very high, and it is often reported in the range of thousands of Ohms per square centimeter (Kim et al., 2020). To be able to maintain intimate contact and to overcome this rigidity, high external fabrication or operating pressures are mostly required. The problem of high pressure, however, leads to internal stress which can cause the destruction of the structural integrity, ionic conductivity, and even the material phase composition of the cell components.
- Dynamic instability and volume change: As an example, during battery cycling, electrode materials such as metallic lithium or alloying anodes, change drastically their volume due to the expansion and contraction. This dynamic volume change puts solid-solid contact under continuous stress, therefore, most of the time it is broken, and resistance becomes higher rapidly, dead lithium is formed, and degradation goes fast (Divakaran et al., 2021). To maintain contact between the current collector and the rigid SEs, which are tightly pressed in an anode-free high-energy-density configuration, a special intervention is necessary. As a result, the bulk material properties remain incompatible with the physical requirements for continuous cycling. To address the poor water contact issue and the worsening side reactions, scientists in the field of mechanical interfaces (SSI) have developed several engineering approaches (Karimi et al., 2025; Liu et al., 2025d). These methods primarily focus on adding an elastic or chemically neutral phase to act as a bridge between the two materials.
- Buffer layers and interlayers: A highly potent chemical passivation and physical contact enhancement measure is the insertion of thin, stable buffer layers between the electrode and the bulk SE. These interlayers are chosen to be stable in terms of chemistry against both the electrode and the SE, while acquiring better wetting properties or producing stable

Table 11.1 Comparative performance and stability metrics of key solid electrolytes (Binninger et al., 2020; Riphaus et al., 2019; Tufail et al., 2023; Paul et al., 2024b).

Electrolyte type	Material example	Room-temp ionic conductivity ($S \cdot cm^{-1}$)	Key advantage	Major drawback/Instability
Sulfide	Argyrodites (Li_6 PS_5Cl)	10^{-3} to 10^{-2}(Superionic)	Highest RT conductivity; mechanical ductility	High moisture sensitivity (H2.S release); interface oxidation
Oxide (Garnet)	LLZOLi7$La_3Zr_2O1_2$)	10^{-4} to 10^{-3}	Excellent chemical stability against Li metal	High rigidity (poor interface contact); reactivity with atmospheric CO_2
Oxide (NASICON)	LATP ($LiAlTiOPO_4)_3$ derivatives)	10^{-4} to 10^{-3}	Good thermal stability, high structural stability	Severe electronic instability against Li metal (reduction of $Ti^{4+}$to Ti^{3+})
Polymer	PEO-based	10^{-8} to 10^{-6} (requires heating)	High flexibility; ease of processing and film formation	Low RT conductivity due to crystallinity; low Li^+ transference number
Hybrid	Polymer + nanoceramic filler	Up to 10^{-4}	Synergistic blend of conductivity and flexibility; improved Li^+ transference number	Increased synthetic complexity; requires careful filler optimization

solid electrolyte interphases (SEIs). For instance, to make LLZO garnets stable against metallic lithium, buffer layers like Li_3PO_4, $LiNbO_3$, or amorphous $Li_3\backslash BO_3$ (LBO) glass are employed (Song et al., 2020). The amorphous nature of the LBO glass structure especially leads to a better wetting with the lithium metal anode. The performance of these layers can be measured in LLZT cells, where the use of an optimized LBO interlayer has been shown to significantly enhance interfacial conductivity. This improvement resulted in interfacial conductivity reaching as high as 1.06×10^{-3} S/cm, compared to 1.32×10^{-4} S/ cm without the LBO interlayer (Tang et al., 2022). Besides, these interlayers are instrumental in halting lithium dendrite penetration and elevating the critical current density of the ASSBs.

- In situ polymerization: The in situ polymerization method is a brilliant way of addressing the mechanical mismatch issue by exploiting the flawless conformity of a liquid precursor (Zhang et al., 2024). A liquid monomer solution is delivered at the electrode-SE interface and then changed to a solid by means such as UV curing, thermal induction, or chemical initiation in this technique. The key advantage of this technique is that the liquid precursor fully wets all surfaces and voids before hardening into a highly conformal polymer electrolyte layer. This process ensures nearly complete contact and significantly reduces interfacial impedance. As a result, in situ polymerized solid polymer electrolytes (e.g., poly (propylene carbonate)-based SPEs) when used in full cells, bring about cycling performances that are much more stable and Coulombic efficiencies that are higher as compared to cells lacking this built-in interface layer (Liu et al., 2025c).
- Nanocomposites and composite electrodes: Composite electrode design means that nanosized SE particles are mixed with the active material and conductive additives. This method is very important for handling the changing of the stress of cycling and for improving ionic transport in the electrode bulk (Kim et al., 2022). Nanocomposites increase the total effective ionic contact area and the ionic percolation pathways in the thick electrode structure. Thus, they minimize the localized stress effects and help maintain stable performance. This designed resilience is a necessary condition for obtaining stable low-pressure operation in ASSBs that use materials with a large volume expansion, such as alloy anodes.

Table 11.2 summarize the main key strategies for reducing interfacial resistance in ASSBs.

11.3 Nanostructured electrodes for enhanced performance

11.3.1 Rationale for nanostructuring

Nanostructuring is a major change in the way electrodes are morphologically designed so as to go beyond the limitations that are naturally present in bulk materials (Das et al., 2025). When the characteristic dimension of the active material is reduced, it has a significant effect on ion transport, charge transfer kinetics, and mechanical robustness.

Kinetic Enhancement: One of the most important benefits of nanostructuring is the drastic reduction of the distance that ions must travel in the solid electrode material. The time for solid-state diffusion (t_d) is dependent on the square of the ion diffusion pathway length (L) and is usually referred to as $t_d \sim L^2/D_{chem}$, where D_{chem} is the

Table 11.2 Key strategies for reducing interfacial resistance in All-Solid-State Batteries (Chen et al., 2025; Zou et al., 2024; Yang et al., 2024).

Strategy	Primary mechanism	Applicable SE type	Material example/ Process	Performance metric improvement
Buffer layers/ Interlayers	Physical wetting enhancement; chemical passivation; dendrite suppression.	Inorganic SEs (garnets, NASICONs)	Amorphous Li_3BO_3 (LBO) glass, Li_3PO_4	Increase interfacial conductivity (e.g., 8-fold increase with LBO); increase critical current density
In situ polymerization	Seamless, conformal contact creation via liquid precursor wetting.	Inorganic/ Composite SEs	UV/Thermally induced polymerization (e.g., PPC-SPEs)	Decrease interfacial impedance; increase Coulombic efficiency and cycling stability
Nanocomposite electrodes	Increased ionic percolation; structural resilience against volume changes.	All SE types	Active material + nano-SE filler + conductive additives	Enables lower operating pressure; improves structural stability during cycling

chemical diffusion coefficient (Gao et al., 2024). By changing L from the micrometer scale to the nanometer scale (usually 10–100 nm), t_d is made several orders smaller. Such a change makes the device able to work at very high charge and discharge rates, thereby power density can be increased significantly. The geometries of highly porous structures, brush-like nanofibers extending from the current collector, or ultra-thin nanosheets (with thicknesses down to approx. 10 nm) are examples that can ensure critically short diffusion pathways. This makes the processes to be closer to intercalation pseudo capacitance, in which ions are rapidly adsorbed or intercalated at the surface or near-surface regions, thus bulk diffusion is not limiting. The definition of the chemical diffusion coefficient, D_{chem}, in these nanostructured semiconductors should also consider nonideal statistics and the kinetic jump diffusion coefficient (Bisquert, 2004). This adjustment reflects the complexity of ion transport kinetics in systems dominated by surface effects and complex geometries.

- Surface area effects and charge transfer efficiency: The downsizing of particles to the nanoscale essentially results in a dramatic increase in the specific surface area (SSA). As an example, Co_3O_4 nanosheets type materials may have SSA values up to 161.9 $m^2\ g^{-1}$ (Umar et al., 2021). Such a high SSA is used to the maximum to expand the interfacial area between the active material and the electrolyte. This enhancement of the contact area is absolutely necessary for the improvement of charge transfer efficiency. Simply, a larger active interface can better distribute the charging current over a more extensive surface, thus decreasing the local current density and consequently lowering the Charge Transfer Resistance (R_{CT}) (Wang et al., 2023b). In fact, the electrode–electrolyte interaction kinetics become quicker, which is extremely important for the specific capacitance and power characteristics to be improved. Besides that, porous structures carefully engineered from either spherical-like particles or interconnected nanofibers will ensure that the electrolyte can easily diffuse throughout the entire structure (Wang et al., 2025a). This design allows for the full utilization of the increased SSA and prevents the formation of inactive regions deep inside the electrode.
- Mechanical tolerance: Nanostructuring is an essential engineering tactic to alleviate the mechanical deterioration that accompanies the materials with high capacity which are liable to have large volume changes during cycling (Cheong et al., 2021). If the volume expansion is left unrestrained, it will result in particle pulverization, loss of electrical contact, and continuous fracturing of the SEI layer. Through modifying the active material to be smaller than a certain size, nanostructuring makes it possible for the particle to share the strain by elastic deformation rather than by cracking severely. Typically, the prevention methods entail fabricating particular electrode architectures such as porous structures, core–shell designs, or materials having internal void spaces. These internal voids are the physical buffers that let the expanding material grow there, thus, the electrode remains intact, and the cycle life is extended.

Table 11.3 presents electrochemical impact of nanostructuring Compared to Bulk Materials.

11.3.2 Nanostructured anodes: mitigation of volume expansion in silicon

The use of silicon (Si) anodes is generally regarded as the primary method for significantly improving the energy density of LIBs (Gonzalez et al., 2017). This is due to

Table 11.3 Electrochemical impact of nanostructuring compared to bulk materials (Xu et al., 2021; Mishra et al., 2024; Péter, 2021).

Property	**Bulk/ Microscale electrode**	**Nanostructured electrode**	**Primary benefit**	**Mechanism**
Specific surface area (SSA)	Low ($<1\ m^2\ g^{-1}$)	High ($10–100\ m^2\ g^{-1}$)	Increased specific capacity, power density	Maximized electrode–electrolyte contact area
Ion diffusion pathway length	Long (micron-scale)	Short (nanometer-scale)	Enhanced rate capability	Reduced solid-state diffusion time ($t_d \propto L^2$)
Tolerance to volume expansion	Low (catastrophic failure)	High	Extended cycle life	Internal void spaces, flexible structure buffering
Rate-limiting process	Solid-state diffusion	Surface charge transfer	Improved kinetics	Shift toward fast surface-limited processes

silicon's outstanding theoretical specific capacity of around 4200 mAh g^{-1}, which is nearly 12 times that of graphite (Gonzalez et al., 2017). But the high specific capacity is inherently associated with silicon's volumetric expansion of around 400% upon lithiation, which has been the reason why it has not been viable commercially for a long time (Jang and Kwon, 2025). Nanostructuring is the necessary tool to get over this severe mechanical problem.

11.3.2.1 The silicon anode challenge

The huge volume change in Si is the main reason for mechanical failure and surface instability in bulk, micron-sized materials (Jang and Kwon, 2025). Thus, the two terminologies for the main degradation mechanisms are from here onwards the ones used. The resulting particle cracking is, thus, the main cause of internal stress, as the crack branches inside particles produce pulverized active material and unconnected electrical current paths. Besides that, the ever-continuing swelling and shrinking destroy the layer of SEI that provides the interface. Thus, the layer is always newly formed from here onwards. This dissipates electrolyte, rapidly increasing resistance, and, consequently, capacity fading is accelerated.

Though, what makes the mechanical problem solvable in nanostructured Si—namely its high SSA—at the same time brings about an electrochemical problem

(Mueller et al., 2023). Nanomaterials are more reactive with the electrolyte, thus the irreversible capacity loss due to SEI formation is more pronounced, leading to a lower first-cycle coulombic efficiency (CE) than in the case of bulk silicon. The research points out that there is a linear relationship between the SSA of the Si material and the irreversible capacity loss (Li et al., 2019). Thus, the success of nanostructuring depends on balancing the mechanical tolerance provided by the small size with the application of protective coatings and specialized binders. These methods are essential to stabilize the resulting high surface area against excessive electrolyte decomposition.

11.3.2.2 Nanostructuring strategies for strain buffering

Nanoscale technologies mainly concentrate on either fabricating controlled void spaces or using flexible, conductive matrices to handle the volume expansion that arises.

- Porous Silicon and Silicon Nanowires (SiNWs): The particle size is significantly reduced so that the nanosized silicon can manage the volume changes much better than a microsized silicon that would break catastrophically (Naseer et al., 2025). Porous silicon structures not only drastically reduce the silicon framework but also create a lot of void spaces, these internal void spaces in turn act as a buffer for the expansion of the active material inside with the overall deformation of the electrode not occurring. An example is the research that has been carried out which states that by combining SiNWs, nanowires can retain almost the same diameter after cycling whereas ones that are nonporous rapidly swell and convert into sheet structures (Mi et al., 2023). One-dimensional silicon nanowires (SiNWs), typically prepared in the 10–100 nm range, exhibit exceptional long-term stability (Mahendran, 2021). Their high aspect ratio allows for anisotropic strain relaxation, making them more efficient than zero-dimensional nanoparticles in managing internal stress.
- Hybrid Si/Carbon (Si/C) Composites: It is very important to combine silicon with carbon as carbon is the one which provides the high conductivity and structural flexibility. Thus, it could alleviate volume change and at the same time, ensuring the electrode integrity throughout the cycling process. The most evolved versions are said to be the core-shell and hollow (yolk-shell) ones in terms of architecture. In a yolk-shell structure, the silicon core is, by design, separated from a stiff carbon shell through a void space (Xie et al., 2017). Such a design guarantees that the expanding Si core does not cause the surrounding carbon structure to be stressed. Moreover, the carbon not only supports electrically but also physically limits the expansion. This could secure the SEI layer. Subsequent enhancements are the incorporation of metal particles such as Ni, Cu, Ag, or Sn into the doping Si/C matrix. Sn is especially interesting because of its similar alloying properties to Si but along with that, it has better conductivity and ductility, thus giving mechanical support and facilitations in conduction pathways.

11.3.2.3 Performance benchmarks and system-level optimization

Through nanostructure engineering and the use of advanced composite formulations, measurable performance milestones have been achieved. In fact, new silicon-carbon anode materials have been able to show a durability level that goes beyond the industry standard of 1000 cycles (Hasan et al., 2025). Hence, they are able to achieve more than 1500 charge cycles while at the same time providing an energy capacity that is almost four times that of standard graphite anodes. The novel materials are also line with existing lithium-ion manufacturing lines; thus, they can be easily integrated without any interruption. Nevertheless, this leading performance is basically a multiparametric system optimization problem. The performance parameters, that is, specific capacity, capacity retention, and rate capability, are not only dependent on the Si nanostructure but are also significantly influenced by the supporting materials. For example, the parameters that determine the formation of the metastable crystalline $Li_{15}Si_4$ phase (which causes a lot of stress) have been shown to vary according to the chosen polymer binder and electrolyte system (Sun et al., 2025). Studies have revealed that the use of an advanced polymer binder such as poly (acrylic acid) (PAA) or poly (vinyl alcohol) (PVA), together with a selected electrolyte (e.g., monofluoroethylene carbonate-based), can bring about a great improvement in cycling stability, especially in the case where the loading of the material is high (Su et al., 2014). This is evidence that the nanostructure plays the role of the base mechanical stability. However, the chemical environment, which is taken care of by the binder and the electrolyte, dictates the kinetic pathway of lithiation and the resulting stress generation. Table 11.4 outlines the volume expansion mitgation strategies and associated cycling stability metrics for nanostructures silicon anode materials.

Table 11.4 Mitigation strategies and performance benchmarks for nanostructured silicon anodes (Rahman et al., 2015; Moin et al., 2025; Khan et al., 2024).

Nanostructure type	Volume expansion mitigation strategy	Primary benefit mechanism	Cycling stability example
Porous silicon	Internal void space buffering	Accommodates expansion inward; maintains stable electrical connectivity	SiNW diameter remains stable after cycling
Si nanowires (SiNWs)	High aspect ratio strain accommodation	Maintains low internal resistance and superior long-term stability	Smallest SiNWs show impressive long-term stability
Si/Carbon composite (yolk-shell)	Flexible carbon shell confinement; intentional void space	Balances high capacity with superior structural integrity and conductivity	Commercial Si–C achieved >1500 cycles
System optimization	PAA/PVA binders and specialized electrolyte	Controls phase formation ($Li_{15}Si_4$) and maintains adhesion	Improved cycling stability at high mass loadings

11.3.3 Nanostructured cathodes: Improving rate capability and long-term stability

Nanoscale engineering is applied to commercial cathode materials (NMC, NCA, LFP) primarily to address structural instability under high operating voltages and to enhance their intrinsic rate capability.

11.3.3.1 Nickel-rich layered oxides (NMC and NCA)

Nickel-rich layered oxides, for example, $LiNixMnyCozO_2$ (NMC) and $LiNixCoyAlzO_2$ (NCA), are vital to energy cells with a high-energy density because of their high operating voltage (Mauger and Julien, 2025). Nevertheless, their power is restricted to phenomena that occur when the device is at a high state of charge (SOC), that is, cation mixing, microcracking, and oxygen release.

- Concentration gradient engineering: it is a significant step in the development of core--shell structures. The full concentration gradient (FCG) material has a high-nickel core to maximize the capacity and the stabilizing elements, for example, manganese (Mn) and cobalt (Co), that are highly concentrated in the outer shell. The Mn/Co-rich surface layer not only increases thermal stability but also isolates the core from the electrolyte (Wu et al., 2024). Importantly, the gradual change in composition through the particle (the gradient) enables the mechanical stress caused by the volume difference between the highly active core and the stable shell to be absorbed, thus preventing microcrack propagation. This structural engineering has been very successful: FCG ($LiNi_{0.78}$ $Co_{0.10}$ $Mn_{0.12}$ O_2) showed an outstanding cycling stability over a long period of time and after 4000 cycles it was able to retain 86.3% of its original capacity (Liu et al., 2025a). On top of that, the concentration grading step frequently leads to the development of primary particles that are radially oriented, thereby, by their nature, helping the force generated during phase transitions to be dispersed, thus stabilizing the structure even more.
- Nanoscale surface coatings: Ultra-thin protective layers ($LiAlO_2$, $LiNbO_3$) are locally put together at the nanoscale to create an artificial SEI, that stabilizes the active material against parasitic side reactions with the electrolyte at high potentials. In general, dual-coating schemes are utilized to obtain multifunctional stability by coupling an ion-conducting phase (e.g., $LiAlO_2$) with an inert, protective phase (e.g., Al_2O_3) (Østli et al., 2021). The dual-layer system such as the $LiNbO^{3+}Li_3BO_3$ coating has been the most effective solid-state battery application to deliver a high discharge capacity of 150.1 mAh g^{-1} at a 5 C rate and a retention rate of 88.4% over 100 cycles (Liu et al., 2025b).

11.3.4 Lithium iron phosphate

$LiFePO_4$ (LFP) is a preferred material because of its long cycle life, low cost, and safety from overheating (Yue et al., 2025). Why this technology is not more widely used, a kind of paradox, is that LFP suffers from inherently low electronic and ionic conductivities which lead to rate capability limitation as well as high polarization. To ensure high power capability the latter must be compensated at nanoscale, which is exactly what is done here.

- Conductive carbon coating for rate capability: Nanostructuring is an absolute requirement for LFP, and this must be accompanied by deposition of an extremely thin, homogeneous, and uniform carbon layer onto LFP. The conductive carbon layer very significantly enhances electron mobility and thus overcomes the kinetic bottleneck. To quantify the efficiency of such a solution, rate capacity can be improved by up to 30% with respect to the pristine LFP where a discharge capacity of 130.9 mAh g^{-1} at a high 3 C rate is attainable (Surace et al., 2015). On top of that, there are the advanced systems that use conductive frameworks like multi-walled carbon nanotubes (MWCNTs) to build a three-dimensional (3D) conductive network. This network can optimize the transport pathways both for electrons and Li + ions. Thus, the main possible causes of the R_{CT} are being significantly reduced resulting in capacity retention higher than 95% after several hundred cycles (Stolz et al., 2024). The fundamental factor in the success of the strategy is the uniform and conformal nature of the nanoscale coating which, among other things. This makes it possible to realize the optimum performance with minimal conductive additive thus the economic benefits are mostly preserved.
- Nanoscale oxide scavengers for chemical protection: Along with the kinetic improvement, nanoscale surface modification was chosen to provide chemical stabilization. Metallic oxide coatings, for example, ZnO or Al_2O_3, employed in this technique are excellent scavengers for HF (hydrogen fluoride) (Tesfamhret et al., 2022). HF attacks the electrolyte and is generated from the salt LiPF6 in the electrolyte. Based on the results for typical equations, oxides like alumina react with HF generating salts and water according to: $Al_2O_3 + 6HF \rightarrow 2AlF_3 + 3H_2O$. Hence, the acid gets neutralized and a stable fluoride protective layer, for example, AlF_3 or ZnF_2, is formed. The process helps prevent surface corrosion by inhibition of metal dissolution and by protecting the cathode material. Thus, it leads to a significant improvement of long-term cycling stability and electrode robustness retention.

The nanoscale engineering approaches for enhancing commercial cathodes are summarized in Table 11.5.

11.4 2D materials for advanced storage

Two-dimensional (2D) materials are changing the advanced energy storage scenario at a very high-speed and crop yield just as they can be used for batteries and supercapacitors. Their ultra-thin atomic structure gives these materials unique characteristics. This is a key reason why they can be used for optimal performance in energy storage devices while simultaneously addressing the issues of conduction and surface area in traditional electrode materials.

11.4.1 Fundamentals of 2D materials

2D materials can be described as crystalline solids that are made of single-layer or few-layer atoms, and the atomic bonds within the plane (usually strong covalent or metallic bonds) are much stronger than the bonds that hold the layers together in the third dimension (e.g., van der Waals forces). In fact, they limit the material to a single atomic plane. The unique properties of 2D materials for energy storage are displayed in Table 11.6.

Table 11.5 Nanoscale engineering approaches for enhancing commercial cathodes (Jayasree et al., 2022; Islam et al., 2025; Zhong et al., 2024; Lim et al., 2025).

Cathode material	Nanoscale engineering technique	Primary goal	Mechanistic benefit	Performance gain example
NMC/NCA (Ni-rich)	Full concentration gradient (FCG)	Structural stability (high voltage)	MnCo rich shell resists phase transition; disperses mechanical stress	86.3% capacity retention after 4000 cycles at 1C
NMC/NCA (Ni-rich)	Nanoscale surface coating ($LiNbO_3$)	High-rate capability, stability	Protective barrier suppresses side reactions; optimizes interface conductivity	High discharge capacity of 150.1 mAh g^{-1} at 5 C
LFP	Conductive Carbon/ MWCNT coating	Rate capability, conductivity	Forms 3D conductive network; reduces charge transfer resistance (R_{CT})	30% increase in rate capacity; 130.9 mAh g^{-1}at 3 C
LFP	Nanoscale oxide scavengers (Al_2O_3)	Chemical stability, cycle life	Neutralizes HF acid, preventing cathode material dissolution and surface damage	Maintained material robustness and extended cycle life

11.4.2 Graphene and its derivatives

Graphene was the first 2D material made of a single layer of carbon atoms in a hexagonal lattice and is the source of derivatives that are extensively used in high-power applications. It plays a crucial role in energy storage including:

- Conductive additive and electrode scaffold: Graphene is mainly employed to generate a 3D most conductive framework of electrode materials (both batteries and supercapacitors) in which the conduction takes place in. Not to mention its superior mechanical strength helps to build a firm scaffold that opens the volume changes of active materials (like silicon or transition metal oxides) during cycling, thus ensuring higher stability and less electrode pulverization.
- Supercapacitors (high power application): With its large surface area and excellent conductivity, graphene is the best active material for electric double-layer capacitors, which allow high storage capacity, energy delivery at high rates, and fast recharging.

Graphene oxide (GO) is a middle material that comes from the oxidation of graphite. Oxygen-containing functional groups (–OH, –COOH, –O, etc.) make it

Table 11.6 Unique properties of 2D materials for energy storage (Abdulrahman et al., 2025; Li et al., 2019; Zhang et al., 2016).

Property	Description	Benefit for storage devices
Ultra-high surface area	Due to their single-layer nature, the theoretical surface area is exceptionally high (e.g., graphene: Approx. 2630 m^2/g).	Maximizes the electrode-electrolyte interface, crucial for rapid ion adsorption/desorption (supercapacitors) and reaction kinetics (batteries).
Exceptional electrical conductivity	Often exhibit metallic-level conductivity (e.g., graphene: Approx 2000 S/cm; MXenes: 6000–8000 S/cm).	Creates a highly efficient conductive network within the electrode, enabling ultra-fast electron transport and high-power density.
Tunable interlayer spacing	The distance between stacked layers can be precisely modified (e.g., by intercalation or functionalization).	Allows for faster and reversible ion intercalation/diffusion, accommodating larger ions (like Na^+ or K^+) and enhancing volumetric capacity.
Mechanical strength and flexibility	Strong in-plane bonds lead to high mechanical resilience.	Provides structural stability to withstand the volume expansion/contraction (swelling/shrinking) during repeated charge/discharge cycles, leading to longer cycle life, and enables flexible device fabrication.

very hydrophilic. Reduced graphene oxide (rGO) is obtained by a chemical or thermal process that removes most oxygen groups from GO and thus partially restores electrical conductivity. Both GO and rGO are indispensable components that facilitate the manufacture of flexible and high-surface area electrodes (mostly composite films or papers) as the functional groups render them water-dispersible, thus enabling scalable solution-based processing. The regulated reduction to rGO is a compromise between high conductivity and good processability.

11.4.3 MXenes (transition metal carbides/nitrides)

MXenes are a newly discovered family of 2D materials (first reported in 2011) that have rapidly evolved to become one of the most important materials for advanced energy storage. MXenes are represented by the general formula $M_{n+1}X_nT_x$, where M is an early transition metal (e.g., Ti, V, Nb), X is carbon and/or nitrogen, n = 1, 2, or 3, and T_x stands for surface functional groups (terminations) such as–OH, –O, and –F (Naguib et al., 2021). The most widely studied example is Ti_3C_2Tx. Additionally, MXenes are obtained through a top-down method from their 3D ceramic precursors,

Table 11.7 The main benefits and applications of MXnes (Pang et al., 2019; Bu et al., 2020; Rahman et al., 2022).

Benefit	Description	Application in storage systems
Metallic conductivity	Exhibits high electron mobility, often faster than graphene (up to approx. 8000 S/cm).	Enables ultra-fast charging/ discharging rates (high power density) in supercapacitors and batteries.
High volumetric capacity	The layered structure allows for efficient intercalation pseudo capacitance, where ions are rapidly stored between the layers through redox reactions.	Leads to high charge storage within a compact volume, crucial for miniaturization and high energy density.
Hydrophilic surface (Tx)	The surface terminations make MXenes highly hydrophilic.	Excellent performance in aqueous-based batteries/ supercapacitors (safer, lower cost electrolytes).
Application	Promising candidates for ultra-fast charging devices (e.g., electric vehicle charging) and hybrid storage systems that seek to combine the high energy density of batteries with the high-power density of supercapacitors.	

called MAX phases (general formula $M_{n+1}AX_n$, where A is an A-group element, typically Al or Si) (Hosseinian et al., 2025). The working principle is the selective removal of the A-layer (e.g., Al) from the MAX phase precursor, most commonly by hydrofluoric acid (HF) or a mixture of salts and HCl. After this etching process, the layers are separated, and the surface of the remaining $M_{n+1}X_n$ structure is functionalized with groups (T_x) from the etchant. The main benefits and applications of MXnes are summarized in Table 11.7.

Table 11.8 summarizes the cutting-edge research and development efforts aimed at transitioning 2D materials from the lab to market.

11.5 Beyond-lithium materials

11.5.1 Motivation for beyond-lithium

Electrification of transportation on a global scale to mitigate climate change and stabilize renewable energy grids has been the main driver of the demand for electrochemical energy storage (EES) at a level that has never been seen before. Although the rechargeable LIB remains the benchmark technology, it has unavoidable limitations

Table 11.8 Cutting-edge research and development efforts aimed at transitioning 2D materials from the lab to market (Ying et al., 2025; Moss, 2025; Ghotia et al., 2025).

Focus area	Key research trends and findings	Impact on advanced storage
Advanced synthesis and stability	Development of fluorine-free etching methods for MXenes to reduce toxicity and improve oxidation resistance, increasing long-term stability in ambient air.	Safer, more scalable production, and longer device lifespan.
Interface engineering (Graphene/MXene)	Computational and experimental studies on creating stable and well-aligned 2D heterostructures (e.g., Graphene-MXene or MXene-Polymer composites). Research found that smooth, well-aligned geometries lead to the most stable interfaces.	Establishing design rules for combining materials to leverage the benefits of each (e.g., a highly flexible, conductive, and structurally robust electrode).
Aqueous/Hybrid devices	MXenes are being heavily explored for high-performance aqueous-based batteries and supercapacitors due to their hydrophilicity and high conductivity, emphasizing safety and cost.	Development of safer, high-power energy storage for grid applications.
High mass loading	Research on scalable electrode fabrication techniques (e.g., doctor blade coating) to incorporate high mass loadings of 2D materials (like Mo1.33 CTx/CNT composites), which is necessary for practical devices.	Moving 2D materials from lab-scale prototypes to commercially viable, high-capacity electrodes.
AI-driven materials discovery	Utilizing machine learning (ML) and density-functional theory (DFT) to rapidly predict and optimize the properties and stability of new 2D material compositions and interfaces.	Accelerating the discovery of next generation 2D materials with tailored properties for specific storage applications.

due to lithium's resource profile. Additionally, the system's dependence on flammable organic electrolytes necessitates an urgent search for high-performance alternatives. Multivalent ion chemistries, involving divalent (Mg^{2+}, Zn^{2+}) and trivalent (Al^{3+}) charge carriers, offer the most realistic way to a strategic resource diversification.

The need for a switch to different battery chemistries is a result of various nontechnical factors that are outside the control of the existing LIB infrastructure and that are threatening its sustained scalability. In essence, one of the main reasons for the scarcity of lithium is its low crustal abundance, which is around 0.0022 wt% in the Earth's crust, and this is what puts a limitation on the expansion of LIB production in the long run. This concern about scarcity is made worse by the high-cost volatility over a long period of time, which can be illustrated by the huge 80% price drop of lithium during 2023 (Wang et al., 2023a).

Besides material economics, the strategic landscape is mainly influenced by geopolitical risk. The red-hot topic of this year, geopolitics, has been recognized as the main factor influencing the mining and metals sector. Thus, as the primary driver of 45% of surveyed industry decision-makers in 2024. The extent of this influence is measured by concerns related to resource nationalism—11% and the impact of industrial and critical minerals policies—7%. Geopolitical tensions contribute directly to supply chain vulnerabilities. For instance, disruptions caused by Houthi attacks on shipping and drought-induced low water levels in the Panama Canal led to increased operating and logistical costs for firms.

These widespread geopolitical and resource-related challenges require companies to make defensive investments to maintain supply continuity and diversity. The move to invest in research and development of multivalent systems despite the lithium price correction is mostly a reflection of a long-term security strategy rather than a short-term commodity arbitrage. The ultimate aim is to phase out the use of critical elements that are concentrated in certain parts of the world and replace them with systems that use elements that are abundant in the earth's crust.

Multivalent systems are highly appealing from an electrochemical perspective mainly due to the charge density advantage that comes with the transfer of multiple electrons per ion. For example, Mg^{2+} and Al^{3+} ions can deliver much higher theoretical volumetric energy density than monovalent Li^{+} (Icaza et al., 2018). In fact, the theoretical volumetric capacity of aluminum is 8046 mAh cm^{-3}, which is about four times that of lithium. As a result of this trivalent efficiency, aluminum also has a theoretical gravimetric capacity of 2980 mAh g^{-1}, which is only lithium's value that is higher (Smajic et al., 2020). Unfortunately, this kinetic bottleneck is introduced by the high charge density of multivalent ions that cause strong electrostatic interaction between the ion and the host lattice during intercalation. This close bonding necessitates a high desolvation energy and causes slow solid-state diffusion kinetics, thus, early multivalent systems fundamentally have limited rate performance and cycle stability. The main technical goal of Mg, Zn, and Al platforms is to surpass this particular kinetic barrier by advanced material and electrolyte engineering. The geopolitical and resource constraints of Lithium are displayed in Table 11.9.

11.5.2 Magnesium-ion batteries

Magnesium-ion batteries (MIBs) are attractive due to the intrinsic properties of magnesium metal: abundant resources, cost-effectiveness, nontoxicity, and a high theoretical specific capacity of 2205 $mAhg^{-1}$ (Zhang et al., 2025). Critically, the use of a

Table 11.9 Geopolitical and resource constraints of lithium (Egbue et al., 2022; Vivoda and Loginova, 2025).

Metric/Risk	Status in 2024	Consequence/Impact
Geopolitical influence	Dominant driver of sector activity (45% of surveyed firms)	Intensified supply chain disruptions, focus on resource nationalism, shift in fundable jurisdictions.
Lithium price volatility	Massive decline (approx. 80% year-on-year in 2023)	Short-term market uncertainty, increased M&A and consolidation potential in battery metals.
Resource scarcity (crustal abundance)	0.0022 wt% in Earth's crust	Fundamental limit on long-term sustainable development of LIBs, driving mandatory diversification.

magnesium metal anode (MMA) mitigates a major safety flaw found in alkali metal batteries: MMAs typically form a smooth passivation layer on their surface without the problematic growth of dendrites that cause internal short circuits. Initial research into MIBs relied on organic electrolytes. These proved problematic due to high cost, corrosiveness, poor cycling performance, and low ionic conductivity. To address both safety (eliminating flammable organic solvents) and cost, researchers have shifted focus toward aqueous magnesium-ion batteries (AMIBs). Aqueous electrolytes (AEs) offer intrinsic safety, environmental friendliness, high ionic conductivity, and simplify manufacturing processes. The fundamental hurdle for AMIBs is managing the narrow ESW of water (1.23 V), which limits working voltage and risks water decomposition (hydrogen evolution reaction, HER, and oxygen evolution reaction, OER). Furthermore, the strong electrostatic effect of the divalent Mg^{2+} ion restricts diffusion kinetics and leads to low reversible specific capacity in cathode materials (Fan et al., 2024).

11.5.2.1 Advanced electrolyte engineering

Recent breakthroughs in material science have largely been concentrated around electrolyte engineering to achieve a metal anode made of Mg stabilized with a wider electrochemical stability window (ESW) which would enable practically operating AMIB.

- Superconcentration Electrolytes: One approach of using superconcentration electrolytes has been the deploying of the highly concentrated salts, such as 4 M $Mg(TFSI)_2$ AE (Wang et al., 2017). This has been proved very effective to expand the electrochemical stability window (ESW) up to 2.0 V, which is a key aspect of making the working voltage of high-potential cathodes compatible.
- Water-in-Salt (WIS) Strategy: WIS concept has been a major turning point in the reversible use of the Mg metal anode in an aqueous solution. This strategy, which is based on a very

concentrated $MgCl_2.6\ H_2O/H_2O$ mixture (25:1 mass ratio), changes the Mg^{2+} solvation structure completely (Qiang et al., 2025). Besides, it changes the passivation layer, usually MgO–Mg $(OH)_2$, formed from the surface and that is impermeable, into an ionically conductive complex interphase. Hence, it offers the reversible removal and deposition of Mg metal, which was limited at the HER-derived surface films.

- Quasi-Solid-State (QSS) Systems: To solve the problem of performance at low temperatures and additionally increase safety, electrolytes in QSSs have been created. The QSS electrolyte consisting of $MgCl_2\ 6H_2O$, polyethylene oxide (PEO), and H_2O (4:1:1 mass ratio) presented an incredibly wide ESW of 4 V along with a high ionic conductivity of 1.24 mS cm^{-1} (Ashfaq et al., 2025). Utilizing this QSS electrolyte, a full cell could be operated continually for more than 900 cycles.

11.5.2.2 Cathode and full cell performance benchmarks

For the cathode, research is mainly directed toward the reduction of electrostatic hindrance for Mg^{2+} diffusion. The most common materials are mainly manganese oxides (such as lambda MnO_2, delta-MnO_2) and polyanionic materials (e.g., $Li_3V_2(PO_4)_3$). The main idea that can be traced out of very recent publications is the structural water playing the positive role. In the case of hydrated birnessite manganese dioxide (delta-MnO_2), the water of crystallization turns out to be the component which screens the strong electrostatic interactions of Mg^{2+} ions. The screening is what provides reversible intercalation kinetics thus performance can go up to high levels. As a matter of fact, a hydrated delta MnO_2 cathode was able to operate at 2.8 V versus Mg Mg^{2+}, and it retained 62.5% of the capacity after 10,000 charging–discharging cycles in AEs (Ling and Zhang, 2017). It is a clear demonstration that advanced material design exploiting the solvation shell structure is not only helpful but necessary in transport acceleration of the divalent ion.

Recent full cell assemblies validate the practicality of these novelties.

- The full cell comprising a lithium vanadium phosphate (LVP) cathode and a poly-pyromellitic dianhydride (PPMDA) anode, along with a superconcentration electrolyte, has gone for a high voltage of 1.9 V and has shown very good cycling stability with an excellent capacity retention ratio of 92% after 6000 cycles at 2 A g^{-1}. The cell was able to provide a desirable energy density of 62.4 Wh kg^{-1}.
- Aqueous Mg metal batteries with the WIS strategy and a copper hexacyanoferrate (CuHCF) cathode achieved ultrahigh voltage plateaus from 2.4 to 2.0 V and a remarkable stability of more than 700 cycles.

11.5.2.3 Zinc-ion batteries

Zinc-ion batteries (ZIBs) are the most mature multivalent chemistry that can be commercially available in the near term, mainly for stationary applications. Zinc is a very abundant, safe, and cheap metal, which makes ZIBs naturally suitable to be operated in AEs, which are the most cost-effective. This combination of low cost and enhanced safety features positions aqueous ZIBs (AZIBs) as a very attractive technology, especially for large-scale stationary grid storage, where they are considered as the most suitable solution. The market for alternative battery technologies, including ZIBs, is expected to be 45.2 Bn by 2034, with this safe, scalable technology being the main driver (Akinyele et al., 2017).

11.5.2.3.1 Anode stability challenges and cutting-edge mitigation (2024)

In general, the use of ZIBs in daily life has been limited due to the unstable Zn metal anode. The problem of the Zn anode spontaneous corrosion and dendrites formation has been at the center of discussion in the last few years; however, the 2024 solution seems to be the stabilization of the Zn surface by the deposition of the scalable coating:

- PVDF-HFP Organic Coating: Scientists used a simple blade-coating method to cast poly (vinylidene fluoride–hexafluoro propylene) (PVDF-HFP) on the Zn anode (P–Zn). Density Functional Theory calculations revealed that this organic coating significantly reduces the energy gap of the adsorption between the P–Zn (002) and (101) planes, which, in turn, leads to the uniform deposition of Zn^{2+} thus dendrite formation and corrosion are effectively alleviated. The symmetric cells with such a protected anode were able to sustain stable cycling performance for 700 h at a current density of 1 mA cm^{-2} (Li et al., 2024b).
- LSO Polymer Coating: To mechanically strengthen the electrode and restrain the dendrite growth the multifunctional zinc silicate polymer, lithium polysilicate (LSO), was suggested by the authors, through the blocking of direct contact between the electrolyte and the Zn anode. The stability of symmetrical cells using the Zn-LSO anode was outstanding, as they were able to cycle for 833 h at 0.5 mA cm^{-2} (Wang et al., 2025b).

11.5.2.3.2 Full cell performance and TRL assessment

When these stabilized anodes are combined with high-performance cathode materials, they have shown incredible durability for a very long time.

- Exceptional Cycle Life: A full cell with the PVDF-HFP coated Zn anode and an MnO_2 cathode at 2C kept almost the whole capacity even after 1400 cycles (Li et al., 2024a). The performance of this device is very far from that of Zn anodes without modification, which normally show only 50% of capacity retention after 600 cycles.
- High-Capacity Reversibility: The full cells using the LSO coating and an MnO_2 cathode have demonstrated the high-capacity reversibility reaching even 234 mAh g^{-1} (Verma et al., 2018).

The main reason for the success of ZIB solutions is their scalable nature or the size building-up is not a problem. Anode stabilization methods feature the application of easily industrialized coating techniques for the polymer/composite materials. This engineering approach that opts for simplicity and good performance in aqueous media leads to a much higher Technology Readiness Level (TRL) for ZIBs in comparison with MIBs which need complicated synthesized electrolyte formulations like WIS or QSS systems for their operation. Therefore, ZIBs are in a very advantageous position to take over the cost-sensitive, large-volume, and low-priced stationary storage market in the near future.

11.5.2.4 Aluminum-ion systems

Aluminum-ion systems (AIBs) are a good example of how one can utilize aluminum, which is the most abundant metal in the Earth's crust, along with the highest charge density of the trivalent Al^{3+} ion. Although it is kinetically hindered to directly intercalate Al^{3+} into solid host materials, the best AIB design is the rechargeable Aluminum-Graphite Dual-Ion Battery (AGDIB). AGDIBs use a special dual-ion mechanism to harness the extremely high volumetric capacity of aluminum of 8046 mAh cm^{-3} theoretically (Dutta and Mary Gladis, 2024). During the charging, the Al anode receives the Al plating, and at the same time, complex chloroaluminate anions ($AlCl_4^-$ or $Al2Cl_7^-$) intercalate the graphitic carbon cathode. The anionic intercalation/deintercalation here is a way to get around the problem of very slow Al^{3+} cation diffusion and thus the system can have very fast kinetics.

The main differentiator for AGDIBs is their ultra-high-power capability. Their recent advancements demonstrate that AGDIBs can achieve ultra-high-power densities as high as 175 kW kg^{-1} (Kravchyk and Kovalenko, 2019). As per the comparative analysis, AGDIBs are a clear distinction of the EES spectrum. Their power density is 10–100 times higher than that of conventional LIBs, sodium-ion batteries (NIBs), and redox flow batteries, thus, making it close to that of the latest electrochemical supercapacitors. This unique performance combination—holding a stable energy density of around 66 Wh kg^{-1} while providing an incredible power—places AGDIBs as a high-performance system that is hierarchically between traditional batteries and supercapacitors (Shi et al., 2025). These characteristics, for instance, are "perfect" for high-power applications that require both rapid energy delivery and high rate cycling, for example, kinetic energy recovery systems in electric vehicles or quick grid frequency stabilization. AGDIBs have two main challenges that limit their use in a broader way. The first one is that the graphite carbon cathode has a relatively low specific capacity, which is the main reason why the overall energy density of the full cell is low. To increase the range of applications of AIBs, it is necessary to find and use new cathode materials with higher intrinsic capacity.

The second one is that the system uses specially designed nonaqueous ionic liquid electrolytes to stabilize the chloroaluminate charge carriers ($AlCl^-_4$). These electrolytes usually cause aggressive corrosion of the parts of the cell. The need for these specially designed, often expensive, electrolytes is currently against the initial reason for AIBs which is only based on low cost and abundance. This technical fact shows that AIBs are currently concentrated on high-value, high-performance niches where fast kinetics are crucial, and where the cost of materials is of less importance. Therefore, future R&D should be directed toward the finding of stable, cheap electrolyte systems that will still have high conductivity in order to be able to realize completely the cost-reduction potential of the Al anode.

11.5.2.5 Comparative performance assessment and strategic outlook

The analysis of multivalent alternatives reveals that Mg, Zn, and Al chemistries are not competitive substitutes but are, rather, strategic compliments, each fulfilling a unique role dictated by its fundamental electrochemistry and current TRL. Table 11.10 provides a comparison of the inherent advantages, key performance indicators, and primary hurdles differentiating the three systems.

The recent breakthroughs in AMIB and AZIB systems are summarized in Table 11.11.

Table 11.10 Comparative performance metrics and challenges of beyond-Lithium-ion systems (Lu et al., 2025; Gupta et l., 2024).

Parameter	Magnesium-ion (MIBs)	Zinc-ion (ZIBs)	Aluminum-ion (AIBs) (AGDIBs)
Charge carrier	Mg^{2+} (divalent)	Zn^{2+} (divalent)	Al^{3+} (trivalent, utilizing $AlCl_4^-$ intercalation)
Crustal abundance rank	High	High	Highest
Theoretical gravimetric capacity (mAh g^{-1})	2205	820 (approximate)	2980
Theoretical volumetric capacity (mAh cm^{-3})	High	Moderate/High	8046 (4× that of Li)
Key operational advantage	Dendrite-free Mg metal potential, exceptional cycle life (up to 10,000 cycles)	Low cost, intrinsic safety (aqueous), high TRL for grid	Ultra-high-power density (up to 175 kW kg^{-1})
Primary technical challenge	Sluggish solid-state diffusion kinetics, complex electrolyte requirements (WIS/QSS)	Dendrite growth and self-corrosion of Zn anode in aqueous systems	Low cathode capacity (graphite), corrosivity of chloroaluminate electrolytes
Representative full cell energy density (Wh kg^{-1})	sim 62.4	Moderate (focus on cost/safety, not peak density)	sim 66
Application niche	Stationary storage, future mobile (volumetric focus)	Large-scale stationary grid storage	High-power, rapid cycling/Hybrid systems

Table 11.11 Recent breakthroughs in aqueous magnesium-ion (AMIB) and zinc-ion (AZIB) systems (Gull and Chen, 2022; Cui et al., 2022).

System	Innovation focus	Mechanism/ Material	Key performance metric
AMIB	Anode protection/ ESW expansion	Water-in-salt (WIS) electrolyte ($MgCl_2 6H_2O$/ H_2O)	Reversible Mg metal plating/ stripping, stable >700 cycles (CuHCF/Mg)
AMIB	Low-temperature/ ESW expansion	Quasi-solid-state (QSS) electrolyte ($MgCl_2$ PEO)	Wide ESW (4 V), stable >900 cycles (CuHCF parallel Mg)
AMIB	High cycle life	Superconcentration Mg TFSI $_2$ electrolytes	62.4 Wh kg-1, 92% retention after 6000 cycles (LVP//PPMDA)
AMIB	Cathode stability	Hydrated delta-MnO_2 cathode	2.8 V operating voltage, 62.5% retention after 10,000 cycles
AZIB	Anode dendrite mitigation	PVDF-HFP organic coating (P–Zn anode)	Stable cycling 700 h (symmetric cell); full capacity retention after 1400 cycles (MnO_2 cathode)
AZIB	Anode dendrite mitigation	Zinc silicate polymer (LSO) coating	Long symmetric life (833 h); 234 mAh g^{-1} capacity (full cell)

11.5.2.6 Technology readiness level and strategic outlook

According to the current technical milestones, the three chemistries are heading to different markets:

- Zinc-ion Batteries (AZIBs): AZIBs have the highest TRL for bulk manufacturing and can be immediately deployed in large-scale stationary grid storage. This is because they utilize scalable engineering solutions to stabilize the anode, such as the simple blade-coating of polymer films (Zhou et al., 2023). This method produces robust, industrial-scale cycle life (1400 cycles) with a slight increase in manufacturing complexity or cost. The present work is geared toward safety and cost rather than energy density maximization.
- Magnesium-ion Batteries (AMIBs): MIBs are behind ZIBs in terms of TRL as they require complex and often expensive electrolyte formulations (WIS, QSS) that are necessary to enable Mg metal reversibility (Liu et al., 2024). However, MIBs have a high potential for ultra-long life (up to 10,000 cycles) and higher volumetric density, thus, they can become a premium source for long-life stationary storage or eventually volumetric-critical mobile applications. The core issue for MIBs is to lower the chemical complexity to reach ZIB-level manufacturing simplicity.
- Aluminum-ion Systems (AGDIBs): AGDIBs are at the moment functioning beyond the cost limits of the mass-market, thus they are concentrating on the high-power niche. Largely, the singular accomplishment of 175 kW kg^{-1} power density is the reason why specialized, expensive ionic liquid electrolytes are used (Kravchyk and Kovalenko, 2019). The system

is now a fully fledged technology for high-rate/supercapacitor-replacement applications; however, it still needs a breakthrough in materials—namely, higher capacity cathodes and more straightforward, less corrosive electrolytes—to be able to move to the energy storage of the general market.

11.6 Trade-offs in performance, stability, and safety considerations

The evolution of EES systems is limited by a basic set of natural, frequently opposing, trade-offs. The aim of having a battery that can deliver high performance is safe by nature, and is profitable at the same time, is still impossible to reach within the existing technology frameworks. Consequently, the design of EES involves making trade-offs that have led to a market characterized by different optimized chemistries for various application areas.

Advances in battery technology are not just a question of simple metric improvement one by one, but rather about figuring out the complex interrelations of these metrics. As a matter of fact, creating a high-energy density battery is the primary cause of an increased possibility of a fire or explosion if the released energy is too fast or if the battery is kept in a volatile environment. Therefore, it is frequently the case that one very important parameter can be improved at the cost of another. Hence, it implies the necessity of a portfolio approach to battery technology, which acknowledges that the market will become diversified to meet different needs instead of a single, universally optimal solution.

11.6.1 The Performance Triangle

The Performance Triangle defines the core electrochemical tension that is at the root of all rechargeable devices. This trilemma revolves around balancing energy density (ED), power density (PD), and cycle life (CL) (Yang and Wang, 2018):

- Energy density (Wh/kg) is the figure of merit for an electrical battery that shows the amount of charge it can store relative to the weight of the battery and is thus the main factor determining the run time or range of a portable device.
- Power density (W/kg) is the rate at which the stored energy can be given to or taken from the system and is thus the most important factor for rapid acceleration or fast charging.
- Cycle life talks about the device's long-term viability and indicates the number of charge/discharge cycles that can be performed before the capacity drops to a certain level, which is generally 80% of its original value.

The trilemma is what conditions the battery system whether it is appropriate essentially for high-energy, sustained discharge applications (e.g., long-range electric vehicles or laptops) or high-power, short-burst applications (e.g., power tools or frequency regulation systems). The inverse relationship between energy density and power density is a fundamental law in battery design, rooted in the kinetics of ion transport within the cell architecture.

11.6.2 Fundamental conflict: Kinetics versus capacity

Energy storage capacity (ED) can be maximized by packing the electrodes with the highest amount of active material, which usually means that the electrodes become thicker and less porous. This, in turn, leads to longer diffusion paths for lithium ions, thus slowing down the transport kinetics and increasing internal resistance, which, in its turn, limits the speed of the energy that can be extracted or inserted. On the other hand, power density (PD) can be maximized by minimizing impedance and maximizing surface area for fast electrochemical reactions. This includes the design of thinner electrodes with higher porosity and the optimization of the electrolyte conductivity. These kinetic improvements lead to a decrease in the total volume fraction of the active material, thus resulting in a sacrifice of ED. The flux of electrons and ions is the researchers' challenge, which is closely related to the electrode architecture problem they are facing.

The most prominent example of such a dilemma is the comparison of conventional LIBs and SCs. SCs hold energy electrostatically, a non-Faradaic process, thus enabling an extremely fast charge transfer. As a result, they show an exceptionally high specific power (usually more than 10,000 W/kg) and an incredibly long cycle life (over one million cycles). 11 However, this kinematic speed is achieved by sacrificing almost all energy density, with typical values of around 5–10 Wh/kg. Hybrid devices, such as lithium-ion capacitors (LiC), attempt to bridge this gap by combining capacitive cathodes (like activated carbon) with battery anodes. But the low capacity of the activated carbon cathode is still the main factor that limits the specific energy of LiCs from reaching that of their battery counterparts.

11.6.3 Visualizing the trade-off: The Ragone plot analysis

The inverse correlation of specific energy and specific power can be visualized through the Ragone plot. This plot puts specific energy (Wh/kg) on the x-axis and specific power (W/kg) on the y-axis, both in a logarithmic scale. The comparisons of energy storage devices that are radically different in power and energy become feasible by means of this instrument. Different specific discharge durations are represented by the diagonal lines that span the plot.

The analysis of 18,650 cell formats draws the positioning of common Lithium-ion systems:

- High Energy Systems: As an example, the Sanyo UR18650F cell is the one that gives the highest specific energy (2600 mAh) but its discharge rate is limited to a moderate one (5 A), so it is located at the side of the plot where energy is high, and power is low thus making it good for laptops or e-bikes (Lain et al., 2019). In this architecture, the maximum active material loading is the main focus.
- High Power Systems: On the other hand, high power output chemistry such as Lithium Iron Phosphate (LFP) (e.g., A123 APR18650M1) and Lithium Manganese Oxide (LMO) achieve high continuous discharge currents (e.g., 20–30 A) but have less total stored energy (Keil and Jossen, 2016). 14 Besides, the example of Lithium Titanate (LTO) systems can help to understand more about this matter. They are the extreme case of specific power and cycle

life (5000–10,000 cycles) while at the same time they have the lowest energy density amongst the standard Li-ion variants (60–120 Wh/kg), that is why they trade range for speed and longevity.

The huge difference between the maximum theoretical energy (e.g., Lithium–Sulfur at 2654 Wh/kg) and the actual performance (e.g., 420 Wh/kg as of 2022, only 15.8% of the theoretical limit is realized) reveals a critical issue that limits high-energy systems of the future (Akri and Lidong, 2024). That is, this large gap shows that the main factor limiting energy density is not the thermodynamic potential of the active materials but the kinetic limitations of the transport properties, rate improvements, and stability of the system during rapid operation. Therefore, future breakthroughs depend on solving electrode architecture and interfacial engineering problems rather than just finding new materials.

11.6.4 Cycle life constraints and degradation mechanisms

Cycle life, to a large extent, depends on energy and power demands. In general, degradation mechanisms fasten if the battery is frequently operated at a high power, for example, rapid charging. One of the major sources of capacity loss and hazard is lithium plating, which takes place when the kinetics of lithium intercalation into the anode do not match the current demand. Lithium that is metallically plated consumes the active lithium and may also generate metallic dendrites that cause internal shorts and thermal runaway to follow.18 The breakdown process is further intensified if the battery is operated at low temperatures or at high C-rates.

Besides that, the cathode material employed in the battery determines the energy-content versus cycle stability trade-off. In general, high nickel NMC chemistries are selected for their excellent energy density (up to 300 Wh/kg). However, they are less stable and usually have shorter life spans (1000–1500) than lithium iron phosphate (LFP) one. LFP possesses a much longer cycle life (2000–8000 cycles) as well as better thermal stability while conceding an energy density of 90–205 Wh/kg to the other side. When it comes to industrial applications, long-term cost-effectiveness evaluation is to be done by connecting the technical parameter of cycle life with financial viability directly. Highly aggressive operational profiles, such as a very frequent-cycle energy arbitrage, significantly increase degradation and make long-term economic returns at risk. Consequently, the use of Multi-Objective Optimization (MOO) frameworks for performance-technically and economic-outcomes balancing purposes becomes the location for the battery's expected load-induced degradation eventually. Table 11.12 summarizes the performance trade-offs for common and emerging chemistries.

11.6.5 The stability-safety-cost trilemma

The second trilemma explores how to make cells that perform well both robust and economically viable. Stability can be defined as the compatibility of the electrode and electrolyte materials, especially the resistance of the materials to chemical and

Table 11.12 The performance triangle trade-offs in commercial and emerging chemistries (Koech et al., 2024; Piat et al., 2025).

Chemistry type	Specific energy (Wh/kg)	Specific power (W/kg)	Typical cycle life	Dominant trade-off
High-Ni NMC	200–300	Moderate	1000–1500	Energy density versus cycle life and safety
LFP (olivine)	90–205$	Moderate-high	2000–8000 +	Energy density versus cycle life and safety
LTO (spinel)	60–120	Very high	5000–10,000 +	Energy density sacrificed for extreme power and cycle life
Li–S (emerging)	Up to 420 (theoretical 2654)	Low	Low (stability issues)	Power density/ Cycle life traded for high energy density potential
Supercapacitors	5–10	Extreme (>10,000)	1,000,000+	Energy density sacrificed for power and instantaneous charging

mechanical degradation over time. Safety aims at preventing thermal runaway, mainly using nonflammable components, such as SSEs or AEs, and the metal dendrite growth physical inhibition. Cost and scalability is the idea that the expenses of the materials, the complexity of the synthesis, and the capability of the production at a high volume are the factors to consider. For instance, high-performance, experimental materials, like MXenes, may require complex synthesis methods that increase production costs and limit large-scale commercial viability, thus a trade-off between peak performance and industrial readiness is necessary. The transition from cobalt-rich high-energy nickel manganese cobalt (NMC) chemistries to cheaper, safer lithium iron phosphate (LFP) is an example of this market tension. The direct increase in cell voltage and energy density of the battery intensifies the demand for strong safety measures, thus the attention is shifted to electrolyte systems that are inherently nonflammable.

11.6.5.1 Safety baseline and dendrite risk

Commercial LIBs are equipped with LOEs, which are volatile and flammable, thus, are the main causes of the risk of thermal runaway (Bugryniec et al., 2024). Thermal runaway usually starts internally, thus when deposited metallic lithium (dendrites) goes through the separator, it short circuits inside the battery. This danger becomes more potent, if the cell contains large volumes of highly reactive active material. While the engineers have found solutions to the problem of dendrite formation in standard graphite anodes, it is still possible that wrong operating conditions or bad anode and cathode pairing will lead to lithium deposition and later dendrite growth. In addition, the lithiated anode materials are also flammable, hence the need to use flame retardants in polymer binders or conductive frameworks to improve the safety of the overall system.

11.6.5.2 Solid-state electrolytes: Maximizing safety and energy density

SSEs represent a fundamental change in the way batteries can be intrinsically safe and have higher energy density. They are also the major reason why rechargeable lithium batteries are the leading contenders for the electromobility and stationary energy storage markets in Europe and the United States by 2030. SSEs are nonvolatile and nonflammable by nature and thus very close to zero-speed fire risks are practically eliminated. In addition, these SEs feature excellent thermal stability as well, which means they do not get easily degraded even when subjected to very high temperatures, and therefore, the possibility of thermal runaway is lowered. What is more, SSEs have enough mechanical strength to be able to physically block the growth of lithium dendrites. This is the physical blocking, which is the main factor that leads to the utilization of lithium metal anodes that, theoretically, provide the highest energy density that can be achieved in rechargeable batteries. In this way, by alleviating the dendrite problem, SSEs can open the way for full lithium energy chemistry potential.

11.6.5.3 The stability trade-offs in SSE implementation

While solid-state batteries (SSEs) undoubtedly offer improved safety, the drawback is that significant stability and performance issues at the SSI have emerged, thus the risk has been shifted from chemical volatility to physicochemical impedance (Boaretto et al., 2021). High interfacial resistance (Ri) is the main factor that limits the performance of solid-state batteries at present. This resistance is due to a number of sources:

- Poor Physical Contact: In general, intercalated cathode materials undergo volume changes by about 5% during their cycling. The repeated mechanical stresses caused by these changes result in microscopic cracking and loss of close contact between the solid active material particles and the SE, thus the impedance is rapidly increased as the cycling proceeds.
- Space Charge Layer Formation: The presence of a large chemical potential difference at the interface can result in the generation of a lithium-depleted space charge layer, which causes a considerable increase in the interface resistance.
- Element Cross-Diffusion: Chemical instability is a factor that should not be neglected either. High-voltage oxide cathodes may oxidize sulfide-based SSEs due to a narrow electrochemical window, which causes element cross-diffusion and the decomposition of the interface. The leading SSE candidates, for example, argyrodites, are still having a hard time handling this decomposition and interface resistance, which indicates that stability is still the biggest challenge.

11.6.5.4 Aqueous electrolyte batteries: Maximizing safety and cost

Aqueous electrolyte batteries (AEBs) are next generation intrinsically safe batteries that use water-based electrolytes. The main benefits are high intrinsic safety, low cost due to the abundance of raw materials, and long cycle life, with a few systems even showing capacity retention above 90% after 1000 cycles. Nevertheless, the design limitation of AEBs is a fundamental and thermodynamic one: the stability of water. Water is electrochemically stable only within a very small voltage range, roughly 1.23 V. Anything beyond this window leads to water splitting, a reaction that lowers battery efficiency and accelerates degradation. This thermodynamic limitation sets an absolute maximum cell voltage that cannot be overcome, thus limiting AEBs to a lower energy density level as compared to conventional LIBs. Consequently, the technology is a perfect fit for the needs of the stationary storage, grid balancing, and short-range mobility sectors where the increased safety and lower cost make the heavy energy density penalty acceptable. The fixed stability limit of water makes AEBs unable to compete in very high specific energy applications, for example, long-range electric vehicles. Table 11.13 summarizes the key safety risks, dendrite control, electrochmical stability windows, and primary limitations of conventional and next generation electrolyte types.

11.6.5.5 The cost, sustainability, and scalability trilemma

Commercial success in the EES sector is less determined by peak performance, but rather by factors related to supply chain resilience, environmental stewardship, and manufacturing cost. As a result, a strategic shift has occurred in which Total Cost of Ownership (TCO) and low environmental, social, and governance (ESG) risk are given more importance than just maximizing specific energy density.

Current market conditions have led to a split between NMC with a high nickel content and LFP-based chemistries:

- High Performance (NMC): NMC batteries, especially those with a high nickel content, are the ones that can offer the highest energy density (up to 300 Wh/kg), which is the most suitable for increasing the range of premium electric vehicles (Hammed et al., 2025). Nevertheless, such a performance is accompanied by a higher cell cost (100–130 kWh) and significant sustainability risks are implied by the reliance on cobalt and nickel (Hammed et al., 2025). These two minerals are subject to supply chain issues and environmental problems and therefore their usage requires safety management systems to be in place due to their moderate thermal stability.
- Sustainability and Cost (LFP): LFP batteries stand for a deliberate strategic compromise. Their energy density is lower (90–205 Wh/kg) (Tan and Paul, 2024). However, they provide better safety due to their high thermal stability, and they also have a considerably longer lifespan (2000–8000+ cycles). Besides that, LFP is much more environmentally friendly and less expensive than the alternative is by the fact that it does not use nickel or cobalt and instead is based on iron and phosphate, which are both plentiful. The cost of cells is usually on the lower side (70–100 kWh) and it is predicted that the prices will be dropping at a fast

Table 11.13 Comparative safety and stability trade-offs in electrolyte systems (Phogat et al., 2025; Hsieh and Tuan, 2024).

Electrolyte type	Flammability risk	Dendrite control	Electrochemical stability window (V)	Primary stability challenge	Energy density limitation
Liquid organic	High (volatile, flammable)	Poor (initiates thermal runaway)	Wide (sim 4.0–5.0)	Low thermal stability, side reactions	Safety mitigation required
Solid-state (SSE)	Inherently none	Excellent (physical barrier)	Wide (enables Li metal)	High interfacial resistance (chemical/mechanical)	Performance limited by impedance
Aqueous (AEB)	Inherently none	Moderate (anode dependent)	Narrow (approx. 1.23)	Water splitting, rapid degradation at high voltage	Low energy density (voltage constrained)

rate. The acceptance of this trade-off by the market can be observed in the moves of major manufacturers who are shifting their focus to LFP not only for ESSs but also for electric vehicles with a standard range. The prevalence of LFP means that the energy density metric has reached such a level that further increments bring very small commercial advantages in most mass-market and stationary sectors and. Thus, the decision to reduce TCO and supply chain risks is justified.

11.6.5.6 Beyond-lithium systems: The sodium-ion cost roadmap

Sodium-ion batteries (Na-ion) contribute to the energy trend which is mainly about cost and sustainability rather than energy density. Na-ion chemistry is a game-changer in the energy industry as it doesn't require lithium or any other metal.

Performance versus Cost: The main compromise of a Na-ion system is a less energy density (75–160 Wh/kg) than an LFP equivalent one. The cost trajectory, however, is very aggressive (Nekahi et al., 2024). The Na-ion cell cost is estimated to be slightly more than that of LFP in 2025 (59 kWh vs. 52 kWh for LFP), but the industry projections anticipate the cost to go down to around 40 kWh with a large production scale (Lamb, 2025). This speedy cost drop caused by a huge supply of sodium raw materials and their low price is what makes Na-ion systems naturally cheaper to scale and less vulnerable to geopolitical supply shocks. The acceptance of a lower energy density makes Na-ion the perfect nonmobile applications, like grid energy storage and urban electric vehicles, where weight and volume are not so important but cost and supply resilience are. Table 11.14 summarizes a comparison between different chemistries in terms of cost, sustainability, and performance.

Table 11.14 Cost, sustainability, and performance comparison of key chemistries (Gharehghani et al., 2024; Sanusi et al., 2025).

Chemistry	Energy density (Wh/kg)	Critical mineral dependence	Safety (thermal stability)	Estimated cost (2025, $/kWh)	Ideal application
High-Ni NMC	250–300	High (Ni, Co)	Moderate	100–130	Long-range EV
LFP	90–205	Low (Fe, P abundant)	Very high	52–100	Stationary storage, standard EV
Sodium-ion	75–160	None (Na abundant)	High	40–59	Grid balancing, urban mobility

Table 11.15 Scalability constraints of advanced materials (MXenes) (Subramanyam et al., 2024).

Advanced material	Performance advantage	Synthesis method baseline	Material cost estimate	Primary scalability constraint
MXenes	High power density, fast charge kinetics	Acid etching (HF) or molten salt	(Estimated $12.3 million/ton)	Complex chemical processing, reproducibility, and automation

11.6.5.7 High-performance, low scalability: The challenge of advanced materials (MXenes)

The advent of new materials like MXenes (two-dimensional transition metal carbides and nitrides) presents the possibility of breakthroughs in power density and charge kinetics (Fotius and Oliveira, 2025). However, these substances demonstrate the most extreme trade-off in terms of performance potential versus scalability for commercial applications. The limitation is not due to the scarcity of the material-fact is, titanium is plentiful-but rather than the complexity of the synthesis procedure. Generally, the fabrication of MXenes involves complicated multi-step routes, and one of the steps is selective acid etching, which requires the use of a very corrosive and toxic agent, hydrofluoric acid (HF). The post-etch cleaning and control needed for purity usually also require a specially designed oxygen-free environment, which makes the manufacturing process even more difficult. The complexity, thus, results in a very high material cost, which is figured to be close to $12.3 million per ton (Zaed et al., 2023). The enormous cost gap serves as evidence that the main challenge of materials like MXenes is the engineering of the industrial process. Several initiatives are in place to produce more eco-friendly and less energy-consuming routes, for example, by adopting electrochemical techniques and automated, high-throughput synthesis. However, the materials will be confined to expensive, niche applications or the research stage until the gap between lab-scale, high-purity synthesis, and cost-effective continuous industrial production is narrowed. Therefore, the cost of innovation is not just the material but rather the process. The scalability constraints of MXenes are summarized in Table 11.15.

11.7 Second-life batteries and circular economy applications

The rapid increase in the use of electric vehicles (EV) and grid ESS has made LIBs the key elements of the global energy transition. At the same time, the expansion of this market has led to concerns about resource security, environmental sustainability, and waste management. To ensure sustainability in the long run, it is necessary to separate energy transition objectives from resource depletion, thus the deployment of a strong circular economy for LIBs, including second-life applications (reuse) and sustainable materials recovery (recycling), is a compulsory strategic priority. The study herewith provides an in-depth understanding of the technical routes and economic factors that are shaping this circular transition.

11.7.1 The full life cycle of lithium-ion batteries: A stage-gate analysis

The lifespan of a LIB is divided into five different phases, each with their own specific resource requirements, technological challenges, and negative environmental impacts (Paul et al., 2024a). A complete understanding of the battery life is the basis for a circular economy, which is possible by the LCA framework. Such knowledge is vital to determine the locations in the system where actions aimed at creating a closed loop—like recycling and repurposing—have the highest overall benefit.

11.7.1.1 Stage 1: Resource extraction and primary processing

The first part of the life cycle of a LIB includes the hard and energy-intensive processes of sourcing and refining critical minerals. Global demand for LIBs has crossed the threshold of 1 terawatt-hour per year already by late 2024, thus, putting extreme pressure on mineral supply chains (Liu et al., 2022). And it is not just a huge resource requirement by current standards, but this demand is expected to shoot up quite drastically; worldwide demand for lithium is expected to grow by as much as 40 times by 2040 in order to meet the requirements of electric vehicles (EVs) and ESSs. The reliance on these critical materials, especially on lithium (Li), cobalt (Co), and nickel (Ni), imposes significant resource limitations and strategic risks of a deep nature. For instance, the risk of geopolitics is a major concern around cobalt production, with almost 70% of total production coming from the Democratic Republic of Congo. Such extreme concentration of production in one region coupled with rapid demand growth not only leads to the circular economy being a means of environmental protection but also a mechanism for granting mineral independence and lessening supply chain resilience risks. The global commitment to developing domestic "urban mining" infrastructure is largely driven by this imperative. Major investments, such as Redwood Materials' $3.5 billion facility aimed at supporting the domestic supply of lithium, nickel, cobalt, and copper, are paving the way. Typical extraction methods such as hard rock mining and brine evaporation come at a heavy cost to the environment. These techniques generally involve areas of water scarcity and tend to produce more dust that may contain harmful metals, for example, nickel, copper, cobalt, and chromium. The environmental burden embedded in this first stage (extraction and refining) is considerable; therefore, the entire concept of circularity is about not doing these high-impact steps again but rather extending the life of the manufactured asset via second-life use and proper recycling. Research on alternative sources such as deep-sea mining of manganese nodules is intended to provide copper, nickel, and cobalt in a less environmentally harmful way, thus, lessening waste and environmental impact than land-based extraction methods.

11.7.1.2 Stage 2: Cell and module manufacturing (embedded energy costs)

The production stage is the moment when raw materials are converted into usable energy assets with a notable amount of energy and processing costs being "embedded" in the final product. The fabrication of a battery cell is a very exact, multi-phased endeavor that is the main factor which sets the performance parameters of the cell, that is, service life, charging capacity, and overall quality. The first step is mixing, in which anode and cathode materials are mixed with conductive additives, solvents, and binders to create a homogeneous slurry (Rezaei et al., 2025). This step is also held under tight quality control, particularly with respect to the homogeneity of the mix and ensuring that the water content is extremely low at parts-per-million (ppm) level. The prepared slurry is then taken to the coating stage where it is coated onto current collectors (e.g., cathode material, generally a metal oxide, is coated on an aluminum electrode). Directly after coating, the electrodes are subjected to Drying by convective air dryers, and then Calendaring, a rolling process that determines the final thickness and porosity of the electrode, is performed. Preparation of the electrodes for cell assembly follows (Rezaei et al., 2025). This is done by slitting the electrodes to standard widths, and then Final Drying stage is performed to ensure removal of all solvent content and further reduction of free water ppm. 10 Cutting not only finalizes the shapes but also creates the necessary tabs for connection. The assembly of cells is done by winding (for cylindrical cells) or by stacking the anode, cathode, and separator layers. Accuracy is very important during the assembly process to avoid punctures in the separators, to make sure that the layers are exactly aligned and to prevent the contamination of the humidity (Degen and Schütte, 2021). In the end, terminal welding is used to connect the electrodes to the cell terminals, which is then followed by Canning or Enclosing, and the battery is filled with liquid electrolyte mixture. The performance of the battery before the first use is in a large part determined by these accurate operations.

11.7.1.3 Stage 3: First life (EV deployment) and degradation mechanics

The first life of a LIB generally is the active utilization period in an electric vehicle. In this phase, the battery capacity will automatically degrade, which is accompanied by a drop in the State of Health (SoH). The degradation curve is nonlinear, that is, the capacity does not decrease in a constant manner over time, hence, it is difficult to precisely forecast the future performance and calculate the Remaining Useful Life (RUL). The major mechanisms that cause the battery to age and lower its energy storage capacity include (Gonzalez-Rodriguez et al., 2025):

- SEI Layer Dynamics: The layer that the SEI forms, which is the anode surface, is the most stable one. Still, the continuous growth and decay of this layer consume active lithium ions, thereby the lifetime is shortened directly, and the total capacity goes down.

- Thermal Stress and Heat Generation: High discharge rate or deep DoD operation leads to increased internal resistance and, therefore, the production of heat to the cell is high. The consequent thermal stress aggravates the battery aging phenomena.
- Phase Transitions and Strain: The metal backing tissues in the cathode are, for instance, the same ones that will endure structural changes in the lattice of the crystal when the charge and discharge of the cycle progress, thus strain which is the mechanical one accumulates with loss of performance.

These processes together determine the battery second most demanding life application.

11.7.1.4 Stage 4: Second life (repurposing)

Stage 4 is the transitional phase, a pivotal circular strategy that moves around the stages. When the performance of a battery pack is less than that required for high-power automotive applications by a threshold—generally it is a capacity of around 70%–80% of the original one that is kept—it still has a considerable residual value (Zhu et al., 2021). By repurposing, the economic and functional life of the asset is prolonged because of the move to less consuming, nonmobile applications, thus solving the problem of how to effectively exploit the asset's intrinsic value prior to the need for materials recycling.

11.7.1.5 Stage 5: End-of-life management (recycling)

The last phase in the circular economy is end-of-life management that aims at the safe recovery of the constituent materials. Local energy is usually discharged through passive methods like a resistor-based short circuit or, more commonly, immersion in an ionically conducting salt solution. The next objective is to recycle the most critical materials, for example, lithium, nickel, and cobalt, that is, to let them re-enter the supply chain and thus reduce the need for primary resource extraction. Such a recycling process is indispensable for the implementation of a closed-loop system and for the reduction of the cradle-to-gate environmental impact of new battery production.

11.7.2 Second-life battery applications

Second life batteries (SLBs) deployment is mainly motivated by the twofold economic and environmental reasons of maximizing energy and material utilization that are already embedded. The growth of this segment, therefore, depends on how effectively these technical issues arising from the diversity of the old assets and the difficulties in setting up firm standardization protocols can be resolved.

11.7.2.1 Defining the SLB transition threshold and applications

SLBs are essentially batteries that have been removed from electric vehicles (EVs) and their state of health (SoH) has been decreased to around 70%–80% of the original capacity (Gharebaghi et al., 2024). The performance of high-power, high-range EV applications is severely limited by the loss of this capacity; however, the leftover capacity is in most cases enough for stationary ESS where the cycling demands are lower and the volumetric or gravimetric constraints are less. The main applications of SLBs can be divided into two broad categories (Costa et al., 2022):

- Residential/commercial storage: Such systems usually mean the combination of SLBs with renewable energy generators, for instance, photovoltaic (PV) systems, for home energy storage or commercial backup. Evaluations, like those of the systems in the United Kingdom with Nissan Leaf batteries, can serve as proof of the technical realization of this integration (Gao et al., 2024). Although it is pointed out that the stability may be deteriorated in the case of high-demand loads (e.g., EV charging stations) in off-grid systems unless further battery packs are connected.
- Grid-level support: On the utility scale, SLBs are allowed to be pooled together to offer grid stability services such as capacity firming, frequency regulation, and peak shaving. These uses call for the leftover capacity from the battery for less-demanding, lower-power discharge requirements, and thus, they keep the economic life of the asset very much extended.

11.7.2.2 Market overview and economic drivers

The SLB market holds strategic significance as it acts as an essential conduit for not only increasing the asset value but also for reducing the environmental impact that precedes the final recycling. Such a strategic significance is echoed in the market's strong growth projections, which are primarily sustained by the growing number of the out-of-life EV packs that are the source of the pipeline. According to the projections, over the next 10 years, the market for second-life electric vehicle batteries will be growing rapidly, with a compound annual growth rate (CAGR) of more than 25%. The global second life battery market size and growth forecast in displayed in Table 11.16.

The expected CAGR of 25.61% during the period from 2025 to 2034 is the main economic reason for the strong positive perception of the market, which anticipates a large volume of EV batteries at the end of their life that will result in a plentiful supply of second-life assets in the near future. Such a growth path is crucial as a successful development of the SLB market gives a needed time buffer, thus it moves the material "stock" that is technically complex and still in the process of scaling (recycling

Table 11.16 Global second-life EV battery market size and growth forecast (Jegham and Sahlin Jankert, 2024; Lieskoski et al., 2024).

Report coverage detail	Value	Timeframe/Metric
Market size (base year)	USD 1.27 billion	2024
Market size (2025 projection)	USD 1.60 billion	2025
Market size (2034 forecast)	USD 12.42 billion	2034
Compound annual growth rate (CAGR)	25.61%	2025 to 2034
Dominating region	Asia Pacific	N/A
Fastest growing region	Europe	N/A

infrastructure mentioned in 11.7.3) further down the line. This postponement gives the recycling industry a great deal of time for the necessary technologies to be perfected and for the capacity to be increased.

11.7.2.3 Material and logistics challenges: Sorting, grading, and integration

The main difficulty in the creation of a scalable SLB industry is the control of the highly heterogeneous retired EV assets. After all, each battery pack has gone through a different operational history. This history involves different temperatures, charging routines, and depths of discharge. As a result, each battery module has a different SoH and a different set of degradation mechanisms.

11.7.2.3.1 Challenges in characterization

The condition of the batteries varies, which means that a standardized process for grading has not been established universally, and complex individual evaluation is required. Proper characterization and grading are not only complicated but also important steps to safe and efficient repurposing of such assets:

- SoH and RUL Assessment: Very strict measures should be taken to accurately determine the current State of Health (SoH) and to be able to forecast the RUL, which is the expected future SoH, the latter being predicted (Zhou et al., 2016). The process of degradation is not linear; therefore, one cannot simply extrapolate the old data for future results. To solve this problem, advanced predictive models such as support vector regression should be used to determine SoH precisely during the second life application and to give accurate RUL predictions (Yin et al., 2024).
- Safety and Reliability: Safety assurance requires extensive testing and modification of the battery management system (BMS) (Rahmani et al., 2025). To ensure that there is no overheating, strong thermal management systems must be in place and the BMS should be changed for second-life situations. For example, the BMS must be linked with each slave BMS in every module to get exact cell information and it must be connected in series with the modules to allow circuit disconnection when the SOC is low or it reaches its maximum charge limit.

Although the heterogeneity of end-of-life (EoL) EV batteries is the primary factor causing the nonlinearity of degradation, this issue leads directly to high operational costs related to the grading and sorting stage. Additionally, the technical challenges associated with sorting these batteries can impact the economic viability of Second life batteries (SLBs). Without the timely development and deployment of advanced, automated, and high-throughput diagnostic systems, the attractiveness of SLBs may diminish.

11.7.2.3.2 Standardization and regulatory protocols

To control the risks of liability and safety that come with the use of batteries of unknown quality for stationary systems, industry standardization is a must. In the United States and Canada, the UL 1974 Standard for Evaluation for Repurposing Batteries has become the main reference document. The standard examines the condition, safety, and energy capacity of each battery pack individually before the integration of the ESS. Certification is about the audit and validation of a facility's sorting and grading operations, thus offering the required control to allow commercial scalability and the acceptance of insurers and utility providers. The company 4R Energy Corp (a Nissan/Sumitomo joint venture) is one of the companies that already categorizes EoL Nissan LEAF batteries based on capacity and predicted life (Chung, 2021).

11.7.3 Materials for a circular economy: Advancements in recycling and design

Although second life uses can prolong the battery asset's value; top-efficiency recycling (Stage 5) is the absolute requirement for closing the material loop and getting real circularity. The emphasis is moving away from just metal recovery to obtaining very pure cathode active materials (CAMs) with almost no energy consumption.

11.7.3.1 The circular economy recycling market context

The recycling industry is growing substantially as a result of regulatory pressure, the need for supply chain decarbonization, and the increase in the volume of retired batteries. The battery recycling circular economy market that was worth US$ 26.54 billion in 2024 is expected to reach US$ 56.07 billion in 2032, recording a strong CAGR of 9.80% (Intelligence, 2025). Likewise, the US Circular Battery Economy Market is expected to elevate from USD 8.8 billion in 2024 to USD 29.58 billion in 2032 with a CAGR of 16.28% (Intelligence, 2025).

The global capacity for recycling is not enough at the moment; established facilities were approximately at 1.6 million tons per year in 2025, with the forecast of more than 3 million tons per year with the planned additions (Kresse et al., 2025). Nevertheless, to meet the anticipated LIB recycling demand resulting from the EV adoption, worldwide LIB recycling capacity needs to be increased by 50 times within the next decade (Li et al., 2023a). Such a huge gap guarantees the continuation of governmental intervention and investment, for instance, the strategic partnerships that Li-cycle is forming with major automakers and the EPA approval that the American Battery Technology Company has received for recycling damaged batteries. Also, the strategic decision of recycling and manufacturing site selection plays an important role in minimizing the environmental impact. The studies show that recycling and then manufacturing in North America instead of China can lower the carbon footprint of NMC811 CAM by 16% and freshwater toxicity by 30% (Fikru et al., 2025). This geographic benefit is a clear indication of how important localized circular supply chains are.

11.7.3.2 Comparative analysis of pyrometallurgy and hydrometallurgy

Historically, two major methods were used for industrial LIB recycling: pyrometallurgy and hydrometallurgy. The industry is gradually moving away from the first one due to its failure to result in a complete material circularity.

11.7.3.2.1 Pyrometallurgy (high-temperature smelting)

Pyrometallurgy or high-temperature smelting is a process of heating spent LIBs at very high temperatures, usually in the range of 1200 °C–1600 °C (Reinhart et al., 2023). The process is very efficient and quite easy to operate. The high temperature decomposes the organics and polymers, thus allowing the metal elements to be extracted. On the other hand, pyrometallurgy is not a circularity-compliant technology:

- Energy and Emissions: The procedure is very energy-intensive, and it emits harmful gases and, hence, it is much less eco-friendly than chemical methods.
- Material Loss: Most importantly, the only raw materials that can be recovered in a metallic alloy form are mainly cobalt (Co) and nickel (Ni). Besides that, lithium (Li), aluminum (Al), and manganese (Mn) are the elements that end up in the slag and are not recovered economically (Zanoletti et al., 2025). The nonrecovery of lithium is what limits pyrometallurgy to be a very small share of the recycling market in the future of constrained resources.

11.7.3.2.2 Hydrometallurgy

Hydrometallurgy relies on chemical leaching (using solvents and acids) to extract valuable metals from the "black mass" (the interiors of batteries that have been shredded and discharged) (Lee et al., 2025):

- Environment and Resource Advantages: Hydrometallurgy is the most environmentally friendly and resource-efficient method of recycling. It results in a higher recovery of elements, requires less energy, and emits substantially lower amounts of GHGs and other pollutants.
- Recovery Efficiency: In contrast to pyrometallurgy, hydrometallurgy makes it possible to recover lithium, aluminum, and manganese. The advanced technologies introduced by the industry leaders, for example, Li-cycle's patented process, are designed to achieve over 95% recovery rates for materials such as cobalt and nickel (in the form of sulfate salts) and lithium (in the form of a carbonate) (Neumann et al., 2022).

The industry's clear evolutionary pivot from pyrometallurgy to hydrometallurgy is dictated by the mandatory necessity of recovering lithium and improving the overall environmental profile of the process. A comparison of conventional LIB recycling technologies-specifically pyrometallurgy and hydrometallurgy -along their operational metrics and strategic implications, is presented in Table 11.17.

11.7.3.3 The evolution to direct recycling and design for recycling

Although hydrometallurgy is a big improvement, it still depends on the destruction of the CAM structure by chemicals, and re-synthesis steps that are energy-intensive are necessary. The following technological horizon is Direct Recycling.

11.7.3.3.1 Direct recycling and environmental superiority

Direct recycling is a process by which components of batteries are recovered, regenerated, and reused directly, with a special emphasis on the preservation of the

Table 11.17 Comparison of conventional LIB recycling technologies (Reinhart et al., 2023; Zanoletti et al., 2025; Lee et al., 2025; Neumann et al., 2022).

Metric	Pyrometallurgy (smelting)	Hydrometallurgy (leaching)	Strategic implication
Operating temperature	High (1200 °C–1600 degrees C)	Low (ambient/ moderate)	Significant reduction in energy consumption.
Lithium (Li) recovery	Negligible (lost to slag)	High (often >95 % efficiency)	Hydrometallurgy is essential for Li resource security.
Cobalt/Nickel recovery	High (as metallic alloys)	High (as salts/ precursors)	Both methods are effective for primary critical metals.
Environmental burden (GHG)	High (energy-intensive, harmful gases)	Lower (requires careful chemical management)	Hydrometallurgy is the environmental intermediary step.
Primary limitation	Material loss (Li, Al, Mn); high CO_2 emissions	High chemical/solvent use; complex wastewater treatment	Defines the necessity of the direct recycling evolution.

chemical structure of the CAM (Mancini et al., 2024). In this method, performance recovery to the cathode by re-lithiation along with electrolyte recovery, electrode separation based on unique properties (density, magnetic susceptibility), binder removal are some of the stages involved (Hu and Xu, 2021). By saving the material from degradation, the production of the new battery cell is fundamentally less complicated because fewer processes must be done. A comparative modeling method shows that direct recycling leads to significantly lower environmental impacts than the traditional hydrometallurgical method. The path is therefore obvious: Pyrometallurgy (simple, low Li recovery) → Hydrometallurgy (complex, high Li recovery) → Direct Recycling (structural preservation, lowest overall impact) (Wang et al., 2023b). Hence, direct recycling is considered the final strategic option for material circularity.

11.7.3.3.2 Addressing the chemical impact challenge

Advanced recycling methods, environmentally preferred, are still faced with a serious chemical impact challenge that limits their advantage over smelting. In fact, the production of chemicals and the use of solvents (e.g., organophosphates, acids) for both hydrometallurgical and current direct recycling processes contribute significantly to the environmental footprint. Modeling, for example, indicated that a high solvent use, which was based on the scaling up of a lab-scale direct recycling, was the main cause of an unexpected increase in ozone depletion impacts. Such a result implies that technological optimization should turn to chemical engineering innovations instead of merely increasing the physical throughput. The next R&D program should be dedicated to finding ways of reducing the usage of high impact reagents.

11.7.3.3.3 Design for Recycling

The overwhelming chemical load is the main reason for the need of Design for Recycling (DfR) (Wu et al., 2025). DfR mandates that producers redesign material chemistries and battery structures with simpler, more easily dissolvable components that facilitate material separation and recovery operations in the recycling department. Innovative research is in line with this perspective. Researchers have created the flash Joule heating–chlorination and oxidation (FJH-ClO) method, which can separate lithium and transition metals without producing the harsh acids or high energy requirements of traditional commercial systems (Xu et al., 2025). This shortened sequence can produce core materials of a high standard of purity in a few cases, and it is estimated that approximately 95% fewer chemicals and half the energy of the already known processes can be used. Thus, setting a feasible technological route toward really sustainable and chemically efficient circular systems.

11.8 Conclusion and future outlook

The LIB landscape is shifting radically, going beyond mere incremental changes to fundamental architectural and material changes. The electrification of the global economy is happening at a very fast pace, and therefore, the question of how sustainable and resilient the battery supply chain will has become a very important one:

- Solid-State Electrolytes (SSEs): The move from liquid to SEs is a major change in safety and energy density. The use of flammable organic solvents is eliminated in solid-state batteries (ASSBs) thus reducing the risk of fire considerably, at the same time allowing high-capacity lithium metal anodes to be used, which in turn makes it possible to double the specific energy of the devices to more than 500 Wh kg^{-1}.
- Nanostructured Silicon Anodes: Among the materials currently used for LIBs, one is silicon. Graphite can be best replaced by silicon due to its outstanding theoretical specific capacity of about 3590 mAh g^{-1}. significant developments in nanostructuring and in new binders have been made, which in turn are successful in alleviating the mechanical strain that accompanies the volume expansion of silicon.
- Beyond-Li Systems: In order to alleviate the shortage and high price of lithium in the very long term, sodium-ion (Na-ion) systems are becoming the most viable successors to stationary storage in terms of cost. The systems take advantage of the high natural abundance of sodium and can deliver reversible capacities close to 108 mAh g^{-1} which makes them perfect for large-scale grid applications where volumetric density plays a secondary role to cost.

To harness the full power of these inventions, it is necessary to address several critical research frontiers:

- Interface Engineering: The most significant obstacle for ASSBs and beyond-Li systems, interface engineering, is still the main point of concern. "Dynamically adaptive interphases" and ductile inorganic-rich solid-electrolyte interphases (SEIs) made of materials like silver

sulfide (Ag_2S) that can manage interfacial cracking and maintain contact under high current densities are the main directions of the current research efforts.
- Multifunctional Materials: The creation of structural batteries is just one step away from the "massless" concept of energy storage. By utilizing carbon fiber composites capable of both bearing mechanical loads and storing energy, researchers have reached elastic moduli over 70 GPa—similar to aluminum—with energy densities up to 30 Wh kg^{-1}.
- Advanced Characterization and AI Discovery: The pace at which new materials are discovered is set by high-throughput Materials Acceleration Platforms and in situ/operando characterization methods. Additionally, AI-supported smart grading systems that shorten the testing duration for second-life applications by 30% are the main contributors to the EV battery retirement sector, which is becoming safer and more efficient in the classification for repurposing.

These material improvements are the key elements that open the way for a tightly carbon-free energy system. It is highly necessary for the climate change mitigation and for the purpose of mineral independence, a safe and circular battery supply chain—which is estimated to continue its annual growth at over 30% and thus have a value of more than 400 billion by 2030—should be at the core of the efforts tackling the issue. Ultimately, they will fundamentally help to accomplish the goal of global decarbonization and thus, through the generation of equitable clean-energy manufacturing jobs, they will contribute to the worldwide economy.

Declaration of AI assistance

The authors acknowledge the use of generative AI tools in the preparation of this manuscript. The tools were used solely to assist with language refinement, grammar correction, and improving clarity. All scientific content, analyses, interpretations, and conclusions presented in this work were developed independently by the authors.

References

Abdulrahman, G.A.Q., Aziz, A., Qasem, N.A.A., Alazzam, A., 2025. 2D Materials: synthesis, properties, and energy-related applications. Coord. Chem. Rev. 544, 216950.

Ai, S., Wu, X., Wang, J., Li, X., Hao, X., Meng, Y., 2024. Research progress on solid-state electrolytes in solid-state lithium batteries: classification, ionic conductive mechanism, interfacial challenges. Nanomaterials 14, 1773.

Akinyele, D., Belikov, J., Levron, Y., 2017. Battery storage technologies for electrical applications: impact in stand-alone photovoltaic systems. Energies 10.

Akri, A., Lidong, A., 2024. Review of performance analysis of lithium-sulfur batteries for electric vehicles: challenges and future prospects. Indonesian Journal of Green Chemistry 1, 12–17.

Arya, A., Sharma, A., 2020. Hybrid polymer nanocomposites for energy storage/conversion devices: from synthesis to applications. In: Electrochemical Energy Conversion and Storage Systems for Future Sustainability. Apple Academic Press.

Ashfaq, M.M., Salam, A., Zholobko, O., Wu, X.-F., 2025. Electrolysis of aqueous polymer solutions for low-cost hydrogen production: an experimental case study of polyethylene oxide (PEO) electrolysis. Int. J. Hydrogen Energy 151, 150198.

Banerjee, A., Wang, X., Fang, C., Wu, E.A., Meng, Y.S., 2020. Interfaces and interphases in all-solid-state batteries with inorganic solid electrolytes. Chem. Rev. 120, 6878–6933.

Binninger, T., Marcolongo, A., Mottet, M., Weber, V., Laino, T., 2020. Comparison of computational methods for the electrochemical stability window of solid-state electrolyte materials. J. Mater. Chem. A 8, 1347–1359.

Bisquert, J., 2004. Chemical diffusion coefficient of electrons in nanostructured semiconductor electrodes and dye-sensitized solar cells. J. Phys. Chem. B 108, 2323–2332.

Boaretto, N., Garbayo, I., Valiyaveettil-Sobhanraj, S., Quintela, A., Li, C., Casas-Cabanas, M., Aguesse, F., 2021. Lithium solid-state batteries: state-of-the-art and challenges for materials, interfaces and processing. J. Power Sources 502, 229919.

Bu, F., Zagho, M.M., Ibrahim, Y., Ma, B., Elzatahry, A., Zhao, D., 2020. Porous MXenes: synthesis, structures, and applications. Nano Today 30, 100803.

Bugryniec, P.J., Resendiz, E.G., Nwophoke, S.M., Khanna, S., James, C., Brown, S.F., 2024. Review of gas emissions from lithium-ion battery thermal runaway failure — considering toxic and flammable compounds. J. Energy Storage 87, 111288.

Chen, L., Shi, P., Gu, T., Mi, J., Yang, K., Zhao, L., Lv, J., Liu, M., He, Y.-B., Kang, F., 2025. Strategies of constructing highly stable interfaces with low resistance in inorganic oxide-based solid-state lithium batteries. eScience 5, 100277.

Cheong, J.Y., Venkateshaiah, A., Yun, T.G., Shin, S.-H., Černík, M., Padil, V.V.T., Kim, I.-D., Varma, R.S., 2021. Transforming gum wastes into high tap density micron-sized carbon with ultra-stable high-rate Li storage. Electrochim. Acta 367, 137419.

Chung, H.-C., 2021. Charge and discharge profiles of repurposed LiFePO4 batteries based on the UL 1974 standard. Sci. Data 8.

Connell, J.G., Fuchs, T., Hartmann, H., Krauskopf, T., Zhu, Y., Sann, J., Garcia-Mendez, R., Sakamoto, J., Tepavcevic, S., Janek, J.R., 2020. Kinetic versus thermodynamic stability of LLZO in contact with lithium metal. Chem. Mater. 32, 10207–10215.

Costa, E., Wells, P., Wang, L., Costa, G., 2022. The electric vehicle and renewable energy: changes in boundary conditions that enhance business model innovations. J. Clean. Prod. 333, 130034.

Cui, X., Zhang, Y., Cheng, S., Liu, Y., Shao, Z., Sun, Z., Wu, Y., Guo, H., Fu, J., Xie, E., 2022. Achieving high-rate and durable aqueous rechargeable Zn-Ion batteries by enhancing the successive electrochemical conversion reactions. J. Colloid Interface Sci. 620, 127–134.

Dai, B., Zhou, M., Liu, K., He, B., Xiang, B., Kong, L., 2023. The molding of the ceramic solid electrolyte sheet prepared by tape casting. J. Phys. Conf. 2566, 012102.

Das, H.S., Mishra, S., Roymahapatra, G., 2025. Advanced Nano-Structured Materials for Energy Storage Devices. Design, Fabrication, and Significance of Advanced Nano-structured Materials. IGI Global.

Degen, F., Schütte, M., 2021. Life cycle assessment of the energy consumption and GHG emissions of state-of-the-art automotive battery cell production. J. Clean. Prod. 330, 129798.

Divakaran, A.M., Minakshi, M., Bahri, P.A., Paul, S., Kumari, P., Divakaran, A.M., Manjunatha, K.N., 2021. Rational design on materials for developing next generation lithium-ion secondary battery. Prog. Solid State Chem. 62, 100298.

Dutta, T., Mary Gladis, J., 2024. Recent developments on electrode materials and electrolytes for aluminium-ion batteries. J. Energy Storage 86, 111287.

Egbue, O., Long, S., Kim, S.D., 2022. Resource availability and implications for the development of plug-in electric vehicles. Sustainability 14.

Famprikis, T., Canepa, P., Dawson, J.A., Islam, M.S., Masquelier, C., 2019. Fundamentals of inorganic solid-state electrolytes for batteries. Nat. Mater. 18, 1278–1291.

Fan, Z., Li, R., Zhang, X., Zhao, W., Pan, Z., Yang, X., 2024. Defect engineering: can it mitigate strong coulomb effect of Mg^{2+} in cathode materials for rechargeable magnesium batteries? Nano-Micro Lett. 17, 4.

Fikru, M.G., Shen, T., Abegaz, M., 2025. Co-location of carbon capture projects with pollution-intensive manufacturing: evidence from US zip codes. Appl. Econ. 1–15.

Fotius, J.A., Oliveira, H.P., 2025. MXene-based composites for energy harvesting and energy storage devices. Solids 6.

Gao, W., Cao, Z., Vosoughi Kurdkandi, N., Fu, Y., Mi, C., 2024. Evaluation of the second-life potential of the first-generation Nissan Leaf battery packs in energy storage systems. eTransportation 20, 100313.

Gao, Y., Yan, L., Zhao, C., Chen, M., Yang, S., Shao, G., Mao, J., 2024. Accurate calculation of solid-state Li+ diffusion coefficient and kinetic activation energies for an artificial graphite anode. J. Electrochem. Soc. 171, 020558.

Gharebaghi, M., Rezaei, O., Li, C., Wang, Z., Tang, Y., 2024. A survey on using second-life batteries in stationary energy storage applications. Energies 18, 42.

Gharehghani, A., Rabiei, M., Mehranfar, S., Saeedipour, S., Mahmoudzadeh Andwari, A., García, A., Reche, C.M., 2024. Progress in battery thermal management systems technologies for electric vehicles. Renew. Sustain. Energy Rev. 202, 114654.

Ghotia, S., Kumar, P., Srivastava, A.K., 2025. A review on 2D materials: unveiling next-generation hydrogen storage solutions, advancements and prospects. J. Mater. Sci. 60, 1071–1097.

Gonzalez, A., Yang, N.-H., Liu, R.-S., 2017. Silicon anode design for lithium-ion batteries: progress and perspectives. J. Phys. Chem. C 121.

Gonzalez-Rodriguez, P.S., Lozoya-Santos, J.D.J., Gonzalez-Hernandez, H.G., Felix-Herran, L. C., Tudon-Martinez, J.C., 2025. Challenges and opportunities for extending battery pack life using new algorithms and techniques for battery electric vehicles. World Electr. Vehic. J. 16.

Gull, S., Chen, H.-Y., 2022. Recent advances in cathode materials for aqueous zinc-ion batteries: mechanisms, materials, challenges, and opportunities. MRS Energy Sustain. 9.

Gupta, S.K., Vishwakarma, J., Srivastava, A.K., Dhand, C., Dwivedi, N., 2024. Aluminum batteries: opportunities and challenges. Energy Storage Mater. 70, 103538.

Hammed, V.O., Salako, E.W., Edet, D., Ederhion, J., Keshinro, B.I., Uwaoma, I.A., Adeleke, O.J., Odetoran, A., Adedokun, O.J., Makinde, P.F., Alli, Y.A., 2025. Next-generation lithium-ion batteries for electric vehicles: advanced materials, AI driven performance optimization, and circular economy strategies. Measure. Energy 7, 100060.

Hasan, M.M., Haque, R., Jahirul, M.I., Rasul, M.G., Fattah, I.M.R., Hassan, N.M.S., Mofijur, M., 2025. Advancing energy storage: the future trajectory of lithium-ion battery technologies. J. Energy Storage 120, 116511.

Hoang, H.A., Kim, D., 2022. High voltage stable solid-state lithium battery based on the nano-conductor imbedded flexible hybrid solid electrolyte with hyper-ion conductivity and thermal, mechanical, and adhesive stability. Chem. Eng. J. 435, 135092.

Hosseinian, S.B., Ghani, M., Raoof, J.B., 2025. Recent trends in synthesis and application of 2D mxene-based nanomaterials for microextraction purposes. Microchem. J. 213, 113788.

Hsieh, Y.-Y., Tuan, H.-Y., 2024. Emerging trends and prospects in aqueous electrolyte design: elevating energy density and power density of multivalent metal-ion batteries. Energy Storage Mater. 68, 103361.

Hu, Q., Xu, L., 2021. An overview on Lithium-ion batteries recycling processes. J. Phys. Conf. 1885, 032031.

Icaza, J.C., Haasch, R.T., Guduru, R.K., 2018. Effect of ion size and charge density on the electrochemical characteristics of α-MoO_3 using aqueous Be^{2+} and Mg^{2+} sulfate electrolytes. J. Alloys Compd. 740, 88–95.

Intelligence, D., 2025. Circular Economy in Battery Recycling Market is Expected to Reach US $ 56.07 Billion by 2032 | Top Companies - Umicore, Li-Cycle, Redwood Materials, Accurec Recycling. Germany.

Islam, M., Ahmed, M.S., Yun, S., Ali, B., Kim, H.-Y., Nam, K.-W., 2025. Recent advances in nanostructured conversion-type cathodes: fluorides and sulfides. Nanomaterials 15.

Jang, J., Kwon, T., 2025. Various technologies to mitigate volume expansion of silicon anode materials in lithium-ion batteries. Batteries 11.

Jayasree, S.S., Nair, S., Santhanagopalan, D., 2022. Nanoscale-engineered $LiCoO_2$ as a high energy cathode for wide temperature lithium-ion battery applications–role of coating chemistry and thickness. Nanotechnology 33, 275403.

Jegham, J., Sahlin Jankert, H., 2024. Exploring Challenges and Opportunities for Circularity within the Second Life EV Battery Market.

Johari, S.N.A.M., Tajuddin, N.A., Hanibah, H., Deraman, S.K., 2021. A review: ionic conductivity of solid polymer electrolyte based polyethylene oxide. Int. J. Electrochem. Sci. 16, 211049.

Karimi, E., Iglauer, S., Azhar, M.R., 2025. Interface engineering and safety in solid-state batteries: advancing from human-centered insights to AI-driven innovations. Adv. Colloid Interface Sci. 346, 103686.

Keil, P., Jossen, A., 2016. Charging protocols for lithium-ion batteries and their impact on cycle life—an experimental study with different 18650 high-power cells. J. Energy Storage 6, 125–141.

Khan, M., Yan, S., Ali, M., Mahmood, F., Zheng, Y., Li, G., Liu, J., Song, X., Wang, Y., 2024. Innovative solutions for high-performance silicon anodes in lithium-ion batteries: overcoming challenges and real-world applications. Nano-Micro Lett. 16, 179.

Kim, A., Woo, S., Kang, M., Park, H., Kang, B., 2020. Research progresses of garnet-type solid electrolytes for developing all-solid-state Li batteries. Front. Chem. 8, 468.

Kim, J.Y., Jung, S., Kang, S.H., Lee, M.J., Jin, D., Shin, D.O., Lee, Y.-G., Lee, Y.M., 2022. All-solid-state hybrid electrode configuration for high-performance all-solid-state batteries: comparative study with composite electrode and diffusion-dependent electrode. J. Power Sources 518, 230736.

Koech, A., Mwandila, G., Mulolani, F., Mwaanga, P., 2024. Lithium-ion battery fundamentals and exploration of cathode materials: a review. South Afr. J. Chem. Eng. 50.

Kravchyk, K.V., Kovalenko, M., 2019. Rechargeable dual-ion batteries with graphite as a cathode: key challenges and opportunities. Adv. Energy Mater. 9, 1901749.

Kresse, C., Bookhagen, B., Buarque Andrade, L., Frenzel, M., 2025. Global supply of secondary lithium from lithium-ion battery recycling. Recycling 10.

Lain, M.J., Brandon, J., Kendrick, E., 2019. Design strategies for high power vs. High energy lithium ion cells. Batteries 5.

Lamb, J., 2025. (Invited) cost is king: an industrial perspective on sodium-ion batteries. In: ECS Meeting Abstracts, MA2025-02, 750-750.

Lee, C., Arby, D.S., Kim, C., Lim, J., Kwon, K., Chung, E., 2025. Hydrometallurgical process of spent lithium-ion battery recycling Part. 1 Chemical leaching of valuable metals from cathode active materials: review and case study. Hydrometallurgy 235, 106494.

Lee, D., Lee, H., Song, T., Paik, U., 2022. Toward high rate performance solid-state batteries. Adv. Energy Mater. 12, 2200948.

Li, B., Xu, H., Ma, Y., Yang, S., 2019. Harnessing the unique properties of 2D materials for advanced lithium–sulfur batteries. Nanoscale Horiz. 4, 77–98.

Li, F., Zhang, H., Liu, X., Xu, B., Wang, C., 2024a. Dendrite-free Zn anode modified by organic coating for stable aqueous zinc ion batteries. Batteries 10, 420.

Li, H., Berbille, A., Zhao, X., Wang, Z., Tang, W., Wang, Z., 2023a. A contact-electro-catalytic cathode recycling method for spent lithium-ion batteries. Nat. Energy 8.

Li, J., Yang, J.-Y., Wang, J.-T., Lu, S.-G., 2019. A scalable synthesis of silicon nanoparticles as high-performance anode material for lithium-ion batteries. Rare Met. 38, 199–205.

Li, M., Hicks, R.P., Chen, Z., Luo, C., Guo, J., Wang, C., Xu, Y., 2023b. Electrolytes in organic batteries. Chem. Rev. 123, 1712–1773.

Li, W., Song, B., Manthiram, A., 2017. High-voltage positive electrode materials for lithium-ion batteries. Chem. Soc. Rev. 46, 3006–3059.

Li, X.-Y., Zhao, M., Song, Y.-W., Bi, C.-X., Li, Z., Chen, Z.-X., Zhang, X.-Q., Li, B.-Q., Huang, J.-Q., 2025. Polysulfide chemistry in metal–sulfur batteries. Chem. Soc. Rev. 54, 4822–4873.

Li, Y., Zhao, L., Dang, H., Dou, P., Wu, Y., Ran, F., 2024b. Designing macromolecular modifiers for zinc metal batteries. Mater. Sci. Eng. R Rep. 161, 100844.

Lieskoski, S., Tuuf, J., Björklund-Sänkiaho, M., 2024. Techno-economic analysis of the business potential of second-life batteries in ostrobothnia, Finland. Batteries 10, 36.

Lim, J.-H., Kim, J., Oh, J., Kwon, J., Lee, K.E., Lee, Y., Park, S., Lim, J., Shin, D., Jo, C., Kim, Y.-T., Moon, J., Hersam, M.C., Park, K.-Y., 2025. Enhancing mechanical resilience in Li-ion battery cathodes with nanoscale elastic framework coatings. ACS Nano 19, 1588–1599.

Ling, C., Zhang, R., 2017. Manganese dioxide as rechargeable magnesium battery cathode. Front. Energy Res. 5, 2017.

Liu, D., Ge, H., Song, M., Jiang, Y., Gong, X., You, T., Fu, L., Fu, Z., Zhang, Y., 2025a. Machine learning driven optimization of electrolyte for highly reversible Zn-air batteries with superior long-term cycling performance. Adv. Mater. 37, 2417161.

Liu, G., Li, Z., Zeng, L., Lin, J., Zheng, B., Liu, H., Chen, L., Wu, F., 2025b. Single-crystal Ni-rich layered oxide cathodes with LiNbO3-Li3BO3 coating for sulfide all-solid-state batteries. Nano Energy 137, 110798.

Liu, H., Luo, S., Yang, Y., Zhao, X., Huang, G., Jin, X., Zhong, T., Guan, M., Liu, J., Li, Y., 2025c. In-situ polymerization formed self-healing quasi-solid electrolyte for high-loading lithium batteries. Energy Storage Mater. 78, 104250.

Liu, K., Wang, Y., Lai, X., 2022. Key stages for battery full-lifespan management. In: Liu, K., Wang, Y., Lai, X. (Eds.), Data Science-Based Full-Lifespan Management of Lithium-Ion Battery: Manufacturing, Operation and Reutilization. Springer International Publishing, Cham.

Liu, M., Song, A., Zhang, X., Wang, J., Fan, Y., Wang, G., Tian, H., Ma, Z., Shao, G., 2025d. Interfacial lithium-ion transportation in solid-state batteries: challenges and prospects. Nano Energy 136, 110749.

Liu, S., Zhang, R., Wang, C., Mao, J., Chao, D., Zhang, C., Zhang, S., Guo, Z., 2024. Zinc ion batteries: bridging the gap from academia to industry for grid-scale energy storage. Angew. Chem. Int. Ed. 63 e202400045.

Lu, C., Wei, L., Li, J., 2025. Aluminum ion batteries: electrolyte and anode innovations and outlook. Energy Storage Mater. 78, 104274.

Lv, J.-S., Guo, S.-K., He, Y.-B., 2021. Modification strategies of Li7La3Zr2O12 ceramic electrolyte for high-performance solid-state batteries. Tungsten 3, 260–278.

Mahendran, S., 2021. Progress in one-dimensional nanostructures. Mater. Charact. 179, 111373.

Mancini, M., Hoffmann, M.F., Martin, J., Weirather-Köstner, D., Axmann, P., Wohlfahrt-Mehrens, M., 2024. A proof-of-concept of direct recycling of anode and cathode active materials: from spent batteries to performance in new Li-ion cells. J. Power Sources 595, 233997.

Mauger, A., Julien, C.M., 2025. Design principles and engineering strategies for stabilizing Ni-rich layered oxides in lithium-ion batteries. Batteries 11.

Mi, C., Luo, C., Wang, Z., Zhang, Y., Yang, S., Wang, Z., 2023. Cu and Ni Co-doped porous Si nanowire networks as high-performance anode materials for lithium-ion batteries. Materials 16.

Mishra, R., Anne, M., Das, S., Chavva, T., Shelke, M.V., Pol, V.G., 2024. Glory of fire retardants in Li-ion batteries: could they Be intrinsically safer? Adv. Sustain. Syst. 8, 2400273.

Moin, M., Moin, M., Zhao, H., Dong, L., Anwar, A.W., Ahsan, Z., Thumu, U., 2025. Mitigating volume expansion in silicon anodes: recent developments and future prospects. J. Mol. Liq. 440, 128768.

Moss, D., 2025. Integrated Photonics Incorporating 2D Materials Designed for Industrial and Commercial Applications.

Mueller, J., Michalowski, P., Kwade, A., 2023. Impact of silicon content and particle size in lithium-ion battery anodes on particulate properties and electrochemical performance. Batteries 9, 377.

Naguib, M., Barsoum, M., Gogotsi, Y., 2021. Ten years of progress in the synthesis and development of MXenes. Adv. Mater. 33, 2103393.

Narla, A., 2023. Borohydride-Based Solid Electrolytes and Polymer Composite Separators for Lithium-Ion Batteries. Georgia Institute of Technology.

Naseer, M.N., Serrano-Sevillano, J., Fehse, M., Bobrikov, I., Saurel, D., 2025. Silicon anodes in lithium-ion batteries: a deep dive into research trends and global collaborations. J. Energy Storage 111, 115334.

Nekahi, A., Dorri, M., Rezaei, M., Bouguern, M.D., Madikere Raghunatha Reddy, A.K., Li, X., Deng, S., Zaghib, K., 2024. Comparative issues of metal-ion batteries toward sustainable energy storage: lithium vs. Sodium. Batteries 10.

Neumann, J., Petranikova, M., Meeus, M., Gamarra, J.D., Younesi, R., Winter, M., Nowak, S., 2022. Recycling of lithium-ion batteries—current state of the art, circular economy, and next generation recycling. Adv. Energy Mater. 12, 2102917.

Østli, E.R., Tesfamhret, Y., Wenner, S., Lacey, M.J., Brandell, D., Svensson, A.M., Selbach, S. M., Wagner, N.P., 2021. Limitations of ultrathin Al_2O_3 coatings on LNMO cathodes. ACS Omega 6, 30644–30655.

Pan, K., Zhang, L., Qian, W., Wu, X., Dong, K., Zhang, H., Zhang, S., 2020. A flexible ceramic/polymer hybrid solid electrolyte for solid-state lithium metal batteries. Adv. Mater. 32, 2000399.

Pang, J., Mendes, R.G., Bachmatiuk, A., Zhao, L., Ta, H.Q., Gemming, T., Liu, H., Liu, Z., Rummeli, M.H., 2019. Applications of 2D MXenes in energy conversion and storage systems. Chem. Soc. Rev. 48, 72–133.

Paul, D., Pechancová, V., Saha, N., Pavelková, D., Saha, N., Motiei, M., Jamatia, T., Chaudhuri, M., Ivanichenko, A., Venher, M., Hrbáčková, L., Sáha, P., 2024a. Life cycle assessment of lithium-based batteries: review of sustainability dimensions. Renew. Sustain. Energy Rev. 206, 114860.

Paul, S., Acharyya, D., Punetha, D., 2024b. Benchmarking the performance of lithium and sodium-ion batteries with different electrode and electrolyte materials. Energy Storage 6, e70068.

Péter, L., 2021. Electrochemical Methods of Nanostructure Preparation. Springer.

Phogat, P., Rawat, S., Chand, B., Shreya, 2025. Next-generation electrolytes for advanced battery systems: materials, properties, and applications. J. Energy Storage 130, 117314.

Piat, C., Sari, A., Viton, C., 2025. Emerging battery technologies: the main aging mechanisms and challenges. Batteries 11.

Placke, T., Kloepsch, R., Dühnen, S., Winter, M., 2017. Lithium ion, lithium metal, and alternative rechargeable battery technologies: the odyssey for high energy density. J. Solid State Electrochem. 21, 1939–1964.

Qiang, W., Jianchun, S., Xue, L., Yusong, Z., Jiaxin, B., Yiqiang, Y., Mingliang, Q., Jie, T., Wenhong, L., Zhong, Z., Zhiqiang, Z., 2025. A comprehensive investigation of salt concentration effects on electrochemical behavior of magnesium anodes. J. Colloid Interface Sci. 692, 137497.

Rahman, M.A., Song, G., Bhatt, A., Wong, Y., Wen, C., 2015. Nanostructured silicon anodes for high-performance lithium-ion batteries. Adv. Funct. Mater. 26.

Rahman, U.U., Humayun, M., Ghani, U., Usman, M., Ullah, H., Khan, A., El-Metwaly, N.M., Khan, A., 2022. MXenes as emerging materials: synthesis, properties, and applications. Molecules 27, 4909.

Rahmani, P., Chakraborty, S., Mele, I., Katrašnik, T., Bernhard, S., Pruefling, S., Wilkins, S., Hegazy, O., 2025. Driving the future: a comprehensive review of automotive battery management system technologies, and future trends. J. Power Sources 629, 235827.

Reinhart, L., Vrucak, D., Woeste, R., Lucas, H., Rombach, E., Friedrich, B., Letmathe, P., 2023. Pyrometallurgical recycling of different lithium-ion battery cell systems: economic and technical analysis. J. Clean. Prod. 416, 137834.

Rezaei, M., Nekahi, A., Kumar, M.R.,A., Nizami, A., Li, X., Deng, S., Nanda, J., Zaghib, K., 2025. A review of lithium-ion battery recycling for enabling a circular economy. J. Power Sources 630, 236157.

Riphaus, N., Stiaszny, B., Beyer, H., Indris, S., Gasteiger, H.A., Sedlmaier, S.J., 2019. Understanding chemical stability issues between different solid electrolytes in all-solid-state batteries. J. Electrochem. Soc. 166, A975–A983.

Sada, K., Darga, J., Manthiram, A., 2023. Challenges and prospects of sodium-ion and potassium-ion batteries for mass production. Adv. Energy Mater. 13, 2302321.

Sanusi, Y., Tresna, F., Adnan, F., Adnan, F., Ginting, D., Radityan, F., Tresna, Adnan, F., Anis, F., Rhee, J., Soo, Dianta, G., Radityan, F., Adnan, F., Rhee, J.-S., 2025. Performance Evaluation and Economic Analysis of Lithium Iron Phosphate (LFP) and Nickel Manganese Cobalt (NMC) Batteries in Electric Vehicles in Indonesia, vol 43, pp. 187–201.

Shi, H., Teck, A.A., Lander, L., Titirici, M.-M., 2025. Aluminum–graphite dual-ion batteries: recent advances and challenges. Green Chem. 27, 15374–15402.

Smajic, J., Wee, S., Simoes, F.R.F., Hedhili, M.N., Wehbe, N., Abou-Hamad, E., Costa, P.M.F. J., 2020. Capacity retention analysis in aluminum-sulfur batteries. ACS Appl. Energy Mater. 3, 6805–6814.

Song, S., Qin, X., Ruan, Y., Li, W., Xu, Y., Zhang, D., Thokchom, J., 2020. Enhanced performance of solid-state lithium-air batteries with continuous 3D garnet network added composite polymer electrolyte. J. Power Sources 461, 228146.

Stolz, L., Winter, M., Kasnatscheew, J., 2024. Practical relevance of charge transfer resistance at the Li metal electrode|electrolyte interface in batteries? J. Solid State Electrochem. 29, 4181–4186.

Su, X., Wu, Q., Li, J., Xiao, X., Lott, A., Lu, W., Sheldon, B., Wu, J., 2014. Silicon-based nanomaterials for lithium-ion batteries: a review. Adv. Energy Mater. 4.

Su, X., Xu, X.-P., Ji, Z.-Q., Wu, J., Ma, F., Fan, L.-Z., 2024. Polyethylene oxide-based composite solid electrolytes for lithium batteries: current progress, low-temperature and high-voltage limitations, and prospects. Electrochem. Energy Rev. 7, 2.

Subramanyam, S., Trabelsi, Y., Mahajan, P., Khanna, V., Thakur, A., Chahal, S., Singh, S., Singh, A., Kumar, S., 2024. Advancing sustainable energy storage with MXenes: a comprehensive analysis of fabrication techniques and applications. J. Energy Storage 101, 114067.

Sun, F., Zurita-Garcia, L., Wheeler, D.R., 2025. Effect of short-chain polymer binders on the mechanical and electrochemical performance of silicon anodes. Batteries 11.

Surace, Y., Simões, M., Karvonen, L., Yoon, S., Pokrant, S., Weidenkaff, A., 2015. Freeze drying synthesis of Li_3MnO_4 cathode material for Li-ion batteries: a physico-electrochemical study. J. Alloys Compd. 644, 297–303.

Tan, A.K.X., Paul, S., 2024. Beyond lithium: future battery technologies for sustainable energy storage. Energies 17.

Tang, Y., Zhang, L., Chen, J., Sun, H., Yang, T., Liu, Q., Huang, Q., Zhu, T., Huang, J., 2021. Electro-chemo-mechanics of lithium in solid state lithium metal batteries. Energy Environ. Sci. 14, 602–642.

Tang, Z., Choi, J., Lorie Lopez, J.L., Co, A.C., Brooks, C.J., Sayre, J.R., Kim, J.-H., 2022. Optimization of the Li3BO3 glass interlayer for garnet-based all-solid-state lithium–metal batteries. ACS Appl. Energy Mater. 5, 12132–12142.

Tesfamhret, Y., Younesi, R., Berg, E., 2022. Influence of Al_2O_3 coatings on HF induced transition metal dissolution from lithium-ion cathodes. J. Electrochem. Soc. 169.

Tufail, M.K., Zhai, P., Khokar, W., Jia, M., Zhao, N., Guo, X., 2023. Evaluation of solid electrolytes: development of conventional and interdisciplinary approaches. Interdiscip. Mater. 2, 529–568.

Umar, A., Raut, S.D., Ibrahim, A.A., Algadi, H., Albargi, H., Alsaiari, M.A., Akhtar, M.S., Qamar, M., Baskoutas, S., 2021. Perforated Co3O4 nanosheets as high-performing supercapacitor material. Electrochim. Acta 389, 138661.

Verma, V., Kumar, S., Manalastas, W., Satish, R., Srinivasan, M., 2018. Progress in rechargeable aqueous zinc- and aluminum-ion battery electrodes: challenges and outlook. Adv. Sustain. Syst. 3.

Vivoda, V., Loginova, J., 2025. Lithium at the crossroads: geopolitical, economic, and socio-environmental complexities of the Jadar project in Serbia. Mineral Economics 38, 665–681.

Wang, D., Sun, S., Ma, J., He, J., Sun, J., Wang, Q., Jia, Z., Zhao, W., 2025a. Zwitterionic polymers in energy storage applications. Mater. Today Energy 54, 102129.

Wang, F., Fan, X., Gao, T., Sun, W., Ma, Z., Yang, C., Han, F., Xu, K., Wang, C., 2017. High-voltage aqueous magnesium ion batteries. ACS Cent. Sci. 3, 1121–1128.

Wang, K., Baule, N., Jin, H., Qiao, H., Hardy, A., Schuelke, T., Fan, Q., 2025b. Multifunctional zinc silicate coating layer for high-performance aqueous zinc-ion batteries. Energy Materials 5.

Wang, X.-Q., Qin, M., Moldovan, N.-C., Su, C.-W., 2023a. Bubble behaviors in lithium price and the contagion effect: an industry chain perspective. Resour. Policy 83, 103725.
Wang, Y., Feng, X., Huang, W., He, X., Wang, L., Ouyang, M., 2023b. Challenges and opportunities to mitigate the catastrophic thermal runaway of high-energy batteries. Adv. Energy Mater. 13, 2203841.
Wu, X., Gao, Q., Li, W., 2025. Design for recycling: a systematic review of approaches for enhancing product recyclability. Sustainability 17.
Wu, Z., Zhang, C., Yuan, F., Lyu, M., Yang, P., Zhang, L., Zhou, M., Wang, L., Zhang, S., Wang, L., 2024. Ni-rich cathode materials for stable high-energy lithium-ion batteries. Nano Energy 126, 109620.
Xiang, J., Wei, Y., Zhong, Y., Yang, Y., Cheng, H., Yuan, L., Xu, H., Huang, Y., 2022. Building practical high-voltage cathode materials for lithium-ion batteries. Adv. Mater. 34, 2200912.
Xie, J., Tong, L., Su, L., Xu, Y., Wang, L., Wang, Y., 2017. Core-shell yolk-shell Si@C@Void@C nanohybrids as advanced lithium ion battery anodes with good electronic conductivity and corrosion resistance. J. Power Sources 342, 529–536.
Xu, R., Du, L., Adekoya, D., Zhang, G., Zhang, S., Sun, S., Lei, Y., 2021. Well-defined nanostructures for electrochemical energy conversion and storage. Adv. Energy Mater. 11, 2001537.
Xu, S., Sharp, J., Liu, Q., Shin, J., Ye, H., Yang, K., Chen, S., Silva, K., Nour, R., Kittrell, C., Zhu, H., Li, B., Jebailey, K., Teng, C., Yakobson, B., Zhao, Y., Tour, J., 2025. Holistic recovery of spent lithium-ion batteries by flash Joule heating. Adv. Mater. 38.
Xu, Z., Chen, R., Zhu, H., 2019. A Li_2 $CuPS_4$ superionic conductor: a new sulfide-based solid-state electrolyte. J. Mater. Chem. A 7, 12645–12653.
Yang, S., Liang, J., He, J., Fang, D., Zhao, Y., Ding, Y., Yin, A., Li, Y., Wang, C., Li, J., Jin, H., 2024. Regulating the interfacial performance of all-solid-state sodium batteries using lanthanum substitution in a $Na_3Zr_2Si_2PO_{12}$ solid electrolyte. ACS Appl. Energy Mater. 7, 9863–9871.
Yang, X.-G., Wang, C.-Y., 2018. Understanding the trilemma of fast charging, energy density and cycle life of lithium-ion batteries. J. Power Sources 402, 489–498.
Yin, D., Zhu, X., Zhang, W., Zheng, J., 2024. Health state prediction of lithium-ion battery based on improved sparrow search algorithm and support vector regression. Energies 17.
Yin, J.-H., Zhu, H., Yu, S.-J., Dong, Y.-B., Wei, Q.-Y., Xu, G.-Q., Xiong, Y., Qian, Y., 2023. Recent advances of LATP and their NASICON structure as a solid-state electrolyte for lithium-ion batteries. Adv. Eng. Mater. 25, 2300566.
Ying, Y., Fan, K., Lin, Z., Huang, H., 2025. Facing the "cutting edge:" edge site engineering on 2D materials for electrocatalysis and photocatalysis. Adv. Mater. 37, 2418757.
Yue, M., Lumsden, M.D., Deshmukh, J., Zsoldos, E.S., Black, W., Azam, S.M., Dahn, J.R., 2025. Comprehensive study of the degradation of LiFePO4/graphite cells at elevated temperatures. J. Electrochem. Soc. 172, 050502.
Zaed, M., Tan, K., Abdullah, N., Saidur, R., Pandey, D.A., Saleque, A., 2023. Cost analysis of MXene for low-cost production, and pinpointing of its economic footprint. Open Ceram. 17, 100526.
Zanoletti, A., Mannu, A., Cornelio, A., 2025. Solvometallurgy as alternative to pyro- and hydrometallurgy for lithium, cobalt, nickel, and manganese extraction from black mass processing: state of the art. Materials 18.
Zardalidis, G., Ioannou, E., Pispas, S., Floudas, G., 2013. Relating structure, viscoelasticity, and local mobility to conductivity in PEO/LiTf electrolytes. Macromolecules 46, 2705–2714.

Zhai, Y., Hou, W., Tao, M., Wang, Z., Chen, Z., Zeng, Z., Liang, X., Paoprasert, P., Yang, Y., Hu, N., 2022. Enabling high-voltage "superconcentrated ionogel-in-ceramic" hybrid electrolyte with ultrahigh ionic conductivity and single Li+-ion transference number. Adv. Mater. 34, 2205560.

Zhang, D., Li, L., Wu, X., Wang, J., Li, Q., Pan, K., He, J., 2021. Research progress and application of PEO-based solid state polymer composite electrolytes. Front. Energy Res. 9, 726738.

Zhang, L., Liu, Y., You, Y., Vinu, A., Mai, L., 2023. NASICONs-type solid-state electrolytes: the history, physicochemical properties, and challenges. Interdiscip. Mater. 2, 91–110.

Zhang, Q., Cao, D., Ma, Y., Natan, A., Aurora, P., Zhu, H., 2019. Sulfide-based solid-state electrolytes: synthesis, stability, and potential for all-solid-state batteries. Adv. Mater. 31, 1901131.

Zhang, Q., He, L., Liu, A., Han, W., Li, A., Zhao, Z., Zhu, X., Tan, G., 2025. High-performance mg-ion battery materials: recent progress and future perspectives. J. Energy Storage 132, 117900.

Zhang, S., Xie, B., Zhuang, X., Wang, S., Qiao, L., Dong, S., Ma, J., Zhou, Q., Zhang, H., Zhang, J., Ju, J., Xu, G., Cui, Z., Cui, G., 2024. Great challenges and new paradigm of the in situ polymerization technology inside lithium batteries. Adv. Funct. Mater. 34, 2314063.

Zhang, X., Hou, L., Ciesielski, A., Samorì, P., 2016. 2D materials beyond graphene for high-performance energy storage applications. Adv. Energy Mater. 6, 1600671.

Zhang, Y., Liu, N., 2017. Nanostructured electrode materials for high-energy rechargeable Li, Na and Zn batteries. Chem. Mater. 29, 9589–9604.

Zhao, Q., Stalin, S., Zhao, C.-Z., Archer, L.A., 2020. Designing solid-state electrolytes for safe, energy-dense batteries. Nat. Rev. Mater. 5, 229–252.

Zhong, H., Su, Y., Ma, R., Luo, Y., Lin, H., Gu, J., Gong, Z., Yang, Y., 2024. Nano-scale interface engineering of sulfur cathode to enable high-performance all-solid-state Li–S batteries. Adv. Funct. Mater. 34, 2315925.

Zhou, A., Chi, R., Shi, Y., Zhao, X., Li, X., Kou, Z., Zhang, Z., Zhang, X., Nie, G., 2023. Manganese-based cathode materials for aqueous rechargeable zinc-ion batteries: recent advance and future prospects. Mater. Today Chem. 27, 101294.

Zhou, Y., Huang, M., Chen, Y., Tao, Y., 2016. A novel health indicator for on-line lithium-ion batteries remaining useful life prediction. J. Power Sources 321, 1–10.

Zhu, J., Mathews, I., Ren, D., Li, W., Cogswell, D., Xing, B., Sedlatschek, T., Kantareddy, S.N. R., Yi, M., Gao, T., Xia, Y., Zhou, Q., Wierzbicki, T., Bazant, M.Z., 2021. End-of-life or second-life options for retired electric vehicle batteries. Cell Rep. Phys. Sci. 2, 100537.

Zou, J., Zhang, J., Wang, L., Sun, Q., Wang, Y., Zheng, M., Hu, H., Xiao, Y., Liu, Y., Liang, Y., 2024. An all-from-one strategy to flexible solid-state lithium-ion batteries with decreased interfacial resistance. Sci. China Mater. 67, 1445–1454.

Battery energy storage systems for reliability and economic performance in modern power systems

12

Sajeeb Saha[1], *Shamsul Huda*[2] *and Amanullah Maung Than Oo*[1]
[1]School of Science, Technology and Engineering, University of the Sunshine Coast, Moreton Bay, QLD, Australia; [2]Faculty of Science, Engineering and Built Environment, School of Information, Deakin University, Burwood, VIC, Australia

12.1 Introduction

Amid continuous efforts toward decarbonization of power systems around the world, the ongoing energy transition toward low-carbon technologies is facilitating accelerated growth of inverter-based renewable energy sources (RESs)—such as solar photovoltaic (PV) generation and wind generation—into power systems. However, the operational nature of renewable generation—uncertain and variable in output—poses additional complexity and challenge to power system operations. This is due to increased variability of the supply-demand balance, decreased system inertia resulting from replacing synchronous generators with inverter-based resources, and new forms of network constraints (Kundur et al., 2007; Machowski et al., 2020; Ulbig et al., 2014; Milano et al., 2018; Saha et al., 2023).

Unlike traditional synchronous generation, inverter-interfaced RESs are unable to provide rotational inertia natively to the grid. Frequency dynamics occur at a faster rate and increase the severity of disturbances. Decrease in system inertia has been broadly acknowledged to be a severe problem for modern power systems, especially under high renewable penetration cases (Ulbig et al., 2014; Milano et al., 2018; Masood et al., 2021). Furthermore, high penetration of distributed renewables could bring several technical difficulties at distribution level, such as voltage fluctuation, reverse power flow, and congestion on branches and equipment (Turitsyn et al., 2011; Chatzivasileiadis et al., 2016). To cope with these issues, flexible/fast acting resources should be implemented in order to provide services both for system reliability as well as economy-wise operation.

Battery energy storage systems (BESSs) have been considered as one of the promising enabling technologies to mitigate the problems described above. Considering its fast response, bidirectional power capability and operational flexibility, BESSs can provide numerous services from various time-scale perspectives. In terms of reliability objectives, BESSs can support frequency stability using fast active power

Energy Storage Technologies. https://doi.org/10.1016/B978-0-443-45858-3.00009-7

support, fast frequency response and virtual inertia emulation capabilities (Zhong and Weiss, 2010; Markovic et al., 2019; Nguyen et al., 2016). On the other hand, it can provide several services such as energy arbitrage, peak shaving, and better utilization of renewables from economic point of view (Lund et al., 2015; Zakeri and Syri, 2015; Restrepo et al., 2021; Okafor and Folly, 2024).

While there exist clear advantages to this practice, effective planning and operation of BESS to harness these benefits must be considered jointly under uncertainty. Uncertainty-aware planning remains challenging since the extent of support needed from BESS, especially to frequency regulation, varies across operating conditions such as renewable generation, load demand, and disturbance sizes/magnitudes. Planning decisions made using deterministic approaches with fixed assumptions on these parameters could lead to erroneous sizing decisions. Robustness-oriented formulations that explicitly consider the stochastic nature of aforementioned uncertainties have been demonstrated to yield more reliable and practical solutions (Zio, 2013; Masood et al., 2021; Saha et al., 2023).

Uncertainty in operating conditions also complicates the operation of batteries for improved grid reliability. Renewables and load are uncertain and fluctuations in these two components require management strategies that can operate effectively under uncertainty. Numerous energy management techniques exist that leverage forecast information and model predictive control (MPC) frameworks to schedule battery operation under uncertainty (Restrepo et al., 2021; Saleem et al., 2024a). Moreover, degradation behavior of batteries is highly dependent on cycling statistics and operating conditions and must also be considered during the operation of batteries for improved economics (Atalay et al., 2020).

Independently of its operating strategy, deployment of shared battery storage systems can also provide increased flexibility and improved economics through coordinated management of distributed energy resources. Community batteries, which are shared storage systems that allow multiple users to collectively own and dispatch energy, are a growing reality. Such deployments can help reduce overall system variability through aggregation and improve battery utilization (Azaroual et al., 2022). Shared batteries have also demonstrated their efficacy in improving network constraint management such as voltage support and hosting capacity (Parag and Sovacool, 2016; Neda et al., 2025).

In light of the above discussions, a unified framework linking reliability-oriented BESS planning, uncertainty-aware BESS operation, and effective deployment of BESS remains underexplored. Though individual works have investigated various pieces of this puzzle, successful implementation of BESS into power systems calls for joint consideration of all three aspects. To that end, this chapter seeks to provide such a perspective for battery energy storage systems.

In this chapter, we provide a unified perspective on BESSs to improve the economic and reliability metrics of power systems. We discuss the use of BESS for improving power system reliability from the lens of reliability-oriented virtual inertia planning, uncertainty-aware battery operation, and community-scale battery deployment. Leveraging commonalities between each of these problems, we discuss how batteries can be utilized as a versatile resource to enable reliable, efficient, and sustainable power systems. The subsequent sections provide background on BESS and its benefits to modern power systems followed by a discussion on the three areas highlighted above within a cohesive framework.

12.2 Role of BESSs in modern power systems

BESSs play an increasingly important role in modern power systems with high penetration of inverter-based RESs. Their ability to provide fast, controllable, and bidirectional power exchange enables them to support both economic operation and system reliability across multiple time scales (Okafor and Folly, 2024; Azaroual et al., 2022). Unlike conventional generation, BESS operation is not constrained by mechanical inertia or fuel limitations, allowing flexible participation in a wide range of grid services. These include energy balancing, economic optimization, frequency support, and voltage regulation. Importantly, these services must often be delivered simultaneously under uncertain operating conditions, requiring coordinated control strategies (Saleem et al., 2024a; Son et al., 2024).

In addition to frequency-related challenges, voltage stability issues become more pronounced under high penetration of inverter-based renewable energy sources, particularly in weak grid conditions where network impedance is high. Under such conditions, variations in active and reactive power injections can lead to significant voltage deviations, reverse power flow, and reduced hosting capacity. Several studies have shown that the integration of inverter-based resources in weak grids can adversely affect voltage stability margins and system strength (Saleem et al., 2022, 2023, 2024b; Rahman et al., 2022). In this context, BESS can support voltage regulation through coordinated active and reactive power control, complementing its role in frequency stability.

One key function of BESS is to balance generation and demand in the face of variability from renewables and load fluctuations. In systems with high renewable penetration, discrepancies between forecast and actual conditions can cause imbalances that must be corrected in real time. BESS mitigates these deviations through adaptive charging and discharging, effectively smoothing net demand fluctuations. This not only lessens the burden on conventional generators but also enhances overall system stability. Furthermore, the integration of BESS can increase the system's capacity to accommodate higher renewable generation levels (Restrepo et al., 2021; Son et al., 2024).

In addition to the technical support services, BESS also offers economic benefits through the optimization of energy schedules. Due to the time-based changes in electricity prices, BESS can be scheduled to engage in energy arbitrage. This involves charging during periods of low electricity prices and discharging during periods of high prices. In reality, the economic operation of BESS is complicated by factors such as forecast uncertainties and battery degradation. Forecast errors in renewable generation and load demand can have a direct impact on the economic scheduling decisions. Battery degradation, on the other hand, accelerates with increased cycling and reduces battery life. It is therefore important for the economic operation to strike a balance between immediate cost savings and long-term battery health (Saleem et al., 2024a; Okafor and Folly, 2024).

The growing integration of inverter-based RESs has been leading to low system inertia, which results in a faster rate of frequency deviation after a disturbance (Saleem et al., 2025). For such a scenario, it becomes imperative to provide fast frequency support to the power system in order to sustain system stability. BESS can offer fast active power response if proper control strategies are implemented, hence providing inertial as well as primary frequency support to the system to limit the rate of change of frequency and improve frequency recovery after a disturbance (Saha et al., 2025; Masood et al., 2021; Nguyen et al., 2016). The mathematical modeling of the frequency dynamics and inertia provision is described in the next section.

12.3 Dynamic support and reliability-oriented planning of BESS

This section focuses on the quantitative modeling and planning of BESS for dynamic support under uncertain conditions, building on previous discussions. BESS provides services like frequency support and energy balancing, but the key challenge lies in determining the required level of such support to ensure reliable system operation.

12.3.1 *Frequency stability and fast dynamic support*

As mentioned in the previous section, the growing penetration of inverter-based RESs in the power system network has a consequence of reducing the system inertia, which results in a faster rate of frequency deviation when there is a disturbance in the network. Therefore, for a low-inertia system, frequency stability requires a fast and controlled response of active power. As a result, BESSs are one of the most promising flexible resources to deal with the challenges of a modern power system network as it can provide the needed support due to the fast response of active power injection or withdrawal, enabled by power electronic converters used in the system. In this case, fast frequency regulation using a BESS can act almost instantly, as a first-order action to help counteract imbalances between supply and demand and support the stability of the grid. The dynamic response of a BESS can be represented by a mathematical expression as

$$P_{\mathrm{bat}}(t) = -K_f \Delta f(t) - K_d \frac{d\Delta f(t)}{dt}, \tag{12.1}$$

where K_f and K_d are droop and inertial response gains, respectively. The first term, that is, droop response, of this BESS model is the component providing a proportion of the frequency deviation. The second term of the dynamic model provides inertial support based on the rate of change of frequency (RoCoF) and is responsible for responding to frequency changes, hence acting faster to mitigate the first transient frequency excursion. The limits on this response are constrained by the available power and energy capacity of the battery, as well as the state of charge (SoC). More specifically, the fast frequency support is limited by the available power capacity, and the support duration is determined by the available energy.

To determine the required level of dynamic support, system frequency behavior can be described using the swing equation:

$$M\frac{d\Delta f(t)}{dt}+D\Delta f(t)=\Delta P_{\mathrm{Dist}}(t), \tag{12.2}$$

where M represents the equivalent system inertia and D is the damping coefficient. In practical systems, the effective inertia is time-varying and depends on the number and operating status of synchronous generators. This can be expressed as

$$M(t)=\sum_{i\in\mathcal{G}}M_i(t), \tag{12.3}$$

where $\mathcal{G}$ denotes the set of generators contributing inertia at time t. As renewable generation displaces conventional units, the effective inertia decreases, increasing the sensitivity of frequency to disturbances.

Based on the frequency response model, BESS can be utilized to provide virtual inertia through controlled active power injection. The virtual inertia contribution can be expressed as

$$\Delta P_{\mathrm{BESS}}(t)=-2H_{\mathrm{BESS}}\frac{d\Delta f(t)}{dt}, \tag{12.4}$$

where H_{BESS} is the virtual inertia constant. This relationship indicates that the instantaneous power requirement of BESS is directly proportional to RoCoF. Therefore, systems experiencing faster frequency dynamics require higher power capacity for inertia support. In contrast, the associated energy requirement is relatively small, as the response is typically required only over a short duration following a disturbance.

As discussed in Section 12.2, renewable generation and load demand exhibit inherent variability. These uncertainties directly influence the required level of dynamic support by affecting both disturbance magnitude and available system inertia. The system operating condition can be represented as

$$\mathbf{x}(t)=[P_{\mathrm{RG}}(t),P_{\mathrm{load}}(t),\Delta P_{\mathrm{Dist}}(t)]^{\top}, \tag{12.5}$$

where $\mathbf{x}(t)$ follows a stochastic process. Due to the stochasticity of $\mathbf{x}(t)$, there is a range of possible frequency responses, rather than a deterministic trajectory, and the system design needs to be conducted in a probabilistic manner.

In order to account for performance under uncertainty, the required inertia is obtained via a reliability criterion. Let β be a desired reliability level. The required inertia is determined such that

$$\mathbb{P}\left(\Delta f_{\text{min}} \geq \Delta f_{\text{limit}}, \left|\frac{d\Delta f}{dt}\right| \leq \text{RoCoF}_{\text{max}}\right) \geq \beta. \tag{12.6}$$

where Δf_{min} represents the frequency nadir, Δf_{limit} denotes the allowable frequency deviation limit, and RoCoF denotes the rate of change of frequency.

A scenario-based approach is adopted to model uncertainty. A set of operating scenarios is generated as

$$\{\mathbf{x}^{(1)}, \mathbf{x}^{(2)}, \ldots, \mathbf{x}^{(N)}\}. \tag{12.7}$$

The required inertia $H_{\text{req}}^{(n)}$ is calculated in each scenario n to ensure frequency constraints are met. Finally, the percentile-based approach is utilized to obtain the inertia requirement:

$$H_{\text{BESS}} = \text{Percentile}_{\beta}\left(H_{\text{req}}^{(1)}, \ldots, H_{\text{req}}^{(N)}\right). \tag{12.8}$$

After obtaining the required inertia, the associated BESS capacity can be calculated. The required power capacity is

$$P_{\text{BESS}}^{\text{max}} = 2H_{\text{BESS}} \cdot \left(\frac{df}{dt}\right)_{\text{max}}, \tag{12.9}$$

while the energy requirement is

$$E_{\text{BESS}} = \int_0^{T_{\text{resp}}} P_{\text{BESS}}(t)\, \mathrm{d}t, \tag{12.10}$$

where T_{resp} denotes the duration of inertial response.

The above formulation builds upon the dynamic support capabilities of BESS described in Section 12.2 and provides a systematic approach for determining the required level of support under uncertainty. By integrating frequency dynamics, stochastic modeling, and reliability constraints, the framework enables informed sizing of BESS for maintaining system stability. Importantly, allocating battery capacity for dynamic support introduces coupling with economic operation. As highlighted earlier, reserving capacity for reliability services reduces the flexibility available for energy arbitrage and other applications, emphasizing the need for coordinated planning and operation strategies.

12.4 Integrated economics–reliability framework for BESS

12.4.1 Need for an integrated perspective

Motivated by the insights from the sections on dynamic support capabilities and reliability-oriented planning for BESS, this section discusses their integration with economic operation and deployment strategies. Despite the multi-functional roles of BESS, including virtual inertia, ramping support, and demand charge mitigation, their effective utilization requires coordinated consideration of these services.

In modern power systems, decisions related to BESS affect multiple aspects of its utilization. For example, the sizing of BESS for virtual inertia determines how much capacity is available for economic operation, while the operational strategy affects its ability to provide dynamic support during disturbances. Similarly, the choice of deployment and utilization configuration influences overall efficiency and system performance. These interdependencies highlight the need for a unified approach that considers planning, operation, and deployment together.

12.4.2 Framework for BESS utilization

Building on the discussion so far, the integrated utilization of BESS can be organized into three layers, as illustrated in Fig. 12.1. These layers are briefly described below:

- *Planning layer (reliability-oriented sizing)*: This layer focuses on determining the required BESS capacity to provide virtual inertia and other dynamic support functions while ensuring system reliability under uncertain operating conditions. The objective is to

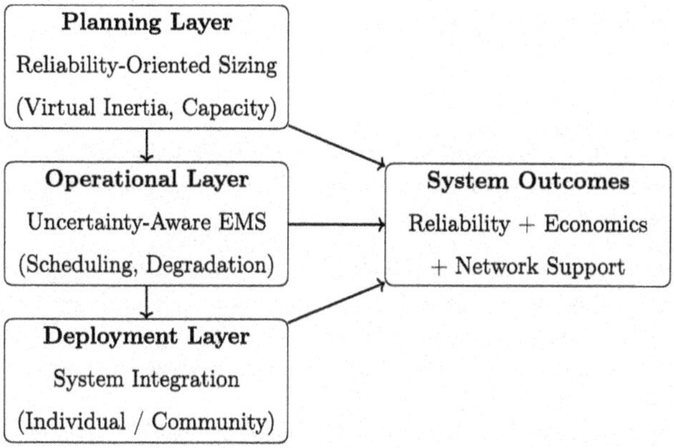

Figure 12.1 Integrated framework for BESS utilization linking reliability-oriented planning, uncertainty-aware operation, and deployment strategies.

determine the minimum required capacity that satisfies frequency stability requirements with a desired level of reliability. This can be expressed as:

$$E_{vi}^{*} = \arg\min_{E_{vi}} E_{vi} \quad \text{s.t.} \quad \mathbb{P}(f_{vi} \geq f_{th}) \geq \beta, \tag{12.11}$$

- where E_{vi} represents the energy capacity associated with virtual inertia provision, f_{vi} denotes the resulting system frequency, f_{th} is the minimum allowable frequency threshold, and β is the desired reliability level.

  This formulation ensures that the frequency remains within acceptable limits with a probability of at least β, accounting for uncertainties in renewable generation, load demand, and disturbances. A detailed development of this reliability-constrained framework and its application to BESS sizing can be found in Saha et al.(2026a). The resulting optimal virtual inertia is then translated into the corresponding BESS power and energy capacity required to deliver the necessary dynamic support.
- *Operational layer (economic and uncertainty-aware management)*: This layer pertains to the optimal charging and discharging schedules of BESS for economic operation under uncertainty in renewable generation, load demand, and price signals. Given the available BESS capacity from the planning layer, this layer involves solving the following optimization problem. In practice, this is typically implemented using a model predictive control (MPC) framework to account for time-varying system conditions and uncertainty, as detailed in (Saha et al., 2025):

$$\min_{\boldsymbol{u}_{bat}} c(\boldsymbol{u}_{bat}) + C_{res} \tag{12.12}$$

$$\text{s.t. } E_{res} + E_{op} = E_{bat}$$

$$0 \leq P_{bat}(t) \leq P_{\max} - P_{res}$$

$$0 \leq P_{ch}(t) \leq \overline{P}_{ch}, P_{dis}(t) \leq \overline{P}_{dis}$$

$$0 \leq E_{op}(t) \leq E_{\max} - E_{\min}.$$

The objective is to minimize the operational cost, while the constraints ensure that the power and energy capacity of BESS are not exceeded and that the required reserve is set aside for reliability services.

- *Deployment layer (system integration)*: This layer focuses on the configuration and placement of BESS within the power system, including both individual and shared deployment models. Given the capacity and operational strategies determined in the previous layers, this layer determines how the available storage resources are utilized across the network to enhance overall system performance.

  In particular, aggregated deployment configurations, such as community battery systems, reduce the variability in net demand through diversity in user consumption and generation patterns. This reduction in variability can be conceptually expressed as:

$$\sigma_{agg} < \sum_{i=1}^{N} \sigma_i, \tag{12.13}$$

where σ_{agg} represents the variability of the aggregated demand and σ_i denotes the variability of individual users.

This reduction in variability improves predictability, reduces extreme operating conditions, and enhances both utilization efficiency and reliability outcomes.

12.4.3 Coupling between reliability and economic objectives

A key feature of the integrated BESS model is the coupling between reliability and economic operation. Specifically, guaranteeing the provision of dynamic support is achieved by ensuring a certain reserve of battery capacity for fast power response, which is inherently related to its utilization for economic operation.

Denoting the overall energy capacity of the battery by E_{total}, a certain amount of energy must be reserved for dynamic support. This leaves an operating energy capacity E_{op} for economic operation:

$$E_{total} = E_{op} + E_{res}, \tag{12.14}$$

where E_{res} is the amount of energy required to deliver dynamic support. In addition to energy capacity, there is also a need to ensure power capacity for headroom of fast response. This leads to a constraint of the form:

$$P_{bat}(t) \leq P_{max} - P_{res}, \tag{12.15}$$

where P_{res} is the amount of power capacity that must be reserved to provide fast dynamic support in the event of a disturbance.

These constraints arise from the need to ensure reliable system operation under uncertain conditions and directly influence the operational flexibility of the BESS. In particular, reserving energy and power for reliability reduces the available capacity for economic activities such as energy arbitrage and peak shaving. Consequently, a fundamental trade-off emerges between reliability and economic objectives. Higher reliability requirements necessitate larger reserve allocations, which may limit economic benefits, while aggressive economic operation can reduce the ability of the system to respond effectively to disturbances. This trade-off is further influenced by uncertainty in renewable generation, load demand, and system operating conditions, which affects both the required level of dynamic support and the achievable economic performance.

12.4.4 Impact of uncertainty across layers

Uncertainty in renewable generation, as well as in the load demand and the disturbances, impacts every aspect of BESS utilization. The consequences of this uncertainty are carried over from planning to operation and deployment layers. It influences both the reliability criteria the BESS system is designed to satisfy, and its economic performance, as describe below:

- *Impact on reliability-oriented planning*: Fluctuations in renewable generation and load demand simultaneously impact the amplitude of the input disturbance and the amount of inertia available to the system. The combination of these influences results in a distribution of the virtual inertia requirement, rather than a deterministic value.
- *Influence on BESS capacity allocation*: A higher level of uncertainty in the system dynamics requires a higher fraction of the BESS capacity to be reserved for reliability services. The system constraints need to be satisfied over a broader range of operating conditions. The relative energy and power capacity allocation to dynamic support is thereby increased.
- *Effect on operational flexibility and economic performance*: The reduction of the capacity available for economic operation limits the flexibility of the charging and discharging schedule. Furthermore, operation under uncertain conditions can lead to constraint violations due to forecast deviations. Optimal operation is thereby hindered, and economic gains are lost.
- *Mitigation through aggregation and deployment strategies*: The aggregation of multiple users and loads smooths out the aggregated net demand due to diversity of user behavior. The smaller amplitude of the aggregated load net of generation thus leads to a better predictability of the disturbance. A lower reserve capacity is needed, which has a positive impact on reliability and economic performance.

12.4.5 Trade-off analysis

The integrated framework presented above highlights several key trade-offs that arise due to the coupling between dynamic support and economic objectives. These include:

- Reliability versus economic efficiency: A higher reliability requirement leads to a larger fraction of the BESS capacity being reserved for dynamic support, which means a lower energy capacity is available for economic operation.
- Short-term gains versus long-term sustainability: Operating BESS more aggressively leads to higher short-term economic benefits but increases battery degradation, which reduces long-term gains.
- Centralized versus distributed deployment: Centralized systems allow for coordinated control for reliability services while distributed or shared systems can have higher utilization and lower uncertainty due to aggregation.

The integrated economics-reliability framework for BESS presented in this section provides a unified perspective by extending the concepts of dynamic support and reliability-oriented planning presented in the previous section with operation and deployment considerations. It recognizes the multifunctional nature of BESS and provides a systematic approach for jointly considering planning, operation, and deployment aspects, as summarized in the layered framework. This integrated view allows for more effective utilization of BESS by taking into account both the need for dynamic support and the economic objectives in the face of uncertainty in operating conditions.

12.5 Case study and validation

The results in this section are based on the reliability-constrained virtual inertia planning framework developed in Saha et al.(2026a).

12.5.1 System description and uncertainty modeling

The framework is validated on a modified IEEE 39-bus system with high penetration of solar PV generation. A 24-hour simulation horizon is considered to capture temporal variations in generation and demand. As solar output increases during daytime hours, synchronous generation is displaced, leading to reduced system inertia and enabling evaluation under low-inertia operating conditions. The system comprises conventional synchronous generators, distributed loads, and network parameters based on the standard IEEE 39-bus test system. Renewable generation is incorporated by replacing a portion of conventional generation with solar PV units at selected buses while maintaining overall power balance.

Time-varying profiles for solar generation and load demand are constructed to reflect realistic daily operating conditions. Solar generation follows a diurnal pattern, while load demand varies according to typical consumption behavior. Uncertainty in renewable generation, load demand, and disturbance magnitude is incorporated through stochastic variations around these nominal profiles. Monte Carlo simulation is employed to generate multiple realizations of system conditions for each hour, resulting in a set of operating scenarios that capture the probabilistic nature of the system. These scenarios are used to evaluate frequency stability and determine the required virtual inertia under uncertainty. The detailed modeling of the stochastic scenario generation process is presented in Saha et al. (2026b).

12.5.2 Frequency stability without virtual inertia support

The system performance is first assessed without BESS-based virtual inertia support. In this context, a *frequency violation* is defined as the condition where the system frequency exceeds acceptable limits following a disturbance. Specifically, a violation occurs if either the frequency nadir falls below the minimum allowable threshold or the rate of change of frequency (RoCoF) exceeds its permissible bound.

The overall system performance is quantified using two key metrics: the *violation rate*, defined as the percentage of operating scenarios in which frequency limits are violated, and the *reliability*, defined as the percentage of scenarios that satisfy all frequency performance criteria following a credible contingency (loss of the largest generation unit).

Table 12.1 shows that during early morning hours, when synchronous generation is relatively high, the system maintains moderate reliability. However, as solar

Table 12.1 Frequency stability performance without BESS-based virtual inertia.

Time period	Violation rate (%)	Reliability (%)
Early morning	36–38	62–64
Midday (11–15 h)	≈100	≈0
Evening	55–67	33–45

Table 12.2 BESS energy and power requirements at different reliability levels.

Reliability level (%)	Energy capacity (MWh)	Power capacity (MW)
60	50.26	191
70	56.97	216.5
80	66.42	252.2
90	79.47	302

generation increases and displaces conventional units, system inertia decreases significantly, leading to severe frequency instability. During midday hours, the violation rate approaches 100%, resulting in near-zero reliability.

12.5.3 Reliability-oriented BESS sizing

In this work, the reliability-constrained framework is utilized to determine the BESS size required to provide virtual inertia. The design is based on a prespecified *reliability level*, which indicates the percentage of operating scenarios that must satisfy the frequency stability constraints.

As shown in Table 12.2, increasing the reliability level results in higher BESS energy and power requirements. This reflects the need to maintain acceptable frequency performance across a broader range of uncertain operating conditions. The nonlinear increase in capacity highlights the growing marginal cost associated with achieving higher reliability levels.

12.5.4 Frequency response improvement

The performance of BESS-based virtual inertia is measured by *frequency compliance*, the percentage of cases for which frequency stays within the limits after loss of the largest generation unit.

Table 12.3 shows that the addition of the BESS based virtual inertia significantly improves the system performance, with even moderate reliability targets leading to a major improvement, and high reliability levels effectively eliminating frequency violations.

Table 12.3 Improvement in frequency stability with BESS-based virtual inertia.

Reliability level	Frequency compliance (%)
No support	≈0 (midday)
60%	63–70
80%	82–89
90%	91–96

Table 12.4 Investment performance of BESS under different reliability levels.

Reliability (%)	CapEx (M AUD)	NPV (M AUD)	ROI (%)
60	20.10	4.10	20.4
70	22.79	3.05	13.4
80	26.57	1.25	4.7
90	31.79	−1.75	−5.5

12.5.5 Economic performance and trade-off analysis

The economic performance of the system is evaluated by considering both capital investment and the reduction in *energy not served (ENS)*, which represents the amount of load curtailed due to insufficient system stability.

From Table 12.4, a trade-off between reliability and economic performance is observed, where improving the reliability results in a lower ENS (load curtailment) and a more resilient system at the expense of higher capital cost and less economic return at very high reliability levels.

12.5.6 Summary of insights

The case study yields several insights into the integration of BESSs in a low-inertia, high renewable penetration system.

- The frequency stability issue becomes increasingly sensitive to renewable penetration due to the reduced system inertia. In particular, lower inertia leads to higher frequency deviations during disturbances and a higher risk of violating frequency constraints.
- BESS-based virtual inertia provides substantial benefits in terms of frequency compliance, particularly for the under secure operation point, due to its capability of offering fast dynamic support.
- Reliability-constrained planning allows for the systematic determination of the required BESS capacity considering the uncertainty in renewable generation, load demand, and disturbances.
- The required BESS capacity is higher at higher reliability levels to accommodate the wider range of operating scenarios.
- There is an apparent trade-off between reliability and economic performance due to the reserved flexibility for dynamic support.
- The utilization of BESS depends on the coordinated consideration of planning, operation, and deployment, which is achieved in the reliability-constrained framework.

In summary, the case study validates the effectiveness of the framework in addressing the challenges arising from high renewable penetration and low-inertia operation while offering a structured approach for balancing the reliability and economic objectives.

12.6 Conclusion

The evolving power system landscape, characterized by increasing penetration of inverter-based renewable energy sources, has presented several operational challenges such as reduced system inertia, increased variability, and new network constraints. BESSs have gained significant attention as an enabling technology to mitigate these challenges, due to their fast response capability, operational flexibility, and provision of multiple support services.

In this chapter, an integrated view of BESS utilization for both reliability and economic performance is presented. Reliability-oriented planning approaches are first used to determine the required level of virtual inertia under uncertain operating conditions. Probabilistic methods are emphasized for capturing the variability in renewable generation, load demand, and other disturbances. In conjunction with this, uncertainty-aware operational strategies are discussed to enable cost-effective scheduling that considers battery degradation and forecast deviations. The influence of deployment is then examined to understand how aggregated configurations, such as community battery systems, can mitigate variability and improve utilization efficiency. A structured framework is developed to connect the planning, operation and deployment aspects of BESS utilization in a coherent manner.

A key insight from this work is that the effective utilization of BESS needs to balance between reliability and economic objectives. The level of capacity reserved for providing dynamic support has a direct impact on operational flexibility and deployment strategies further influence the system performance through variability reduction and aggregation effects. These interdependencies highlight the importance of coordinated decision-making across all three aspects of BESS utilization.

The illustrative case study shows that the reliability-constrained framework offers a systematic approach for determining BESS requirements and assessing system performance under low-inertia conditions. As power systems continue to evolve with higher levels of renewable penetration, future research will likely focus on enhanced uncertainty modeling, advanced control strategies, and appropriate market mechanisms to fully realize the role of BESS in enabling reliable and efficient power system operation.

Nomenclature

D	Load damping coefficient
ENS	Energy not served
M	Equivalent inertia constant
β	Reliability level
$\Delta f(t)$	Frequency deviation
$\frac{d\Delta f}{dt}$	RoCoF
$\Delta P_{\text{Dist}}(t)$	Disturbance power
H_{BESS}	Virtual inertia constant
$\Delta P_{\text{BESS}}(t)$	BESS power

$\boldsymbol{P_{\mathbf{bat}}(t)}$	Battery power
$\boldsymbol{P_{\mathbf{ch}}(t)}$	Charging power
$\boldsymbol{P_{\mathbf{dis}}(t)}$	Discharging power
$\boldsymbol{P_{\mathbf{res}}}$	Reserved power
$\boldsymbol{P}_{\max}$	Max battery power
$\boldsymbol{\overline{P}_{\mathbf{ch}}}$	Max charging power
$\boldsymbol{\overline{P}_{\mathbf{dis}}}$	Max discharging power
$\boldsymbol{E_{\mathbf{bat}}}$	Total battery energy capacity
$\boldsymbol{E_{\mathbf{op}}}$	Energy for operation
$\boldsymbol{E_{\mathbf{res}}}$	Reserved energy
$\boldsymbol{E}_{\min}$	Minimum energy limit
$\boldsymbol{E}_{\max}$	Maximum energy limit
$\Delta f_{\min}$	Frequency nadir
$\Delta f_{\mathbf{limit}}$	Frequency limit
$\mathbf{RoCoF}_{\max}$	Max RoCoF
$c(u_{\mathbf{bat}})$	Cost function
$\boldsymbol{C_{\mathbf{res}}}$	Reserve cost
$\boldsymbol{C}_{\mathrm{deg}}$	Degradation cost
$\boldsymbol{C_{\mathbf{total}}}$	Total cost

References

Atalay, S., Sheikh, M., Mariani, A., Merla, Y., Bower, E., Widanage, W.D., 2020. Theory of battery ageing in a lithium-ion battery: capacity fade, nonlinear ageing and lifetime prediction. J. Power Sources 478, 229026.

Azaroual, M., Mbungu, N.T., Ouassaid, M., Siti, M.W., Maaroufi, M., 2022. Toward an intelligent community microgrid energy management system based on optimal control schemes. Int. J. Energy Res. 46 (15), 21234–21256. https://doi.org/10.1002/er.8343.

Chatzivasileiadis, S., Vu, T.L., Turitsyn, K., 2016. Remedial actions to enhance stability of low-inertia systems. In: 2016 IEEE Power and Energy Society General Meeting (PESGM). IEEE, pp. 1–5.

Kundur, P., et al., 2007. Power system stability, Power system stability and control. Power Sys. Eng.Ser. 10 (1), 7–1.

Lund, P.D., Lindgren, J., Mikkola, J., Salpakari, J., 2015. Review of energy system flexibility measures to enable high levels of variable renewable electricity. Renew. Sustain. Energy Rev. 45, 785–807.

Machowski, J., Lubosny, Z., Bialek, J.W., Bumby, J.R., 2020. *Power System Dynamics: Stability and Control*. John Wiley & Sons.

Markovic, U., Chu, Z., Aristidou, P., Hug, G., 2019. Lqr-based adaptive virtual synchronous machine for power systems with high inverter penetration. IEEE Trans. Sustain. Energy 10 (3), 1501–1512. https://doi.org/10.1109/TSTE.2018.2887147.

Masood, N.A, Shazon, M.N.H., Deeba, S.R., Modak, S.R., et al., 2021. A frequency and voltage stability-based load shedding technique for low inertia power systems. IEEE Access 9, 78947–78961.

Milano, F., Dörfler, F., Hug, G., Hill, D.J., Verbič, G., 2018. Foundations and challenges of low-inertia systems. In: 2018 Power Systems Computation Conference (PSCC). IEEE, pp. 1–25.

Neda, O.M., Adabi, J., Gholinezhadomran, H., Marzband, M., 2025. Machine learning-enhanced coordination of home-microgrids for resource compensation in large-scale energy systems. Sustain. Cities Soc., 106744

Nguyen, H.T., Yang, G., Nielsen, A.H., Jensen, P.H., 2016. Frequency stability improvement of low inertia systems using synchronous condensers. In: 2016 IEEE International Conference on Smart Grid Communications (SmartGridComm). IEEE, pp. 650–655.

Okafor, C.E., Folly, K.A., 2024. Design and implementation of a control system for multi-functional applications of a battery energy storage system (BESS) in a power system network. Clean.Energy Syst. 9, 100153.

Parag, Y., Sovacool, B.K., 2016. Electricity market design for the prosumer era. Nat. Energy 1 (4), 16032.

Rahman, S., Saha, S., Haque, M., Islam, S., Arif, M., Mosadeghy, M., Oo, A., 2022. A framework to assess voltage stability of power grids with high penetration of solar pv systems. Int. J. Electr. Power Energy Syst. 139, 107815.

Restrepo, M., Cañizares, C.A., Simpson-Porco, J.W., Su, P., Taruc, J., 2021. Optimization-and rule-based energy management systems at the canadian renewable energy laboratory microgrid facility. Appl. Energy 290, 116760.

Saha, S., Saleem, M., Roy, T., 2023. Impact of high penetration of renewable energy sources on grid frequency behaviour. Int. J. Electr. Power Energy Syst. 145, 108701.

Saha, S., Saleem, M.I., Roy, T.K., Oo, A.M.T., 2025. Virtual inertia planning for enhancing grid stability in low-inertia systems. IET Energy Syst. Integr 7 (1).

Saha, S., Elliott, M., Roy, T.K., Oo, A.M.T., 2026a. Reliability constrained optimization of virtual inertia and battery energy storage system sizing for frequency stability in low inertia power systems. Sustain. Energy Grids Netw. 45, 102130.

Saha, S., Elliott, M., Roy, T.K., Oo, A.M.T., 2026b. Quantifying hourly virtual inertia and battery energy storage system requirements for frequency stability in low-inertia power systems. J. Energy Storage 153, 120853.

Saleem, M.I., Saha, S., Izhar, U., Ang, L.-m., 2022. Integration challenges of inverter based renewable energy sources in weak grids. In: 2022 IEEE Industry Applications Society Annual Meeting (IAS), pp. 1–18.

Saleem, M., Saha, S., Ang, L., Izhar, U., Roy, T., 2023. Factors affecting voltage stability while integrating inverter based renewable energy sources into weak power grids. IET Gener. Transm. Distrib. 17 (6), 1216–1231.

Saleem, M., Saha, S., Izhar, U., Ang, L., 2024a. A stochastic mpc-based energy management system for integrating solar pv, battery storage, and ev charging in residential complexes. Energy Build. 325, 114993. https://doi.org/10.1016/j.enbuild.2024.114993.

Saleem, M., Saha, S., Izhar, U., Ang, L., 2024b. Stability assessment of inverter-based renewable energy sources integrated to weak grids. IET Energy Syst. Integr. 6 (4), 525–544.

Saleem, M.I., Saha, S., Roy, T.K., 2025. Inertia sensitivity analysis of power grids with high penetration of renewable energy sources. IEEE Trans. Ind. Appl. 61 (4), 5362–5380.

Son, Y.-G., Kim, S.-Y., Bae, I.-S., 2024. Optimal coordination of energy coupling system considering uncertainty of renewable energy sources. Energies 17 (4), 931.

Turitsyn, K., Sulc, P., Backhaus, S., Chertkov, M., 2011. Options for control of reactive power by distributed photovoltaic generators. Proc. IEEE 99 (6), 1063–1073.

Ulbig, A., Borsche, T.S., Andersson, G., 2014. Impact of low rotational inertia on power system stability and operation. IFAC Proc. Vol. 47 (3), 7290–7297.

Zakeri, B., Syri, S., 2015. Electrical energy storage systems: a comparative life cycle cost analysis. Renew. Sustain. Energy Rev. 42, 569–596.

Zhong, Q.-C., Weiss, G., 2010. Synchronverters: Inverters that mimic synchronous generators. IEEE Trans. Ind. Electron. 58 (4), 1259–1267.

Zio, E., 2013. Monte Carlo simulation: the method. In: The Monte Carlo Simulation Method for System Reliability and Risk Analysis. Springer, pp. 19–58.

Cybersecurity and data management challenges in smart energy storage networks

13

Muninathan Kota [1], *H. Leena Catherine* [1], *S. Murugapoopathi* [2] *and Hiren Mewada* [3]
[1]St. Joseph's College of Engineering, Chennai, Tamil Nadu, India; [2]PSNA College of Engineering and Technology, Dindigul, Tamil Nadu, India; [3]Prince Mohammad Bin Fahd University, Al Khobar, Saudi Arabia

13.1 Introduction

The energy storage technologies such as supercapacitors, advanced lithium-ion batteries, and hybrid energy storage systems are becoming increasingly important in the context of coping with the intermittency of renewable power generation, improving the power grid operation reliability, and satisfying peak power demands (Knowles et al., 2015) in the global transition to low-carbon energy sources. Enhancing the efficiency and interconnectivity of contemporary energy storage networks (ESNs) through cloud computing analytics, sophisticated control systems, and high-speed communication networks, digitalization is transforming energy storage systems into smart, efficient, and interconnected assets that are essential for modern integration of renewable energy and grid stabilization (Farhangi, 2010). The fast pace of digital transformation offers many advantages but leaves us open to serious cyber threats which could result in heavy financial, operational, and reputational loss. Also, ESNs produce additional data sets such as operational, environmental, and user information. Confidentiality and integrity are key parts of a holistic data governance framework (Cárdenas et al., 2008). This section discusses the intersection of multidimensional data governance and cyber-physical security, particularly highlighting the vulnerabilities of industrial control systems and battery management systems (BMSs). The study also discusses recent system security techniques such as advanced encryption, AI-based anomaly detection, and role-based access control, as well as benchmarks and the future outlook for sustainable energy systems.

Energy Storage Technologies. https://doi.org/10.1016/B978-0-443-45858-3.00008-5

13.2 Energy storage networks: Architecture and data flows

13.2.1 System architecture

ESNs are interconnected systems where the physical energy infrastructure is integrated with digital control and communication technologies. Their design seamlessly merges the physical and digital to encourage intelligent, secure, and effective energy management. The typical ESN is made up of battery cells and packs, with a BMS which evaluates the battery condition, controls charge allocation, and offers safety. The energy management system (EMS) improves energy distribution (Fig. 13.1), controls power consumption, and facilitates integration with renewable energy sources or the electrical grid. The Supervisory Control and Data Acquisition (SCADA) system also enables real-time monitoring, detects anomalies, and provides diagnostics (Cárdenas et al., 2008). Modern ESNs rely on cloud analytics tools that employ AI and big data to assess performance, anticipate maintenance needs, and enhance system efficiency. These components establish a robust, multilayered framework that improves the overall efficacy, scalability, and reliability of energy storage systems. Table 13.1 details the various components and functions of ESN.

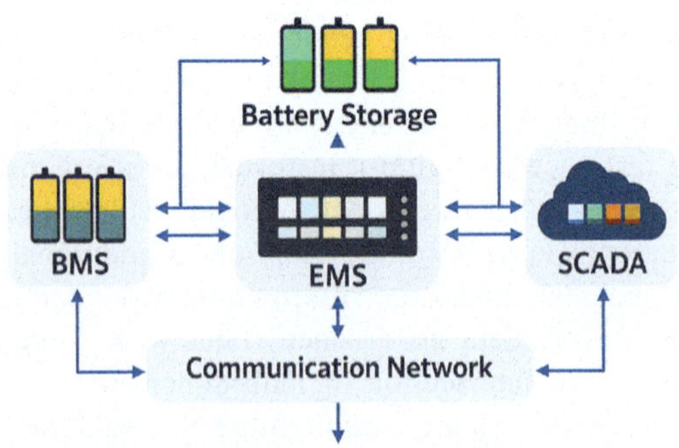

Figure 13.1 Generic architecture of an ESN.

Table 13.1 Components and functions of energy storage networks (Farhangi, 2010).

S. No.	Component	Function	Cybersecurity relevance
1	Battery cells	Energy storage	Safety-critical
2	BMS	Monitoring and protection	Firmware security
3	EMS	System optimization	Access control
4	SCADA	Supervisory control	Network security
5	Cloud platform	Analytics and storage	Data privacy

13.2.2 Data types and characteristics

A large amount of operational and control data is generated by ESNs to ensure system safety and performance. The data streams include electrical parameters such as voltages, currents, and states of charge, thermal and ambient measurements, degradation and health indicators, control instructions, firmware updates, user interactions, and market transactions. All of these provide independent monitoring, diagnosis, and management of storage assets. In many cases, data security means protecting important operational data from being accessed or altered by people who shouldn't have access to it, while still allowing approved analytics, predictive maintenance, and day-to-day operational decisions to proceed without a hitch. The increase in usage of advanced remote systems, cloud-based services, and the Internet of things attracts ESS to implement strong encryption, comprehensive access restrictions, and secure data transfer protocols (ISO/IEC, 2022; He et al., 2024).

13.3 Cybersecurity threat landscape in energy storage networks

13.3.1 Attack surfaces and threat vectors

ESNs are more vulnerable to cyber and cyber-physical threats due to their reliance on communication networks, digital control systems, and cloud-based analytics. Cybercriminals carry out malicious activities such as data manipulation, executing Denial-of-Service (DoS) attacks, and overwhelming systems (Kharlamova et al., 2023). They utilize malicious software such as ransomware, Trojans, and web shells to disseminate malware. This can exfiltrate data, disrupt systems, or create backdoors. In cloud-based and IoT systems, vulnerabilities can be exploited to gain unauthorized access (Öhrström et al., 2025). It is important to protect against data manipulation, DoS attacks, malware distribution, and unauthorized access to the system. The risks of cloud-based analytics and remote interfaces increase when data transmitters and systems are integrated (Naseri et al., 2023). According to more detailed cyber-physical threat models for critical infrastructure, a cyberattack can directly affect the performance and stability of physical systems. Table 13.2 lists the threats and damages arising

Table 13.2 Major cybersecurity threats and their potential impacts on energy storage networks.

S. No.	Threat type	Target component	Potential impact
1	Malware	BMS firmware	Safety risks, battery damage
2	Data spoofing	Sensors	Incorrect control decisions
3	Unauthorized access	Cloud platforms	Data breaches
4	DoS attacks	Communication links	Grid instability

from cybersecurity standards for critical energy infrastructure and industrial control systems (NIST, 2018; IEC, 2018; ISO/IEC, 2022).

13.3.2 Safety implications

In ESNs, both physical operations and digital systems are affected by cyberattacks. Such attacks may lead to serious safety issues, such as heat runaway, accelerated battery discharge, or unexpected system shutdowns, as shown in Fig. 13.2. Cyber security and cyber physical impact: Cyberattacks targeting ESNs differ from those targeting traditional IT systems because they affect complex, interconnected control and electrochemical processes. This suggests that detrimental behaviors can disturb immediate decisions. Therefore, cybersecurity in ESNs must be addressed not merely as an information security issue, but as a critical component of system safety engineering and risk management (NIST, 2018; IEC, 2018; ISO/IEC, 2022).

13.4 Data management challenges in energy storage networks

13.4.1 Data lifecycle management

Effective data management in ESNs requires oversight of the data life cycle. This includes the acquisition of data from different sensors to perform a safe transmission, storage, processing, and analysis and later archiving or deletion. To operate grid-scale storage systems in an optimal manner, it is important to ensure that data are always available, accurate, and retrievable. Strong data governance structures allow for quick identification of issues, better predictive maintenance, and continued compliance with operational and regulatory standards. The increasing combination of other communication networks, cloud infrastructures, and data-centric control methodologies in today's ESNs necessitates comprehensive data lifecycle management to optimize system safety, efficiency, and robustness (Gungor et al., 2013).

13.4.2 Data governance and compliance

A major challenge in ESNs is the data governance as the number of asset owners is high, the data flows are transnational, and the regulations for grid integrated infrastructure are heterogeneous. These include privacy protections, compliance with international data transfer regulations across multiple jurisdictions, and compliance with grid codes and operational reporting standards. Governance and accountability oversight is a complex challenge, especially in distributed systems with multiple stakeholders and heterogeneous hardware and software platforms. ESN ecosystems require strong data governance frameworks to ensure compliance with regulations, maintain transparency in operations, and enable secure data exchange (Weber, 2010). Fig. 13.3 illustrates the data lifecycle and governance framework for ESNs.

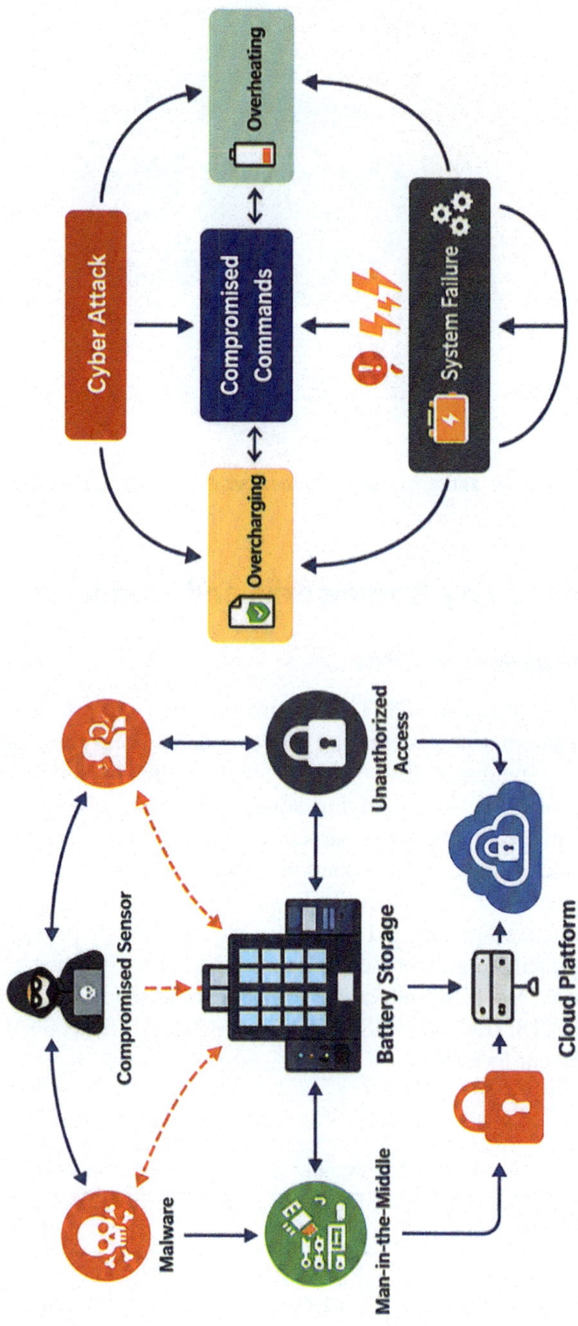

Figure 13.2 Cyber security and cyber physical impact.

Figure 13.3 Data lifecycle and governance framework for energy storage networks.

13.5 Cybersecurity frameworks and standards

13.5.1 Applicable standards

IEC (62443), ISO (27001), and NIST are the international cybersecurity standards for ESNs. Industrial automation systems and grid-connected energy storage solutions are protected by these standards. IEC 62443 emphasizes defense-in-depth security for operational technology (OT) employing zones and conduits with layered security standards. ISO/IEC 27001 provides a structured approach to managing data security, access control, and risk via an Information Security Management System (ISMS). Combined with the NIST Cybersecurity Framework lifecycle model (identification, protection, detection, response, and recovery), these standards promote comprehensive risk management, regulatory compliance, operational resilience, and security across the entire ESN—from individual components to the massive network infrastructure (NIST, 2018; IEC, 2018; ISO/IEC, 2022). Table 13.3 details the cybersecurity standards relevant to ESNs.

Table 13.3 Cybersecurity standards relevant to energy storage networks (IEC, 2018).

S. No.	Standard	Scope	Application
1	IEC 62443	Industrial systems	BMS, EMS
2	ISO/IEC 27001	ISMS	Data governance
3	NIST CSF	Critical infrastructure	Risk management

13.5.2 Layered security architecture

One of the standard approaches to securing cyber-physical systems, for example, ESNs, is to employ a layered defense-in-depth security framework. This approach spans security measures over multiple layers of the system to prevent single points of failure and minimize the effect of successful breaches.

In general, ESNs will apply defense-in-depth by implementing safeguards at the device level (e.g., secure boot protocols and hardware authentication), network level (e.g., encrypted communications and intrusion detection systems), and application level (e.g., access controls on analytics and supervisory interfaces). Continuous monitoring, recordkeeping, and incident management tactics enhance system robustness by enabling rapid risk identification and coordinated resolution. Many argue that multilayered security frameworks are necessary for securing cyber-physical systems, since cyber-attacks may have direct impacts on physical operations (Giraldo et al., 2019). Fig. 13.4 details the layered cybersecurity framework.

13.6 Secure data analytics and emerging technologies

13.6.1 Secure analytics and digital twins

Reliable data is a prerequisite for safe and economical operation of energy storage systems. Malicious data injection remains a risk, despite advancements in data analytics and the increasing real-time nature of digital twins. This persists despite the shift in emphasis toward supervision. Deceptive data injection is a technique that can be used to change sensor readings, which can lead to wrong charging or discharging. This can cause interruptions in energy allocation and speed up battery degradation. So the voltage and the frequency that flood communication channels are the targets of ramp and DoS attacks. Cloud infrastructures and, in particular, BMS sensors are vulnerable to two threats: man-in-the-middle attacks and firmware manipulation. Both sensor drift noise and physical impairment can adversely affect data integrity. Data corruption can lead to control errors, grid instability, and safety risks. Blockchain records, AI anomaly detection, encrypted pipelines, redundant sensors, and edge computing secure data. These strategies help ensure digital twins provide accurate, reliable information, boosting asset safety, performance, and lifespan (Jiang et al., 2013).

13.6.2 Blockchain and distributed ledgers

Some experts argue that blockchain technology can improve the safety and reliability of peer-to-peer ESNs by providing immutable data records and facilitating decentralized coordination. Spreading transaction records across multiple nodes lessens reliance on central authorities and lowers the risk of data manipulation. These benefits are especially important in peer-to-peer energy storage systems, where participants may mistrust one another, requiring transparent energy trading, secure sharing of

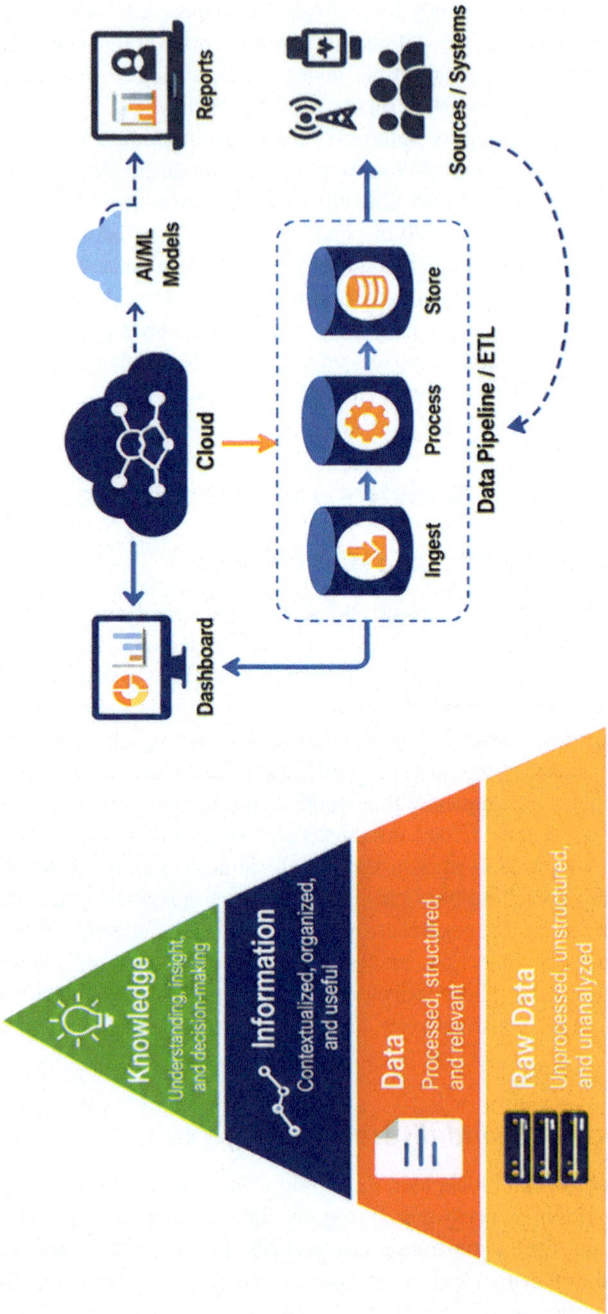

Figure 13.4 Layered cybersecurity architecture.

Table 13.4 Comparison of conventional versus blockchain-based data management (Andoni et al., 2019).

S. No.	Aspect	Conventional	Blockchain-based
1	Data integrity	Centralized	Immutable
2	Trust model	Trusted authority	Decentralized
3	Scalability	High	Moderate

charging information, and trustworthy control decisions (Andoni et al., 2019). Table 13.4 shows the detailed comparison of conventional versus blockchain-based data management.

13.7 Case studies and practical implementations

According to the previous research, the cybersecurity challenges in energy storage systems vary significantly. The security measures (multilayered defenses, risk assessments, intrusion detection) of grid-connected battery energy storage systems are being strengthened, particularly on a large scale and with proper regulations. Small-scale storage systems lack these protective measures and dedicated cybersecurity policies. This results in vulnerable attacks and operational problems. Regulation and ownership are not the only reasons for variation, but the level of investment in security and data-driven (cyber defense) strategies also plays a vital role across the application. Cyber defense explores the gaps and customized security solutions for both distributed systems and utility-scale systems (Kharlamova et al., 2023).

13.8 Future research directions

The future direction of research on ESNs focuses on establishing cybersecurity regulations. Current cybersecurity standards for small utility companies and industrial control systems provide a useful foundation. High cybersecurity requirements are necessary for BMSs to protect ownership and maintain grid connections. Data-driven cybersecurity techniques, such as artificial intelligence and machine learning, appear to have great potential for improving intrusion detection and response. With the emergence of decentralized architectures and advanced analytics, there is a need for privacy-aware data processing techniques that protect sensitive customer and operational information, while improving system efficiency. Furthermore, significant research directions regarding safety and cybersecurity in ESNs are the development of quantitative metrics for cyber-resilience and the creation of coherent certification frameworks. Since cyber and physical threats are interconnected, the accreditation for safety engineering and cybersecurity has to be integrated.

Successfully solving these challenges is crucial for the safety, reliability, and large-scale deployment of the energy storage systems in future power grids.

13.9 Conclusion

For ESNs to work safely, reliably, and last, they need robust cybersecurity and data management. The attack on the cyber-physical interface is increasing significantly as grid and market platforms integrate. This chapter explores key data governance challenges, risk areas, and strategies to mitigate them. Also, this chapter explores the importance of layered security, governance framework, and data lifecycle management. To prevent cascading failures and promote long-term resilience, cybersecurity should be integrated early in the design phase, alongside performance and safety considerations. Emphasizing cybersecurity by design and strong data governance is vital to recognizing the role of energy storage in building secure and sustainable energy systems.

Acknowledgments

The authors acknowledge the use of Grammarly as a language-assistance tool to support linguistic refinement of this chapter. The authors take full responsibility for the technical content, accuracy, originality, and compliance with ethical and publication standards.

Conflict of interest statement

The authors declare no conflict of interest related to this work.

References

Andoni, M., Robu, V., Flynn, D., Abram, S., Geach, D., Jenkins, D., McCallum, P., Peacock, A., 2019. Blockchain technology in the energy sector: a systematic review of challenges and opportunities. Renew. Sustain. Energy Rev. 100, 143–174. https://doi.org/10.1016/J.RSER.2018.10.014.

Cárdenas, A.A., Amin, S., Sastry, S., 2008. Secure control: towards survivable cyber-physical systems. Proc. Int. Conf. Distrib. Comput. Syst. 495–500. https://doi.org/10.1109/ICDCS.WORKSHOPS.2008.40.

Farhangi, H., 2010. The path of the smart grid. IEEE Power Energy Mag. 8 (1), 18–28. https://doi.org/10.1109/MPE.2009.934876.

Giraldo, J., Urbina, D., Cardenas, A., Valente, J., Faisal, M., Ruths, J., Tippenhauer, N.O., Sandberg, H., Candell, R., 2019. A survey of physics-based attack detection in cyber-physical systems. ACM Comput. Surv. 51 (4). https://doi.org/10.1145/3203245. Available from: http://dl.acm.org/citation.cfm?id=J204.

Gungor, V.C., Sahin, D., Kocak, T., Ergut, S., Buccella, C., Cecati, C., Hancke, G.P., 2013. A Survey on smart grid potential applications and communication requirements. IEEE Trans. Ind. Inf. 9 (1), 28–42. https://doi.org/10.1109/TII.2012.2218253.

He, Y., Zhou, Z., Pan, Y., Chong, F., Wu, B., Xiao, K., Li, H., 2024. Review of data security within energy blockchain: a comprehensive analysis of storage, management, and utilization. High-Confid. Comp. 4 (3), 100233. https://doi.org/10.1016/j.hcc.2024.100233.

IEC, 2018. IEC 62443: Industrial Communication Networks – Network and System Security. International Electrotechnical Commission.

ISO/IEC, 2022. ISO/IEC 27001:2022 — Information Security Management Systems — Requirements. International Organization for Standardization.

Jiang, Q., Xue, M., Geng, G., 2013. Energy management of microgrid in grid-connected and stand-alone modes. IEEE Trans. Power Syst. 28, 3380–3389. https://doi.org/10.1109/TPWRS.2013.2244104.

Kharlamova, N., Træhold, C., Hashemi, S., 2023. Cyberattack detection methods for battery energy storage systems. J. Energy Storage 69, 107795. https://doi.org/10.1016/J.EST.2023.107795.

Knowles, W., Prince, D., Hutchison, D., Disso, J.F.P., Jones, K., 2015. A survey of cyber security management in industrial control systems. Int. J. Crit. Infrastruct. Protect. 9, 52–80. https://doi.org/10.1016/j.ijcip.2015.02.002. Available from: http://www.elsevier.com.

Naseri, F., Kazemi, Z., Larsen, P.G., Arefi, M.M., Schaltz, E., 2023. Cyber-physical cloud battery management systems: review of security aspects. Batteries 9 (7). https://doi.org/10.3390/batteries9070382.

NIST, 2018. Framework for Improving Critical Infrastructure Cybersecurity (Version 1.1). National Institute of Standards and Technology.

Öhrström, F., Oscarsson, J., Afzal, Z., Dani, J., Asplund, M., 2025. From balance to breach: cyber threats to battery energy storage systems. Energy Inform. 8 (1). https://doi.org/10.1186/s42162-025-00499-4. Available from: https://www.springer.com/journal/42162.

Weber, R.H., 2010. Internet of Things - new security and privacy challenges. Comp. Law Security Rev. 26 (1), 23–30. https://doi.org/10.1016/j.clsr.2009.11.008.

Future outlook and research directions in energy storage

14

Deepanraj Balakrishnan [1], *Prabhakar Sharma* [2], *Amanullah Maung Than Oo* [3] *and Bachir Benhala* [4]
[1]Department of Mechanical Engineering, College of Engineering, Prince Mohammad Bin Fahd University, Al-Khobar, Saudi Arabia; [2]Delhi Skill and Entrepreneurship University, Delhi, New Delhi, India; [3]School of Science, Technology and Engineering, University of the Sunshine Coast, Moreton Bay, QLD, Australia; [4]Sidi Mohammed Ben Abdellah University, Fez, Morocco

Energy storage technologies (ESTs) are one of the key elements in the transition toward successful implementation of renewable energy. This is necessary for improving the grid stability and efficient distributed energy model. In this chapter, technological, digital, and policy innovations in this domain are reported. This chapter also describes the key research and technological trends that are shaping the future of energy storage (ES). It also explores emerging storage technologies and novel materials, ES digitalization, challenges in grid integration, sustainability, and the regulatory environment.

14.1 Next-generation energy storage technologies

The global transition to a low-carbon energy system has put pressure on the development of new ESTs to help overcome the limitations of the existing systems. While lithium-ion batteries have become dominant in the market owing to their high energy density, falling cost and technological readiness, their long-term sustainability, safety, and material scarcity have driven the search for the next generation of batteries (Njema et al., 2024; He et al., 2021).

One of the most exciting innovations is a solid-state battery, which is a significant advancement of the traditional liquid electrolyte-based lithium-ion batteries. They also increase safety through the prevention of leakage and thermal runaway by replacing combustible liquid electrolytes with solid electrolytes. In addition, solid-state batteries also allow the use of high-energy density lithium metal anodes instead of graphite-based ones. Recent research has focused on sulfide, oxide, and polymer solid electrolytes that each has their own advantages as to ionic conductivity, mechanical strength, and processability (Joshi et al., 2025).

Sodium-ion batteries have recently emerged as an alternative to lithium-ion batteries, particularly for large-scale and stationary ES. These batteries are cost-effective and environmentally friendly due to the low cost and widespread availability of sodium. On top of that, sodium-ion batteries are thermally stable and have a wide range

Energy Storage Technologies. https://doi.org/10.1016/B978-0-443-45858-3.00011-5

of working temperature; hence, they can be applied at the grid scale in different climates. The research now is focused on improving the cycle life, enhancing the stability of the electrodes, and developing electrolyte formulations (Xiong etal., 2025; Liang et al., 2018). Other such next-generation technologies are metal-air batteries (e.g., lithium-air, zinc-air, and aluminum-air batteries). The theoretical energy density of these batteries is very high as the oxygen in air is used as the reactant, thus significantly reducing the mass and volume of the battery. For instance, lithium-air batteries have the potential to achieve the same energy density as fossil fuels and could be a potential alternative to long-range electric vehicles (EVs) (Srinivasarao and Mohamed, 2024). Zinc-air batteries, however, are already being examined for stationery and back-up power sources, as they are relatively stable, cheap, and safe (Li et al., 2023; Olabi et al., 2021).

Besides single chemistries, hybrid ES is also gaining attention as a way to take advantage of the merits of different ESTs and compensate for the limitations of individual technologies (Günther etal., 2025). Hybrid systems are typically combinations of high-energy-density devices, such as batteries, together with high-power-density devices, such as supercapacitors or flywheels. This offers the opportunity for systems to sustain the required power and respond to rapid load changes and is particularly relevant to EVs, renewable energy integration, and grid stabilization solutions (Gopi and Ramesh, 2024; Ciçek et al., 2022).

The flexibility to use new chemistries in hybrid systems also offers new possibilities for system optimization. For instance, hybrid ES systems based on sodium-ion batteries combined with supercapacitors can deliver cost-effective solutions to local ES systems, while solid-state batteries combined with next-generation power electronics can deliver safe and efficient solutions to high-power systems (Vanaraj et al., 2025). These hybrid systems need to be developed with new control strategies and optimization techniques to ensure seamless operation, efficient energy management, and system reliability. Improvements in power electronics, control strategies, and modeling are therefore key to harness the benefits of hybrid ES (Sajedi-Moghaddam etal., 2026; Wen et al., 2025). It is important that future ES system should tackle challenges involved in maintaining sustainability and depleting resources. Alternative chemistries, as for example, sodium-ion and metal-air, can be employed to lower the dependence on scarce resources. It can also promote adopting abundant and nontoxic materials. It is also observed that development of recyclable and biodegradable materials is increasingly day by day and thus leading to ESTs in line with the circular economy (Ghazoui et al., 2026).

Another dimension in the prospects of modern ESTs is the versatility in case of different applications. The higher energy density technologies are required for EVs; on the other hand, we need long-duration discharge technologies in case of grid-scale storage, cost, and lifetime (Ghazoui et al., 2026; Bányai, 2023). The successful implementation and commercialization of renewable energy technologies need newer-generation ESTs. This will also depend on the innovative technologies, cost reduction, scalability, and standardization. The scaling up challenges is also another cause of concern. The researchers, industry and policy makers should come together to accelerate the TRL and market penetration (Sajedi-Moghaddam etal., 2026).

14.2 Advanced materials and performance enhancement

High-performance materials are considered as foundation in the ESTs progress to address the growing needs of efficiency, durability, safety, and sustainability. Although system-level innovations and architectural advancements have an important role in the overall performance, the properties of materials, at the atomic, nano, and micro scales, ultimately determine the electrochemical behavior and long-term stability of storage systems. This implies that current and future studies in ES are highly biased toward the development of advanced materials engineering, especially in the context of customizing the structure-property relationships to attain high-performance indicators (Tawalbeh et al., 2023). A key research focus is on high-capacity electrode materials. In conventional lithium-ion batteries, current anodes (graphite) and cathodes (layered transition metal oxides) are close to the theoretical limits, and new materials with higher energy density are being explored. Silicon anodes, are attractive owing to their higher theoretical capacity (almost 10 times greater to graphite) (Machín etal., 2024; Feyzi et al., 2024).

Greater energy densities and improved thermal properties are being targeted by high-nickel layered oxides, lithium-rich materials and polyanionic materials. These have improved electrochemical properties and the challenge with these materials is that they are less structurally stable, tend to lose oxygen, and their capacity fades over time. These problems are being addressed through surface modification techniques such as coating and doping with a stabilizing element. These can be used to prevent side reactions, improve interface stability, and improve battery lifetime (Yang et al., 2024). The other key component of the ES system safety and performance is electrolyte. Conventional liquid electrolytes have a high ionic conductivity but are flammable and prone to leakage. These materials enable the use of high energy-density metal anodes and prevent thermal runaway. However, challenges in terms of interfacial resistance, mechanical brittleness and mass production should be addressed in order for them to become popular (Kolosnitsyn and Karaseva, 2008; Fotouhi et al., 2016).

The interfacial engineering has become a key strategy in improving the advanced ES materials. Electrochemical reactions generally take place at the interface between electrodes and electrolytes, and the stability of the interface is crucial to the efficiency and lifetime of batteries. The formation of solid electrolyte interphase (SEI) layers, for instance, plays a critical role in the electrode-electrolyte stability. Research is ongoing on development of stable and ionically conductive SEI layers through electrolyte additives and surface treatments. Similarly, artificial interphases and protective layers are being developed to suppress dendrites, particularly in the systems with metal anodes such as Li and Na (Adenusi et al., 2023). Nanotechnology is a breakthrough in designing ES materials as it offers control over the morphology and structure of a material. The nanomaterials have unique properties, such as high surface area, short ion diffusion paths, and flexibility, which translate to excellent electrochemical properties. For instance, the nanoscale of the electrode materials can lead to greater rate of discharge-charge and power capability. However, the larger area of surface in the case of the nanomaterials can also result in an increase of side reactions and a loss of coulombic efficiency. These conflicting issues need to be managed to achieve the right performance by tailoring and fine-tuning the nanostructures (Naraprawatphong et al., 2022; Pérez Mendoza et al., 2024).

In addition to nano-structuring, novel characterization and simulation techniques have also improved the development of high-performance materials. In situ electron microscopy, synchrotron spectroscopy and atomic-scale simulations as tools have provided valuable insights into dynamic processes in ES systems. These techniques enable the observation of structural evolution, ion transport, and degradation mechanisms, which can be applied for materials development with superior performance. Modeling approaches, like molecular dynamics simulations as well as density functional theory (DFT), are getting higher attention to model-predict the material properties and in guiding experimental studies. Environmental sustainability and compatibility are also important aspects of advanced materials. Extraction and processing of the materials that are deemed critical in the construction of conventional batteries, such as cobalt and nickel, has resulted in environmental and ethical concerns. This has resulted in the development of environmental-friendly materials that are nontoxic, cheap, and recyclable. Environmentally friendly options include iron cathodes, organic electrodes, and bio-based polymers. In addition, efforts are made to produce recyclable and reusable materials, which would reduce the environmental footprint of ESTs (Nyangiwe, 2025; Wu et al., 2019).

Thermal stability and mechanical strength are also important aspects in the design of novel ES materials. HPS are often subjected to extreme operating settings, like higher temperatures and charge-discharge currents that can accelerate degradation. Therefore, improved thermal and mechanical properties of the materials are required for safe and stable operation. The usage of multifunctional materials to offer structural, thermal, and electrochemical properties is a future research direction (Li et al., 2021). In addition to electrochemical method, the thermal storage systems and supercapacitors are also being developed with new materials. Similarly, phase change materials (PCMs) and thermochemical materials are also being developed in thermal storage systems to store more energy and have improved thermal conductivity. These advances highlight the multidisciplinary importance of materials innovation in the different ESTs (Ansari et al., 2024; Said et al., 2023).

In the coming years, materials science is expected to be complemented by new research fields such as artificial intelligence and high-throughput experimentation to help in development and discovery of novel materials. Data-based approaches can help shorten the cost and time of the iterative processes of the traditional method of trial-and-error to rapidly uncover new materials candidates. Machine learning can analyze huge data to reveal hidden patterns and relationships that can guide the development of materials with tailor-made properties. And last but not least, advanced materials and strategies for improving performance are used to drive next-generation ESTs. At the same time, the addition of sustainability and advanced analytical tools is shaping a more holistic approach to material design. As material science continues to evolve, it is increasingly feasible to develop high-performance, cheap, and eco-friendly ESTs.

14.3 Smart and digital energy storage systems

The ongoing digital transformation of the energy sector is revolutionizing the design, operation, and management of ESTs. The fact that ES is becoming an integral element of the energy system, particularly with the integration of renewable energy and decentralized power supply, makes smart, dynamic, and data-driven technologies even more important. Smart and digital ES technologies take advantage of recent advances in AI, ML, data analytics and cyber-physical systems to improve the performance, reliability, and efficiency. This digital integration of ES is a move from nondynamic to dynamic and self-optimizing systems (Kadulkar et al., 2022; Senthilkumar and Yuvaperiyasamy, 2025). The primary use of AI and ML in ES is performance prediction and modeling (Mhatre et al., 2026). Conventional modeling methods use simple assumptions and parameters that may not account for the dynamic and nonlinear characteristics of ES systems for different operating conditions (Senthilkumar etal., 2024). These parameters are important for safe and efficient operation, especially for EVs and grid-scale ES systems.

AI diagnosis systems are able to track the evolution of the operating data (voltage, current, temperature, and impedance) to identify issues and predict failures. This allows the development of condition-based maintenance programs, as compared to the traditional time-based maintenance programs. The digital twin technology can be used for real-time monitoring and optimization of ES systems. A digital twin is a model of a physical system that is updated in real time with data from sensors and control systems. Digital twins can be used to better understand system's performance under various operating conditions by combining physical models and data analytics. The ability to simulate different scenarios for performance improvement and to test the effects of control policies on the system's behavior without affecting the real system are enabled by this approach. A digital twin can be applied to ES for optimizing the charge-discharge cycle, thermal management, and to increase the lifespan of the ES (Semeraro et al., 2023).

Smart control systems also transform the relationship of ES with the grid. ES plays an important role in supply and demand balance of intermittent renewable energy systems. Smart energy management systems (EMSs) are complex algorithms to control the operation of storage systems based on the energy market signals, grid, and consumer requirements. These systems can alter the charge and discharge profile, involved in demand response programs and provide the ancillary services like frequency and voltage regulation. These systems are also made more efficient and flexible by reinforcement learning and optimization control strategies. The other critical feature of digital ES systems is that they support distributed energy systems (Guler et al., 2024; Mischos etal., 2023).

The use of blockchain technology is also being proposed to facilitate secure and transparent energy trading in decentralized systems. Blockchain can enable P2P energy trading, in which prosumers (those who both produce and consume energy) can exchange energy. ES systems are vital in such systems as they provide flexibility to store energy and make it available on demand. To facilitate energy markets that are efficient and secure, smart contracts (that are implemented on blockchain platforms) can automate transactions, based on certain criteria (Wu et al., 2022). The smart ES systems depend on data handling and communication technologies. The use of new sensors, IoT devices, and fast communication systems enable the continuous generation and transmission of large volumes of data. To support this, cloud and edge computing technologies are increasingly being used. While the cloud offers flexibility and processing power, edge computing can be applied to enable faster reaction to data by processing it close to the source. It is not uncommon to adopt a hybrid approach of using both concepts (Ramdan et al., 2025).

A significant challenge for digitalizing ES systems is cybersecurity. As they become more interconnected and dependent on digital technologies, these systems are more vulnerable to cyber security threats. Cyber security threats can compromise system integrity, operations, and pose serious safety risks through hacking, data theft, and sabotage. Therefore, strong cybersecurity measures like authentication, encryption, and intrusion detection should be in place for intelligent ES systems (Cavus et al., 2025).

14.4 Grid integration and long-duration storage

The ever-increasing use and adoption of renewable energy sources (RESs) have transformed the operation and management of modern power systems. RES like wind and solar are intermittent in nature, and it remain a big challenge. This creates challenges in terms of RES based grid reliability, stability, and balancing. ESTs play a crucial role in grid integration in this dynamic energy environment as they provide the flexibility required to balance the energy supply and demand (García et al., 2021). Long-duration energy storage (LDES) technologies are also becoming particularly noticeable among other types of ESTs because of their ability to store energy for longer periods (i.e., hours to days or even seasons). ES integration is the seamless integration of ESTs with the grid to enable a wide range of applications such as energy arbitrage, peak-shaving, frequency control, voltage regulation, and back-up power (Mohseni and Brent, 2020; Obakhume and Opatola, 2025).

Hydrogen ES is also attracting a lot of attention as a potential seasonal ES and sector coupling technology. Here, surplus renewable energy is employed to generate hydrogen via electrolysis process, that may be stored and again re-use by converting to electricity either with a fuel cell or a turbine. The hydrogen can be used for other end-use sectors, such as transportation and industry, enabling sector coupling. While the hydrogen storage has significant storage time (nearly infinite) and energy density, its round-trip efficiency is low compared to other ES methods. The ongoing research is concentrated on enhancing the electrolysis efficiency, cost, and developing a robust storage and transport infrastructure (Alzahrani et al., 2022).

The successful integration of ES is also dependent on the transmission and distribution system. The ESTs optimization in terms of location can help in congestion mitigation, and it can also improve power delivery. For example, close (in vicinity) placement of ESTs with RES can help in enhancing demand side management. Modern approach involving ML and optimization methods can help in configuration and location configurations. In recent times, the LDES technologies and grid integration approaches have improved but challenges remain in terms of higher capital cost, TRLs, sustainability challenges, and regulatory barriers (Ramdan et al., 2025).

14.5 Future research roadmap and policy directions

The evolution toward a sustainable and low-carbon energy system heavily depends on the advancement of ESTs, both in terms of technical aspects and in the integration of research, policy, and market. Since ES is no longer a niche element of the contemporary energy system, but a key element, a clear research roadmap and, in parallel, forward-looking policy interventions are needed to promote innovation and accelerate its deployment and to ensure its long-term sustainability. ES systems are complex, involving thermal, electrochemical, and mechanical processes, so it is necessary to develop detailed models which may capture multiscale interactions. Such models can help to better predict, manage and control in real time, and design and optimize systems.

The research in this field should be based on the principles of circular economy and sustainability. This means that new materials being developed should be sustainable and processes should be energy efficient, also the developed materials should be recyclable. It should minimize the use of critical raw materials and improve the material recovery rate, and designing for disassembly and the re-use is also needed. There is potential as well as challenges of ES in smart and distributed systems. Research is needed for better control and optimization approaches, control algorithms, and communication protocols, for efficient integration of storage systems with other grid components. Digital technologies and the AI can enhance flexibility, forecasting, and higher degree of independence, but there are also challenges such as privacy, cybersecurity, and interoperability, these need to be addressed to have safe and secure systems. For the commercialization, right mechanisms and the incentives are required such as subsidies, tax credits, the low-cost finance, and public-private partnerships. Long duration ES systems have high upfront cost and long payback; these need supportive measures to be commercialized. Research and policy should also consider equity and inclusivity, not just tech development but also considerations for fair and equitable use.

14.6 Concluding thoughts on the future of energy storage

Energy storage plays a key role in the process of transforming the world to sustainable and resilient working energy systems. It is underlined that the innovation in the sphere is no longer based on the technological breakthroughs but is shaped by the combination of developments in the field of materials, system engineering, digital technologies, and policies. A shift in the type of focus is one of the changes that can be identified as the shift of the technology to the system focus. The second lesson is the variety in ESTs. There can hardly be a single technology that will be able to substitute all the applications, and a range of technologies will be created, and each is specific to particular applications. The wave of digitalization will continue to be disruptive in the dynamic landscape of ES. The incorporation of data analytics, AI, and digital twins can be accomplished to ensure that the ES systems are effectively managed in real time. The following applications can be made by these approaches: condition monitoring, smart control, and predictive maintenance.

This is because there is growing emphasis on environmentally sustainable ESTs. The ecological price of material resources, production, and recycling should also be considered in the context of the growing use of renewable energy sources. Circular economy principles like recycling, reuse, and efficient use of resources should be adopted. This will help to lower the negative impact on environment. The key to mainstreaming is safety and reliability. It is to be noted there is need of stable materials, efficient design and monitoring of systems to reduce the risk of thermal runaway and system breakage. The combination of advanced diagnostics and old safety mechanisms is crucial to the safe and stable functioning of applications. Policies and regulations also affect ES. The current policies have to be adjusted to the new multifunctional and decentralized character of the storage systems.

AI disclosure

During the preparation of this work the author(s) used Grammerly in order to perform grammar refinement, clarity enhancement, and language editing. After using this tool/service, the author(s) reviewed and edited the content as needed and take(s) full responsibility for the content of the publication.

References

Adenusi, H., Chass, G.A., Passerini, S., Tian, K.V., Chen, G., 2023. Lithium batteries and the solid electrolyte interphase (SEI)—progress and outlook. Adv. Energy Mater. 13 (10). https://doi.org/10.1002/aenm.202203307.

Alzahrani, A., Kumar Ramu, S., Devarajan, G., Vairavasundaram, I., Vairavasundaram, S., 2022. A review on hydrogen-based hybrid microgrid system: topologies for hydrogen energy storage, integration, and energy management with solar and wind energy. Energies 15 (21), 7979. https://doi.org/10.3390/en15217979.

Ansari, S.N., Arya, R., Shah, A., Singhania, T., Mathur, J., 2024. Passive cooling of buildings with phase change materials using fins to increase the effectiveness of thermal storage units. Maejo Int. J. Energy Environ. Commun. 6 (1), 62–69. https://doi.org/10.54279/mijeec.v6i1.255047.

Bányai, Á., 2023. Energy-efficiency and sustainabilityincross-docking supply using e-vehicles. Appl. Eng. Lett. 8 (4), 139–147. https://doi.org/10.18485/aeletters.2023.8.4.1.

Cavus, M., Ayan, H., Bell, M., Dissanayake, D., 2025. Advances in energy storage, AI optimisation, and cybersecurity for electric vehicle grid integration. Energies 18 (17), 4599. https://doi.org/10.3390/en18174599.

Çiçek, M., Gençtürk, M., Balcı, S., Sabancı, K., 2022. The modelling, simulation, and implementation of wireless power transfer for an electric vehicle charging station. J. Eng. 6 (3), 223–229. https://doi.org/10.31127/tuje.930933.

Feyzi, E., Anil Kumar, M.R., Xia, L., Deng, S., Nanda, J., Karim, Z., 2024. A comprehensive review of silicon anodes for high-energy lithium-ion batteries: challenges, latest developments, and perspectives. Next Energy 5, 100176. https://doi.org/10.1016/j.nxener.2024.100176.

Fotouhi, A., Auger, D.J., Propp, K., Longo, S., Wild, M., 2016. A review on electric vehicle battery modelling: from lithium-ion toward lithium–sulphur. Renew. Sustain. Energy Rev. 56, 1008–1021. https://doi.org/10.1016/j.rser.2015.12.009.

García, R.G., Orozco Gamboa, G., Zamora Antuñano, M.A., Ramírez, M.C., García, S.G., López, L.V., Ortiz Pérez, E.L., 2021. Sizing photovoltaic systems interconnected to the grid in the industry. Maejo Int. J. Energy Environ. Commun. 3 (2), 32–37. https://doi.org/10.54279/mijeec.v3i2.245840.

Ghazoui, M., Boudouch, O., Sylla, A.S., Moulakhnif, K., Dabali, S., Elkacmi, R., 2026. Engineering particulate architectures for hybrid energy storage: bridging the gap between intercalation capacity and adsorption kinetics. Particuology 109, 192–210. https://doi.org/10.1016/j.partic.2025.12.019.

Gopi, C.V.M., Ramesh, R., 2024. Review of battery-supercapacitor hybrid energy storage systems for electric vehicles. Results Eng. 24. https://doi.org/10.1016/j.rineng.2024.103598.

Guler, N., Ismail, Z.M., Hazem, Z.B., Naik, N., 2024. Adaptive fuzzy logic controller-based intelligent energy management system scheme for hybrid electric vehicles. IEEE Access 12, 173441–173454. https://doi.org/10.1109/ACCESS.2024.3496897.

Günther, S., Bensmann, A., Hanke-Rauschenbach, R., 2025. Representative energy management strategies for hybrid energy storage systems derived from a meta-review. Renew. Sustain. Energy Rev. 216, 115610. https://doi.org/10.1016/j.rser.2025.115610.

He, G., Michalek, J., Kar, S., Chen, Q., Zhang, D., Whitacre, J.F., 2021. Utility-scale portable energy storage systems. Joule 5 (2), 379–392. https://doi.org/10.1016/j.joule.2020.12.005.

Kolosnitsyn, V.S., Karaseva, E.V., 2008. Lithium-sulfur batteries: problems and solutions. Russ. J. Electrochem. 44 (5), 506–509. https://doi.org/10.1134/s1023193508050029.

Joshi, A., Mishra, D.K., Singh, R., Zhang, J., Ding, Y., 2025. A comprehensive review of solid-state batteries. Appl. Energy 386, 125546. https://doi.org/10.1016/j.apenergy.2025.125546.

Kadulkar, S., Sherman, Z.M., Ganesan, V., Truskett, T.M., 2022. Machine learning-assisted design of material properties. Annu. Rev. Chem. Biomol. Eng. 13, 235–254. https://doi.org/10.1146/annurev-chembioeng-092220-024340.

Li, C., Yang, W., He, W., Zhang, X., Zhu, J., 2021. Multifunctional surfactants for synthesizing high-performance energy storage materials. Energy Storage Mater. 43, 1–19. https://doi.org/10.1016/j.ensm.2021.08.033.

Li, T., Huang, M., Bai, X., Wang, Y.X., 2023. Metal–air batteries: a review on current status and future applications. Prog. Nat. Sci.: Mater. Int. 33 (2), 151–171. https://doi.org/10.1016/j.pnsc.2023.05.007.

Liang, Y., Lai, W.H., Miao, Z., Chou, S.L., 2018. Nanocomposite materials for the sodium–ion battery: a review. Small 14 (5). https://doi.org/10.1002/smll.201702514.
Machín, A., Morant, C., Márquez, F., 2024. Advancements and challenges in solid-state battery technology: an in-depth review of solid electrolytes and anode innovations. Batteries 10 (1), 29. https://doi.org/10.3390/batteries10010029.
Mhatre, S.V., Dongre, V.J., Mande, S., 2026. A hardware-software co-design framework for energy efficient LSTM based SoC estimation on resource-constrained FPGAs. J. Comput. Mech. Manag. 5 (3), 157–174. https://doi.org/10.57159/jcmm.5.3.26806.
Mischos, S., Dalagdi, E., Vrakas, D., 2023. Intelligent energy management systems: a review. Artif. Intell. Rev. 56 (10), 11635–11674. https://doi.org/10.1007/s10462-023-10441-3.
Mohseni, S., Brent, A.C., 2020. Economic viability assessment of sustainable hydrogen production, storage, and utilisation technologies integrated into on- and off-grid micro-grids: a performance comparison of different meta-heuristics. J. Hydrogen Energy 45 (59), 34412–34436. https://doi.org/10.1016/j.ijhydene.2019.11.079.
Naraprawatphong, R., Chokradjaroen, C., Thiangtham, S., Yang, L., Saito, N., 2022. Nanoscale advanced carbons as an anode for lithium-ion battery. Mater. Today Adv. 16, 100290. https://doi.org/10.1016/j.mtadv.2022.100290.
Njema, G.G., Ouma, R.B.O., Kibet, J.K., Ponnusami, V., 2024. A review on the recent advances in battery development and energy storage technologies. J.Renew. Energy 2024, 1–35. https://doi.org/10.1155/2024/2329261.
Nyangiwe, N.N., 2025. Applications of density functional theory and machine learning in nanomaterials: a review. Next Mater. 8. https://doi.org/10.1016/j.nxmate.2025.100683.
Obakhume, K.A., Opatola, F.F., 2025. Enhancing renewable energy integration through energy storage and smart grid innovations: a systematic review. J. Emerg. Sci. Eng. 3 (2). https://doi.org/10.61435/jese.2025.e38.
Olabi, A.G., Sayed, E.T., Wilberforce, T., Jamal, A., Alami, A.H., Elsaid, K., Rahman, S.M.A., Shah, S.K., Abdelkareem, M.A., 2021. Metal-air batteries—a review. Energies 14 (21). https://doi.org/10.3390/en14217373.
Pérez Mendoza, A.E., Schmidt, A., Zarbin, A.J.G., Winnischofer, H., 2024. Review of nanoscale approaches for tailoring electrode materials for advanced energy storage systems. ACS Appl. Nano Mater. 7 (20), 23295–23320. https://doi.org/10.1021/acsanm.4c03493.
Ramdan, M., Hakim, A.L., Ramdani, S., Wicaksana, F.A., 2025. IoT integration for renewable energy storage: a systematic literature approach. Eng.Proc. 107 (1). https://doi.org/10.3390/engproc2025107101.
Said, Z., Sohail, M.A., Pandey, A.K., Sharma, P., Waqas, A., Chen, W.H., Nguyen, P.Q.P., Nguyen, V.N., Pham, N.D.K., Nguyen, X.P., 2023. Nanotechnology-integrated phase change material and nanofluids for solar applications as a potential approach for clean energy strategies: progress, challenges, and opportunities. J. Clean. Prod. 416. https://doi.org/10.1016/j.jclepro.2023.137736.
Sajedi-Moghaddam, A., Rahmanian, E., Aboutalebi, S.H., 2026. Overcoming the energy--power trade-off in fiber electronics: the role of metal-ion hybrid capacitors. Small 22 (9). https://doi.org/10.1002/smll.202512541.
Semeraro, C., Olabi, A.G., Aljaghoub, H., Alami, A.H., Al Radi, M., Dassisti, M., Abdelkareem, M.A., 2023. Digital twin application in energy storage: trends and challenges. J. Energy Storage 58. https://doi.org/10.1016/j.est.2022.106347.
Senthilkumar, N., Yuvaperiyasamy, M., Deepanraj, B., 2024. Application of Taguchi design to optimize the operational parameters of pyramid solar still integrated with TiO 2 nanoparticles. Eng. Res. Express 6 (4), 045540. https://doi.org/10.1088/2631-8695/ad8b0f.

Senthilkumar, N., Yuvaperiyasamy, M., 2025. Experimental investigation and numerical modelling of anaerobic digestion process using de-oiled cakes. Turkish J. Eng. 9 (3), 460–470. https://doi.org/10.31127/tuje.1554884.

Srinivasarao, P., Mohamed, M.R., 2024. A Comprehensive Review of Electric Vehicle Energy Management Techniques, Optimizations, and Controllers. … Communication.

Tawalbeh, M., Khan, H.A., Al-Othman, A., Almomani, F., Ajith, S., 2023. A comprehensive review on the recent advances in materials for thermal energy storage applications. Int. J. Thermofluids 18. https://doi.org/10.1016/j.ijft.2023.100326.

Vanaraj, R., Arumugam, B., Mayakrishnan, G., Kim, S.C., 2025. Advancements in metal-ion capacitors: bridging energy and power density for next-generation energy storage. Energies 18 (5). https://doi.org/10.3390/en18051253.

Wen, Q., Li, C., Chen, Q., Zhao, P., Wu, C., Wu, X., Chou, S.-L., 2025. Carbon engineering for sodium batteries: multi-role architectures bridging material design and hybrid system innovation. Chem. Soc. Rev. 54 (20). https://doi.org/10.1039/d5cs00515a.

Wu, X., Kang, F., Duan, W., Jia, L., 2019. Density functional theory calculations: a powerful tool to simulate and design high-performance energy storage and conversion materials. Prog. Nat. Sci. Mater. Int. 29 (3), 247–255. https://doi.org/10.1016/j.pnsc.2019.04.003.

Wu, Y., Wu, Y., Cimen, H., Vasquez, J.C., Guerrero, J.M., 2022. P2P energy trading: blockchain-enabled P2P energy society with multi-scale flexibility services. Energy Rep. 8, 3614–3628. https://doi.org/10.1016/j.egyr.2022.02.074.

Xiong, K., Qi, T., Zhang, X., 2025. Advancements in graphite anodes for lithium-ion and sodium-ion batteries: a review. Electroanalysis 37 (1). https://doi.org/10.1002/elan.202400318.

Yang, Y., Long, X., Zhang, H., Dong, B., Chen, H., Zhang, L., Shi, S., Zhou, C., 2024. Surface modification of lithium-ion sieves by silane coupling agent for improved adsorption performance. Sep. Purif. Technol. 330, 125422. https://doi.org/10.1016/j.seppur.2023.125422.

Index

Note: Page numbers followed by "f" indicate figures and "t" indicate tables.

F

Printed and bound by CPI Group (UK) Ltd, Croydon, CR0 4YY
19/09/2026
02219501-0004